Pitman £ 2·80 S/72

ISBN 0 273 36148 1

SURVEYING

THE SOUTH GEORGIA SURVEY 1955–56: CAPTAIN A. G. BOMFORD, R.E.,
USES A THEODOLITE NEAR THE HEAD OF THE DRYGALSKI FJORD

Photograph by George B. Spenceley

SURVEYING

BY

A. BANNISTER

M.C., M.Sc., F.I.C.E.
Reader in Civil Engineering
University of Salford

AND

S. RAYMOND

M.Sc., Ph.D., Dip.T.P.(Manchester)
F.I.C.E., M.R.T.P.I., M.Inst.H.E.
Reader in Civil Engineering
University of Salford

Pitman Publishing

Second edition 1965
First paperback edition 1967
Reprinted 1968
Reprinted 1970
Reprinted 1971
Third edition 1972

SIR ISAAC PITMAN AND SONS LTD.
Pitman House, Parker Street, Kingsway, London, WC2B 5PB
P.O. Box 46038, Portal Street, Nairobi, Kenya

SIR ISAAC PITMAN (AUST.) PTY. LTD.
Pitman House, Bouverie Street, Carlton, Victoria 3053, Australia

PITMAN PUBLISHING COMPANY S.A. LTD.
P.O. Box 11231, Johannesburg, S. Africa

PITMAN PUBLISHING CORPORATION
6 East 43rd Street, New York, N.Y. 10017, U.S.A.

SIR ISAAC PITMAN (CANADA) LTD.
495 Wellington Street West, Toronto 135, Canada

THE COPP CLARK PUBLISHING COMPANY
517 Wellington Street, Toronto 135, Canada

Cased edition ISBN: 0 273 36149 X

Paperback edition: ISBN: 0 273 36148 1

Text set in 10 pt. Monotype Modern, printed by letterpress,
and bound in Great Britain at The Pitman Press, Bath
G2—(T.1256/1348:73)

PREFACE

THIS BOOK WAS WRITTEN with the needs of students reading for the following examinations in mind: Engineering Degrees, Ordinary and Higher National Diplomas in Building, and Higher National Certificates in Structural Engineering, Civil Engineering or Building. It also is intended to cater for the needs of those studying for the examination of the Council of Engineering Institutions. Naturally, with such diverse syllabuses to consider, it was not possible to deal exhaustively with every branch of Surveying and still keep the book to reasonable size. We have, in fact, aimed at producing a concise, modern text-book, giving the elements of surveying, to be used in conjunction with the other comprehensive reference works on the subject which are available.

Although it may be possible that the amount of research in Surveying has been less than in, for example, Soil Mechanics and Theory of Structures, nevertheless a surprisingly large number of new developments have taken place since the conclusion of the Second World War. It was the excellent papers and articles by A. Stephenson, O.B.E., M.A., F.R.I.C.S., and J. Clendinning, O.B.E., B.Sc.(Eng.), A.M.I.C.E., on recent trends in surveying which, *inter alia*, inspired us to first write this book, as we are firmly convinced that in the education of technologists accent should be placed, where possible, on latest developments after the fundamental techniques have been absorbed. This extends, of course, to practical surveying, in that modern equipment and techniques should be employed where possible.

Modern instruments embody all the advances in knowledge made by the physicist, optical scientist, instrument maker, etc., and will give at least the same accuracy as the older instruments in much less time. Where the cost of a new technique (e.g. aerial surveying) or instrument (e.g. Watts Microptic theodolite or Watts Autoplumb) can be shown to be cheaper, having regard to the cost of the engineers' time using lengthier methods, the new methods and instruments should normally be adopted by employers, and the trend in this direction will be accentuated if students are made familiar from the start with modern developments.

Acknowledgment is hereby made and our thanks tendered to the following firms, bodies and persons who have supplied material for, or who have granted permission for the use of materials in this book: Cambridge University Press; Technical Press Ltd.; Controller of H.M.S.O.; Royal Geographical Society; Hilger & Watts Ltd.; Vickers Instruments Ltd.; W. A. Stanley & Co. Ltd.; Wild Heerbrugg Ltd.; Kern Co.; Survey & General Instrument Co. Ltd.; Zeiss Co.; Carl Degenhardt Ltd.; C. Z. Scientific Instruments Ltd.; Kelvin Hughes Ltd.; Decca Ltd.; Holmes Bros. (Leyton) Ltd.; Short and Mason Ltd.; R. W. Munro Ltd.; Halden's Ltd.; British Transport Commission; C. F. Casella Ltd.; Fennel Ltd.; Tellurometer (U.K.) Ltd.; Askania

G.m.b.H.; Williamson Manufacturing Co. Ltd.; AGA (UK) Ltd.; Hansa Luftbild; Franke & Co. Optik G.m.b.H.; Institution of Civil Engineers; Institution of Structural Engineers; and University of London.

Finally our thanks are due to the Vice-Chancellor and Senate of the University of Salford for their help and encouragement.

A. BANNISTER
1972   S. RAYMOND

CONTENTS

Inset: FIG. 11.17A, between pages 394 and 395

CHAPTER 1

INTRODUCTORY

Definitions

Surveying may be defined as the art of making measurements of the relative positions of natural and man-made features on the earth's surface, and the plotting of these measurements to some suitable scale to form a map, plan or section.

In practice, however, the term "surveying" is often used in the particular sense of meaning those operations which deal with the making of plans, i.e. working in the two dimensions which form the horizontal plane, and the term "levelling" covers work in the third dimension, namely the dimension normal to the horizontal. Thus we have—

Surveying: operations connected with representation of ground features in plan.

Levelling: operations connected with the representation of relative difference in altitude between various points on the earth's surface.

Surveying is divided primarily into: (i) geodetic surveying, (ii) plane surveying.

(i) In geodetic surveying the curvature of the earth is taken into account, so that a knowledge of spherical geometry is required. All surveys of countries, such as the Ordnance Survey of Great Britain, are geodetic surveys.

(ii) In plane surveying the area under consideration is taken to be a horizontal plane, and the measurements plotted will represent the projection on the horizontal plane of the actual field measurements. For example, if the distance between two points A and B on a hillside is l, the distance to be plotted will be $l \cos \alpha$, where α is the angle line AB makes with the horizontal, assuming a uniform slope.

A horizontal plane is one which is normal to the direction of gravity, as defined by a plumb bob, at a point, but owing to the curvature of the earth such a plane will in fact be tangential to the earth's surface at the point. Thus, if a large enough area is considered on this basis, a discrepancy will become apparent between the area of the horizontal plane and the actual curved area of the earth's surface.

It can be shown that for surveys up to 250 km^2 in area this discrepancy is not serious, and it is obvious therefore that plane surveying will be adequate for all but the very largest surveys.

Uses of Plane Surveying

The uses of plane surveying include—

    (*a*) the measurement of areas;

    (*b*) the making of plans in connexion with legal documents (including land transfer), Parliamentary Bills, etc.;

(*c*) the making of plans in connexion with the work of the civil engineer, architect, builder, structural engineer and town planner; and the reverse process—working from the plan back to the site—involved in setting out works;

(*d*) the making of maps and plans for military, geographical, geological, and other purposes.

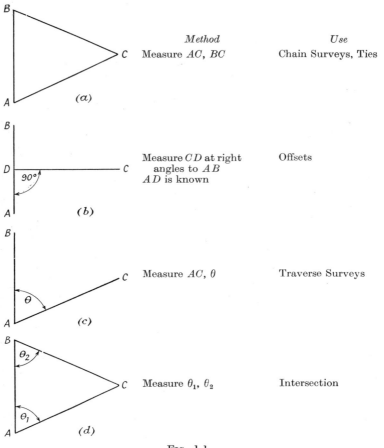

	Method	*Use*
(*a*)	Measure *AC, BC*	Chain Surveys, Ties
(*b*)	Measure *CD* at right angles to *AB* *AD* is known	Offsets
(*c*)	Measure *AC*, θ	Traverse Surveys
(*d*)	Measure θ_1, θ_2	Intersection

FIG. 1.1

There are many other lesser uses too numerous to mention, and the importance of plane surveying, particularly to those connected with constructional work in some form or other, is readily seen.

Basic Principles of Surveying

One might truthfully say that surveying as a subject is not difficult, though at the advanced stages of some branches of the art a fair demand is made on mathematical knowledge. In this respect surveying is similar to other subjects, such as the theory of structures, in that as well as common sense and a feeling for the subject, the possession of a mathematical background is an advantage. The fundamental principles of surveying, however, are few and simple in concept.

On any area of land to be measured, it will always be possible to choose two points and to measure the distance between them. This line AB can be drawn to scale on paper. Other points can be located relative to the line by taking *two* other measurements which can of course be similarly drawn to scale on the paper, and in this way a map is constructed. The two measurements can consist of two measured lines, one line and an angle, or two angles. The principles are illustrated in Fig. 1.1, A and B representing in each case the two original points, and C a point to be located.

Checking Measurements

It is appropriate to say a word here on the prevention of mistakes and the elimination of errors in surveying. There is no such thing as an absolutely exact measurement, and the precision (itself a relative term) achieved depends on the instruments used in making the measurement, the care taken, and the ability of the surveyor to guarantee that his work is free from mistakes. It is axiomatic that surveying work should be self-checking where possible, and in practice some of the measurements given in the previous section, representing as they do theoretical minimum requirements, would require to be supplemented by check measurements. For example, in chain surveying, if only the lines joining the three points of a triangle were measured (Fig. 1.1a), it would still be possible to plot the triangle even though a mistake had been made in the measurement of one of the lines. By chaining one more line, for instance one running from one corner of the triangle to the opposite side, the error would be revealed, for, except in the improbable event of two self-compensating errors, it would not be possible to plot the check line. Also, in Fig. 1.1d, although with absolutely exact and mistake-free measurement it is sufficient to measure angles θ_1 and θ_2 alone, in practice the angle θ_3 at C would be measured, so that, by applying the test that $\theta_1 + \theta_2 + \theta_3$ should equal 180°, a check may be made on mistakes and a distribution made of other errors. (Note that this relationship applies only in plane surveying; in geodetic surveying the laws of spherical trigonometry apply.)

CHAPTER 2

CHAIN SURVEYING

THIS BRANCH OF SURVEYING derives its name from the fact that the principal item of equipment used is the measuring chain. Notwithstanding that this and the allied items of equipment are simple in construction, work of a sufficiently high order of accuracy to cover the requirements of much ordinary engineering work is possible, especially when large-scale plans of relatively small areas are required. Also, in addition to being a complete method of surveying, some of the operations of chain surveying occur in other branches (notably traverse surveying, where the chain and tape are often used to survey detail). A good knowledge of chain surveying is therefore essential to a proper knowledge of surveying as a whole.

HISTORICAL NOTE. The use of chain surveying dates back to the earliest times even though the records are scanty compared with those on other subjects. Only a brief reference to some of them is possible here. The most definite mention of chaining comes from drawings on the walls of some Egyptian tombs showing the "cordmen" stretching their measuring cords of plaited palm strip. The precision with which the pyramids, and later, the Greek cities, were set out, indicates that the ancient civilizations had, even by present standards, good surveying technique. Probably the best record of the first thousand years B.C. is that due to Heron, a Greek who lived in Alexandria in about the first century A.D. He provided the first serious account of surveying techniques, and from this it is clear that the work of Euclid and the other geometers was used in measuring up and setting out. Many of the methods and much of the equipment described by Heron were in fact too complicated for the Romans, who, like present-day engineers, preferred simple methods where possible. As the Roman Empire became stabilized, great use was made of chain surveying in the fixing of boundaries, setting out new cities, aqueducts, roads, etc. The work was carried out by a trained body of men known as the "agrimensores," and many examples survive to show the skill of these Roman surveyors. From Roman times until the present day, the main improvement has been in equipment, e.g. the link chain of wrought iron introduced in the seventeenth century, and the optical squares and steel tapes, etc., introduced in the past century.

Equipment Used in Chain Surveying

The items of equipment required fall under three broad headings: those used for linear measurement, those used for measuring right angles, and other items.

1. Equipment for the Measurement of Lines

THE CHAIN. Chains manufactured according to British Standard 4484:Part 1:1969 are to be 20 m long, but it is also possible to obtain chains 30 m long. They are made of tempered steel wire, 8 or 10 SWG, and are made up of links which measure 200 mm from centre to centre of each middle connecting ring. Swivelling brass handles are fitted at each end and the total length is measured over the handles. On the 20 m chain tally markers, made of plastic, are attached at every whole metre position, and those giving 5 m positions are of a different colour. The 30 m chains may have brass tallies marking every tenth link.

FIG. 2.1. LAND MEASURING CHAIN

FIG. 2.2a. SURVEYOR'S BAND
(*Courtesy of Rabone Chesterman, Ltd.*)

The chain is robust, easily read, and easily repaired in the field if broken. It is liable to vary somewhat in length, however, owing to wear on the metal-to-metal surfaces, bending of the links, mud between the bearing surfaces, etc. Also its weight is a disadvantage when the chain has to be suspended.

THE SURVEYOR'S BAND (OR DRAG TAPE). This is made of steel strip, some six millimetres in width, and is carried on a four-arm open frame winder. A handle is fitted for returning the band into its frame after use, and this also provides a locking device for retaining the band. Rawhide thongs are supplied for attaching to the small loops at the extremities of the bands to allow them to be pulled or straightened. Alternatively handles, similar to those of the chain, can be fitted: lengths of 30 m or 50 m are normal but 100 m bands may be encountered. B.S. 4484:Part 1:1969 requires that metres, tenths and hundredths of metres should be marked, with the first and last metres also subdivided into millimetres. The operating tension and temperature for which it was graduated should preferably be indicated on the band in addition. Other types, made of special steel and used with supplementary equipment to apply a constant pull, are used for more accurate work than we are considering here, and these will be dealt with later in the chapter on Triangulation Surveys.

The steel band is a much more accurate measuring instrument than the chain, and with careful use maintains its length, so that it can be used as a standard measure. Its main disadvantages are its lack of robustness and the difficulty in doing field repairs.

TAPES. These may be made of synthetic material, glass fibre being typical, or coated steel or plain steel. B.S. 4484:Part 1 suggests 10 m, 20 m or 30 m as the desirable lengths, and these lengths are generally available.

For the synthetic types the British Standard requires major graduations at whole metre positions and tenths, with minor graduations at hundredths, and 50 mm intervals indicated. Those manufactured of glass fibre have a PVC coating (Fig. 2.2b) They are graduated every

FIG. 2.2b. TYPICAL MEASURING TAPE
(*Courtesy of Rabone Chesterman, Ltd.*)

10 mm and figured every 100 mm; whole metre figures are shown in red at every metre. These tapes are said to have good length-keeping properties, but it is conventional the use them for relatively short measurements.

Steel tapes may be provided with a vinyl coating or may be plain. The former type has sharp black graduations on a white background. They can be obtained graduated every 5 mm and figured every 100 mm; the first and last metre lengths are also graduated in millimetres. Whole metre figures are again shown in red at every metre.

The latter type have graduations and figures etched on to the steel and they present the same subdivisions as the vinyl-covered types. However, they are generally wider and are contained in a leather case rather than a plastic case.

2. Equipment for Measuring Right Angles

THE CROSS STAFF. A typical cross staff is shown in Fig. 2.3a, and consists essentially of an octagonal brass box with slits cut in each face so that opposite pairs form sight lines. The instrument may be mounted

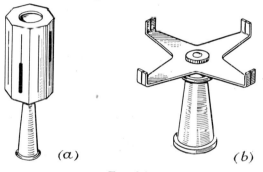

(a)   (b)

FIG. 2.3

on a short ranging rod and, to set out a right angle, sights are taken through any two pairs of slits whose axes are perpendicular. The other two pairs then enable angles of 45° and 135° to be set out. An alternative type of cross staff is shown in Fig. 2.3b.

THE OPTICAL SQUARE. There are two types of optical square, one using two mirrors and the other a prism. The instrument is compact, rarely measuring more than 75 mm in diameter by about 20 mm thick, and is more accurate than the cross staff.

The mirror type makes use of the fact that a ray of light reflected from two mirrors is turned through twice the angle between the mirrors, which in turn is easily derived using the principles of reflection of light. Fig. 2.4 shows the mounting. Mirror A is completely silvered, while mirror B is silvered to half its depth, the other half being left plain.

Thus, the eye looking through the small eye-hole will be able to see half an object at O_1. An object at O_2 is visible in the upper (silvered) half of mirror B, and when $O_1 \hat{X} O_2$ is a right angle (where X is the

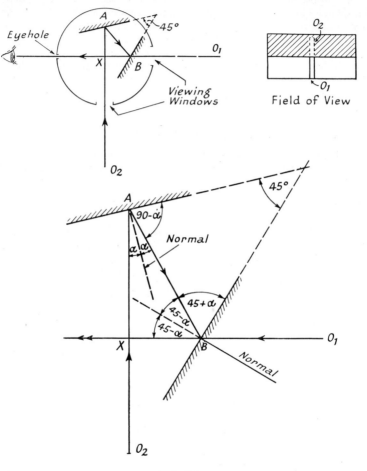

Fig. 2.4

centre of the instrument), the image of O_2 is in line with the bottom half of O_1 seen direct through the plain glass.

The surveyor stands at X, sights O_1, and directs his assistant to move O_2 until the field of view is as shown above. Then $O_1 \hat{X} O_2$ is a right angle for, considering any ray from O_2 incident on mirror A at angle α to the normal, it will emerge at the same angle to the normal.

Therefore $\qquad X\hat{A}B = 2\alpha$

and, from a consideration of the angles,

$$X\hat{B}A = 90° - 2\alpha$$

$$\therefore \qquad A\hat{X}B = O_1\hat{X}O_2$$
$$= 90°$$

i.e. the result is independent of α.

The prismatic type of optical square employs a pentagonal-shaped prism, cut so that two faces contain an angle of exactly 45°. It is used in the same way as the mirror square, but is rather more accurate.

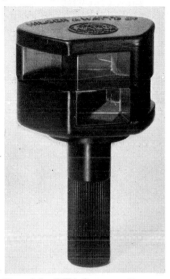

The model shown in Fig. 2.5 (by Messrs. Hilger and Watts) has two such prisms, and for setting out right angles the top one only is used. By using both of them, it is possible to set out two points O_1 and O_2 such that $O_1\hat{X}O_2$ is 180°.

3. Other Equipment

RANGING RODS. These are poles of circular section 2 m, 2·5 m or 3 m long, painted with characteristic red and white bands which are usually 0·5 m long, and tipped with a pointed steel shoe to enable them to be driven into the ground. They are used in the measurement of lines with the chain, and for marking any points which re-

FIG. 2.5. SH4 OPTICAL SQUARE

(Courtesy of Hilger & Watts, Ltd.)

quire to be seen. A sectional tubular type is also made, the short bottom section of which, apart from its robustness, is often useful where headroom is restricted.

In hard, or paved, ground a tripod is used to support the rods.

ARROWS. When chaining a long line, the chain has to be laid down a number of times and the positions of the ends are marked with arrows, which are steel skewers about 40 mm long and 3 mm to 4 mm diameter. A piece of red ribbon at the top enables them to be seen more clearly. An arrow with a lump of lead at its bottom end is occasionally used as a form of plumb-bob. It is known as a drop-arrow, and its principal use is in connexion with chaining on a slope, an operation described later.

PEGS. Points which require to be more permanently marked, such as the intersection points of chain lines, are marked by oak pegs driven into the ground by a mallet. A typical size is 40 mm × 40 mm × 0·4 m long. In very hard or frozen ground, steel dowels are used instead, while in asphalt roads, small 5 or 6 mm square brads are used.

Procedure in Chain Surveying

Chain surveying consists in measuring with the chain the lengths of a series of straight lines, the positions of which are governed by principles given later in this chapter, and then locating points on the ground relative to these lines by the methods given in the Introduction, namely, by measuring two other lines, known as *ties*, or by measuring *offsets* at right angles to the main chain line. Since ties and offsets are short measurements, they can be made with the glass fibre tape and rarely exceed one tape length. Thus the two basic procedures which require to be known at this stage are (1) the chaining or "ranging" of lines, and (2) the setting out of the right angles in connexion with offsets.

1. **Chaining a Line.** The operation is carried out by two assistants known as chainmen, one acting as *leading chainman* and the other as

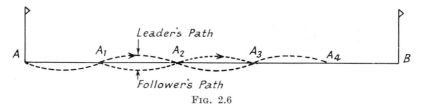

Fig. 2.6

follower. To throw out a chain, initially bundled, the handles are held in the left hand and the chain is thrown out with the right hand. Knots and twists are taken out as the chain lies stretched on the ground in two equal lengths with the brass handles near each other. The chainmen then take one end each and the chain is pulled out full length and examined for defects. The leader equips himself with ten arrows, and a ranging rod, and the follower also takes a ranging rod. Then, (Fig. 2.6) to measure line AB having previously positioned ranging rods at both A and B—

(*a*) The leader drags his end of the chain forward to A_1, and holds his ranging rod about one link short of the end.

(*b*) The follower holds his end of the chain firmly against station A and the surveyor lines in the leader's pole between A and B by closing one eye, sighting poles A and B, and signalling the leader till he brings his pole into line AB. The system of signalling usually adopted is to swing the left arm out to the left as an instruction to the leader to move his pole in that direction: the right arm is similarly used to indicate movement to the right, while both arms extended above the head, then brought down, indicates that the pole is on line.

(*c*) The leader straightens the chain past the rod by sending gentle 'snakes" down the chain.

(*d*) The follower indicates the chain is straight, and the leader puts an arrow at the end, outside and touching the handle, at A_1. (At this stage offsets or ties may be taken from known chain positions to required detail.)

(e) The leader then drags his end to A_2, taking nine arrows and his pole.

(f) The follower moves to A_1, and puts his pole behind the arrow, and the surveyor again lines in from here or from A.

The above procedure is repeated, the follower picking up the first arrow before he moves from A_1. The leader moves to A_3, carrying eight arrows. The follower moves to A_2, carrying arrow from A_1.

If the line measured is more than ten chains long, the leader will exhaust his supply of arrows at the 10-chain point, so that when the 11th chain length is stretched out, the follower will have to hand over the ten arrows back to the leader. This fact is pointed out to the surveyor who notes it in his field book.

The following exercises should be carried out—

(i) Throwing out the 20 m chain and gathering it in again. This latter operation consists in carefully doubling the chain, so that starting from the 10 m tally the links may be gathered in two at a time. Note the characteristic wheatsheaf shape of the gathered-in chain, which is achieved by holding the links firmly at the mid point, and laying each pair as gathered in at a constant angle to the previous pair.

When winding up a 30 m chain made of 8 SWG links the last few links are not easily wrapped in a neat and tidy manner when starting from the middle. It is, however, possible to fold the chain inwards so that the handles lie next to the 74th links from each end. Two men can gather in the chain as mentioned previously by now working inwards; this leaves the handles in the correct position for throwing out.

(ii) Straightening the chain and marking the chain length with arrows. This length can be checked against the steel band.

(iii) Measuring the lengths of two or three lines, chaining each one in both directions.

2. **Setting out Right Angles.** Since this operation is often required in connexion with the measurement of offsets, this is a convenient point at which it may be discussed. There are two cases to consider, (1) dropping a perpendicular from a point to a line, and (2) setting out a line at right angles to the chain from a given point on the chain.

1(a) For short offsets, the end of the tape is held at the point to be located, and the right angle is estimated by eye. Although a usual method in practice, it is not so accurate as the following methods.

(b) Again for offsets, the tape is swung with its zero as centre about the point and the minimum reading at which it crosses the chain is noted. This occurs, of course, when the tape is perpendicular to the chain, but the method can be used only on smooth ground where a free swing of the tape is possible.

(c) Where the above method is not applicable, with the free end of the tape at centre P (the point) strike an arc to cut the chain at

A and B (*see* Fig. 2.7a). Bisect AB at Q. Then $P\hat{Q}A = 90°$.

(*d*) Run the tape from P to any point A on the chain (Fig. 2.7b). Bisect PA at B, and with centre B and radius BA strike an arc to cut the chain at Q. Then $A\hat{Q}P = 90°$, being the angle in a semicircle.

2(*a*) Cross Staff. This is mounted on a short ranging rod which is stuck in the ground at the point at which the right angle is to be set out. The cross staff is turned until a sight is obtained along the chain line and the normal is then set out by sighting through the slits at right angles to the chain.

(*b*) Optical Square. This is used in the manner already described, the instrument being either held in the hand or else propped on a short ranging rod.

(*c*) Pythagoras' Theorem. 3, 4, 5 rule or any multiple of same, say 9, 12, 15. With the zero end of the tape at P take the 24 m

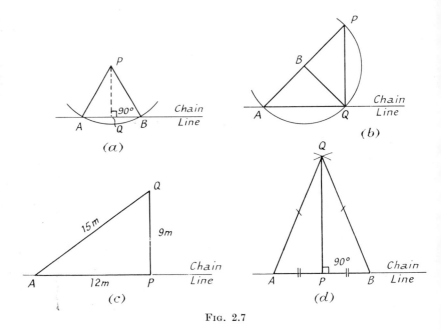

Fig. 2.7

mark of the tape to A, where $AP = 12$ m on the chain. Take the 9 mark on the tape in the hand and ensuring that the tape is securely held at A and P pull both parts of the tape taut to Q. Then $A\hat{P}Q = 90°$ (Fig. 2.7c).

Note: A convenient combination is 8, 8·4, 11·6. If the 8 is set off on the chain then $8·4 + 11·6 = 20$ m can be set off on a 20 m tape.

(*d*) Take *A* and *B* on the chain so that $PA = PB$ (Fig. 2.7*d*). Strike arcs from *A* and *B* with equal radii to intersect at *Q*. Then

$$A\hat{P}Q = 90°.$$

As exercises, the above methods of setting out right angles should be practised.

Errors in Chaining and their Correction

In all surveying operations, as indeed in any operation involving measurement, errors are likely to occur, and so far as is possible they must be guarded against or their effects corrected for. The types of error which can occur may be classed under the following three headings: (i) mistakes, (ii) systematic or cumulative errors, (iii) accidental or compensating errors. The three types will be dealt with briefly, with examples of their occurrence and the remedies for them in chain surveying.

Mistakes. These are due to inexperience, or to carelessness on the part of the surveyor or the chainmen, and are of course quite random in both occurrence and magnitude. If allowed to pass unchecked, mistakes could lead to a faulty plan being produced. By careful work, however, and by taking suitable check measurements, it should be possible to make a survey which is free from mistakes. Typical mistakes in chaining the length of a line are—

(1) Omitting an entire chain length in booking. This is prevented by noting down each chain length, and by the leader keeping careful count of the arrows, as described earlier.

(2) Mis-reading the chainage by confusing the tallies—say the 14 m and 16 m tallies on a 30 m chain. Also the units four and six are capable of being mis-read. Only careful reading can prevent these mistakes; it is best if two people make important readings.

(3) Erroneous booking sometimes occurs; it is prevented by the chainman carefully calling out the result and the surveyor repeating it, paying attention when calling 5 or 9, 7 or 11.

Systematic or Cumulative Errors. These arise from sources which are known, and their effects, therefore, can be eliminated.

The most careful chaining will not produce an accurate survey if, for example, the chain has been damaged and is therefore of incorrect length, because every time the chain is stretched out it will measure not 100 links in the case of a 20 m chain but $100 \pm$ (some constant or systematic error). If uncorrected, such an error could have serious effects. By checking the chain against a standard, which might be a steel band, or two marks measured for the purpose, the exact error per chain length is known. If this error cannot be eliminated, a correction can be applied which will enable the effect of the error to be removed. There are two such errors in chaining to which corrections are applied—the one just mentioned, and the other due to slope of the ground.

(1) Correction of chain for standardization. Checking the chain frequently is necessary for this correction to be effective, because a chain changes in length due to wear and tear. Checking at the beginning of each

day's chaining is a good rule. If the chain has a large error, say a link or more, it should be overhauled and brought to correct length. Otherwise, it is used in its incorrect condition, and each line corrected, thus—

Correct length of line

$$= \text{measured length of line} \times \frac{\text{length of chain used}}{\text{length of standard}} \qquad . \qquad (2.1)$$

The length of the standard is, of course, usually 100 or 150 links, preferably the former in accordance with B.S. 4484.

EXAMPLE. A line is measured with a chain believed to be 20 m long which gives a length of 376·4 m. On checking, the chain is found to measure 20·04 m. What is the correct length of the line?

$$\text{True length of line} = 376·4 \times \frac{20·04}{20}$$

$$= 376·4(1 + 0·002)$$

$$= 376·4 + 0·7$$

$$= 377·1 \text{ m}$$

Note the method of working, which enables a slide rule or four-figure logarithms to be used. The error can be expressed as 0·002 m per metre, and the adjustment has to be added because by using a chain that is too long the length of the line is measured as less than it actually is.

If the survey is carried out to determine an area, it is not necessary to correct each measurement first. Calculate the area from the measurements taken, and correct this with the formula

$$\text{Correct area} = \text{measured area} \times \left(\frac{\text{length of chain used}}{\text{length of standard}}\right)^2 \qquad . \qquad (2.2)$$

This formula derives from (1) because an area is given by one length multiplied by another, and all the lengths are corrected by a constant factor.

EXAMPLE. A field was measured with a 20 m chain 0·3 of a link too long. The area thus found was 1·032 hectares. What is the true area?

From (2.2),

$$\text{True area} = 1·032 \times \left(\frac{100·3}{100}\right)^2$$

$$= 1·032 \times (1 + 0·003)^2$$

Put

$$0·003 = e$$

$$(1 + e)^2 = 1 + 2e + e^2$$

$$\simeq 1 + 2e \quad \text{if } e \text{ is small}$$

$$\therefore \qquad \text{True area} = 1·032(1 + 0·006)$$

$$= 1·038 \text{ hectares}$$

(2) Correction for sloping ground. As was stated in the Introduction, all measurements in surveying must either be in the horizontal plane, or be corrected to give the projection on this plane. Lines chained on sloping land must be longer than lines chained on the flat, and if the

slope is excessive, then a correction must be applied. There are two methods—

(*a*) "Stepping": on ground which is of variable slope this is the best method, and needs no calculation. The chaining is done in short lengths of 30–50 links, the leader holding the length horizontal. (Any attempt to measure in lengths longer than this will lead to errors due to chain sag greater than those caused by slope.) The point on the ground below the free end of the chain is located by plumb bob or drop

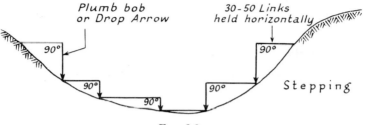

Fig. 2.8

arrow, as shown in Fig. 2.8. It will be seen that it is easier to chain downhill when "stepping" than to chain uphill, the follower then having the difficult job of holding the chain taut, horizontal, and with the end vertically over the previous arrow. The leader has therefore to line himself in.

(*b*) Measuring along the slope, a method applicable where the ground runs in long regular slopes. The slope is measured either by an instru-

Fig. 2.9

ment known as a clinometer, described in a later chapter, or by levelling, a procedure which gives the surface height at points along the slope. In either case the angle of slope can be found, and hence the corrected length from

$$\text{Correct length} = \text{measured length} \times \cos \alpha \qquad . \qquad (2.3)$$

where $\alpha = $ angle of slope   (*see* Fig. 2.9*a*.)

Note : The measured length used in (2.3) might have to be corrected first if a non-standard chain is used.

This method corrects only the total length of the line, and if intermediate measurements are to be correctly made, adjustments must be made during chaining. These can be readily made. In Fig. 2.9*b*, *A B* represents one 20 m chain-length measured along the slope. What we require is the point *C* beyond *B* such that a plumb bob at *C* will cut the horizontal through A at D, where AD is 100 links on the horizontal; i.e. we require the correction BC which is to be added, with a tape, to each chain length measured along the slope. If α is measured with a clinometer, then

$$AC = AD \sec \alpha$$

Therefore

$$BC = AC - AB = AD \sec \alpha - AB$$

$$= AD\left(1 + \frac{\alpha^2}{2} + \frac{5\alpha^4}{24} + \ldots\right) - AD \quad \text{where } \alpha \text{ is in radians}$$

$$= AD\left(1 + \frac{1 \cdot 5\alpha^2}{100}\right) - AD \quad \text{where } \alpha \text{ is in degrees}$$

$\therefore$ correction

$$= \frac{1 \cdot 5\alpha^2}{100} AD$$

i.e. if $\alpha = 10°$, correction is 1·5 links per 100 links.

If the slope is measured by levelling it will be expressed as 1 in n, which means a rise of 1 unit vertically for n units *horizontally*. Then, from previous working

$$\text{Correction per 100 links} = 50\alpha^2 \ (\alpha \text{ in radians})$$

$$= \frac{50}{n^2}$$

Since

$$\alpha = \frac{1}{n}$$

i.e. if slope is 1 in 10, the adjustment is 0·5 links per 100 links.

EXAMPLE. In chaining a line, what is the maximum slope (*a*) in degrees, and (*b*) as 1 in *n* which can be ignored if the error from this source is not to exceed 1 in 1,000.

(*a*) From Fig. 2.9*b* $\qquad$ error $= 1$ in $1,000 = \dfrac{AD}{1,000}$

Therefore $\qquad \dfrac{1 \cdot 5\alpha^2}{100} AD = \dfrac{AD}{1,000}$

and $\qquad \alpha = 2 \cdot 6°$

(*b*) $\qquad$ error $= \dfrac{AD}{1,000}$

Therefore $\qquad \dfrac{AD\alpha^2}{2} = \dfrac{AD}{2n^2} = \dfrac{AD}{1,000}$

$$n = \sqrt{500}$$

$\therefore$ max. slope is 1 in 22·4.

Accidental or Compensating Errors. This third group of errors arises from lack of perfection in the human eye and in the method of using equipment. They are not mistakes and as there is as much chance of their being positive as being negative, the errors from these sources tend to cancel out, i.e. tend to be compensatory. They do not entirely disappear, however, and it can be shown that they are proportional to $\sqrt{L}$, where L is the length of the line. They are, therefore, second-order errors compared with cumulative ones, which are proportional to L. Usually, no attempt is made to correct for them in chaining. This type of error could arise from the leader holding the marking arrow in an inclined position relative to the chain instead of vertically, so that the point of the arrow, from which the next chain length is to be measured, is either in front of or behind the actual end of the chain.

Field Work

The basic principle is that if A and B are fixed (Fig. 2.10), C can be located by measuring AC and BC. The lengths of the three sides

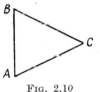

FIG. 2.10

of $\triangle ABC$ being known, the triangle can be plotted. Any area of land can be divided into a series of triangles which form a framework, which may be plotted, and which covers the greater part of the area to be surveyed. To locate topographical and man-made features relative to this framework, measurements are made with the tape from the lines during the course of chaining. The two methods for locating detail are shown in Fig. 2.11.

Such measurements should be as short as possible, and in any case not greater than one tape-length, so the chain lines will normally run

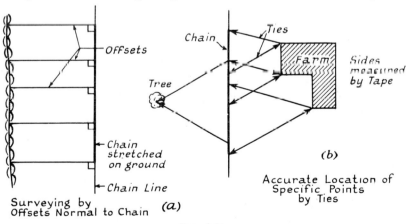

Surveying by Offsets Normal to Chain **(a)**

Accurate Location of Specific Points by Ties **(b)**

FIG. 2.11

as close to the site boundary as possible. The accuracy of the work will be increased if the framework of triangles is founded on a backbone line run through the site to be surveyed, bearing in mind that the fewest number of well-shaped (or "well-conditioned") triangles necessary for the work normally gives the best results. By well-conditioned we mean that all the intersections are clean for plotting purposes, giving nearly equilateral triangles if possible, for where points are plotted by striking arcs, representing the measured tie lengths, the determination of intersections at angles of much less than 30° is difficult. Each triangle should have a check line (Fig. 2.12), so that an error in the chaining of one line will not go undetected.

The intersection points of the lines are called *stations* and these are established first by placing ranging rods (and later, pegs if permanency

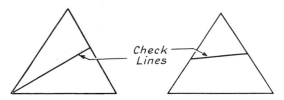

FIG. 2.12. TYPES OF CHECK LINE

is required), after a preliminary reconnaissance survey of the site has been made by the surveyor. The principles governing the location of the stations and the lie of the chain lines may be summarized—

1. Although general topography will dictate to a large extent the actual layout of the triangles, as few chain lines as necessary should be used, and obstacles and steep, uneven slopes avoided as far as possible.

2. There should be at least one long "backbone" line in the survey upon which the surveyor may found the triangles.

3. The triangles should have the angles lying between 30° and 120° so as to give clean intersections, and check lines must be provided for all independent figures.

4. Where possible, avoid having chain lines without offsets unless they are check lines; and keep offsets short, especially to important features.

Fig. 2.13 shows some layouts of chain lines.

The maximum length of line which can be ranged is normally governed by visibility, in this country the limit being about 250 m. Also, too long a line may have excessive errors due to faulty alignment during chaining.

Booking the Survey. Booking is carried out by the surveyor in a field book, which consists of good quality paper, the pages each approximately 150 mm long by 100 mm wide, say. The field book is bound in the same way as a reporter's notebook. Each page is ruled up the centre with either a single coloured line or with two such lines about half an inch apart to represent the chain line, and booking starts at the bottom of the page. The salient points to note when starting a survey are—

1. After a preliminary reconnaissance of the site, bearing in mind the principles enumerated, make a sketch showing the location of the chosen stations and chain lines.

2. Take enough measurements, generally ties from nearby easily

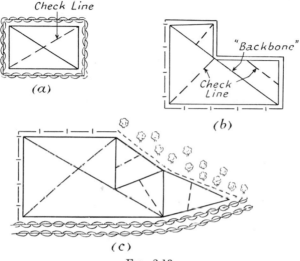

(a)

"Backbone"

Check Line

(b)

(c)

Fig. 2.13

recognizable features, and note enough information to enable each station to be relocated if necessary.

3. Take the bearing from true or magnetic north of at least one of the lines.

The principal features involved in booking the lines, offsets, etc., may be summarized as—

1. Begin each line at the bottom of a fresh page.

2. Take plenty of room and make no attempt to scale the bookings.

3. Exaggerate any small irregularities which are capable of being plotted.

4. Make clear sketches of all detail, inserting explanatory matter in writing where necessary: *do not rely on memory.*

5. Book systematically, proceeding up one side of the chain and then the other, starting with the side having more detail (and hence more offsets).

These and other features involved in booking are best dealt with by considering the following survey in Peel Park, Salford (Figs. 2.14 and 2.15). It will be noted that, to prevent confusion or the risk of mistakes, the decimal point has not been used, i.e. 5·52 m is written as 5/52.

Accuracy of Measurements. To a certain extent this depends on the scale to which the survey is to be plotted, and bearing in mind that the

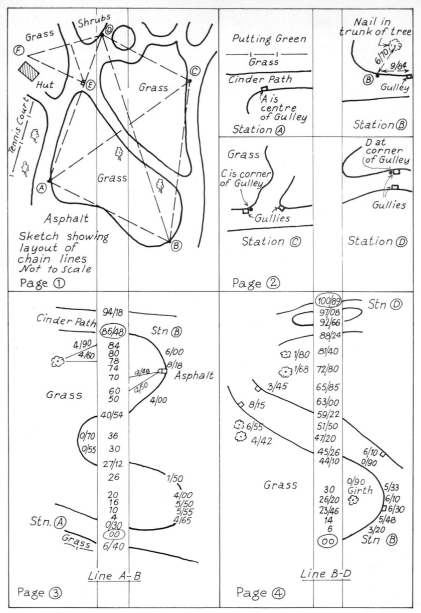

FIG. 2.14. PAGES FROM FIELD BOOK SHOWING SURVEY OF PEEL PARK
(*Continued on pp. 21 and 22*)

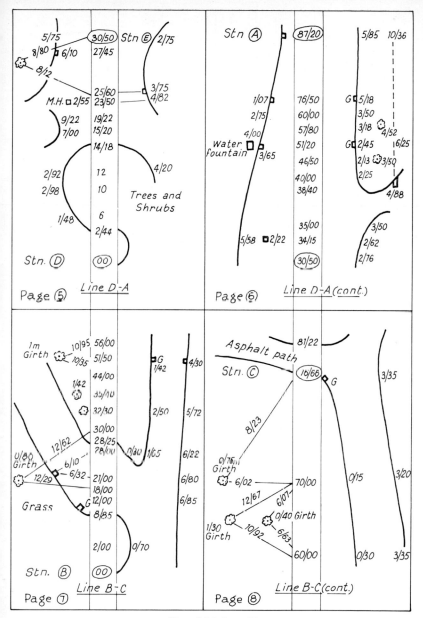

FIG. 2.14 (contd.)

Note: Whole-metre values can be written with or without zero decimal values, e.g. 21/00 or 21.

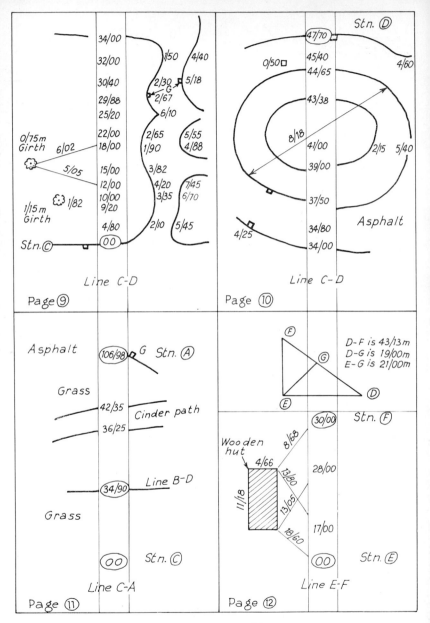

FIG. 2.14 (contd.)

scale might be increased, it is better to be more accurate than may appear to be strictly necessary. Normally we measure—

> chain lines to 1/10 of link, i.e. 20 mm
> offsets to nearest $\frac{1}{4}$ link, i.e. 50 mm.

A good draughtsman can plot a length to within 0·5 mm. Hence if the scale used is 1/500, i.e. 1 mm represents 500 mm on the ground, then

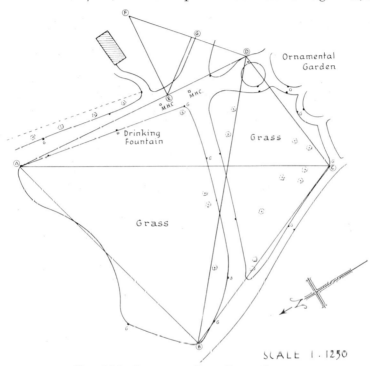

FIG. 2.15. SURVEY IN PEEL PARK, SALFORD

0·5 mm represents 0·25 m on the ground. Thus the order of accuracy given above is sufficient, but since parts of the survey may be required to be plotted to a larger scale later, it is suggested that measurements be made to 10 mm, i.e. 0·01 m. This is easily effected by means of any of the tapes previously described. One can readily estimate to 0·1 link on the chain but this is facilitated by laying down a 0·2 m metal scale alongside the relevant link. If, however, a building survey is made for plotting at 1/100 then 0·5 mm represents 0·05 m on the ground and measurements again should be taken to 10 mm.

The perpendicularity of short offsets is normally judged by eye or by swinging the tape, but with longer offsets one of the methods given earlier (e.g. optical square), or ties, should be used to prevent error in location. Where the offset would be longer than one tape-length, a subsidiary triangle should be employed (Fig. 2.16a). The use of subsidiary lines and triangles to improve accuracy should also not be forgotten when picking up detail where the offsets to the chain line would be at a very acute angle to the detail (Fig. 2.16b).

Office Work

Preliminary calculations and checking must first be carried out before plotting can take place, e.g. if any chain lines have been measured along regular slopes, the projection of these lengths on the horizontal

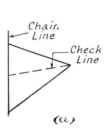

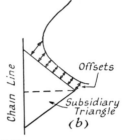

FIG. 2.16

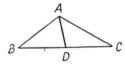

(a)

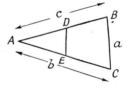

(b)

$$AD^2 = \frac{AB^2.DC - AC^2.BD}{BC} - BD.BC$$

$$\cos A = \frac{b^2 + c^2 - a^2}{-2bc}$$

Calculate DE from

$$DE^2 = AD^2 + AE^2 \\ - 2AD.AE \cos A$$

FIG. 2.17

must be calculated. Check line lengths can be calculated using the appropriate formulae and compared with the measured lengths, or they can be scaled off after the triangles have been plotted. Checking by calculation is preferable; the formulae are given in Fig. 2.17. Any errors revealed at this stage will require rectification by repeat measurements in the field, after which plotting may proceed.

Materials and Instruments. Original surveys should be plotted on good hand-made paper, or better, on paper mounted on holland cloth to reduce shrinkage.

4H pencils are used for drawing the framework, 2H pencils for plotting detail. Waterproof Indian ink is used for inking in the completed survey.

Additional requirements to the normal drawing instruments are—
Beam compasses for striking large-radius arcs.
A steel straight edge, 1 m or 2 m long.
A set of railway curves.
Set squares, protractor, and French curves.
Colour brushes and a set of stick water colours.

Scales for plotting may be wood-based with celluloid facings, 300 mm long and supplied with small offset scales. The common scales are dealt with later.

Scale of Surveys. This is usually decided beforehand and the choice depends on (*a*) the use to which the survey is to be put, (*b*) the professional practice in a particular office. Basic scales may range from 1/10 to 1/1,000,000, say. The former is appropriate for certain detail drawings and the latter for small-scale mapping. Between those limits it is possible to particularize to some extent, i.e.—

Architectural work, building work, location drawings: 1/50, 1/100, 1/200.

Site plans, civil engineering works: 1/500, 1/1,000, 1/1,250, 1/2,000, 1/2,500.

Town surveys, highway surveys: 1/2,000, 1/2,500, 1/5,000, 1/10,000, 1/20,000, 1/50,000.

Mapping: 1/50,000, 1/100,000, 1/200,000, 1/50,000, 1/1,000,000.

Plotting. It is often good practice to rough out the chain lines on tracing paper, so that by overlaying this on the paper to be used the survey may be properly centred on that sheet. It is preferable that the north be towards the top of the sheet, but by no means essential, since the North point must always be shown.

A line to represent the longest chain line is drawn and the length scaled off. By striking arcs the other stations are established and the network of triangles can be drawn. Check lines are scaled off and compared with the measured distances.

Offsets and ties are plotted systematically in the same order in which they were booked, i.e. working from beginning to end of each line up one side and then the other. The right angles for offsets may be set out separately with a set square, or more conveniently an offset scale may be used as shown in Fig. 2.18 in conjunction with a scale. The scale is set parallel to the line. Both chainage and offset are scaled off simply in one operation.

When the points are plotted the detail is drawn in, using the symbols which have by now become more or less standardized. Some of the symbols used vary according to the scale of the plan: e.g. a feature of width 400 mm on a 1/500 scale plan can be shown by two fine parallel lines about 0·8 mm apart; on any smaller scale than this, however, a single line will be used. A list of some typical conventional symbols is given in Fig. 2.19, and to supplement this reference may be made to a list supplied by the Ordnance Survey* for Large-scale Plans.

After the detail has been drawn in, the plan should be taken to the

* Ordnance Survey Large-scale Plans—H.M.S.O.

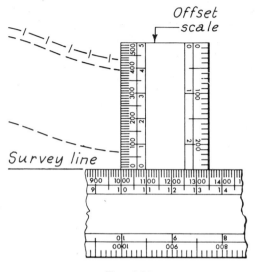

FIG. 2.18

site and checked. If nothing has been omitted, or there is no error, the detail is inked in, the North point is drawn, and any necessary lettering and titling carried out. It is usual nowadays for UNO stencils to be used, giving uniform and neat printing even in the largest letters, although it undoubtedly lacks the character of good hand lettering. Fig. 2.20 gives some of the range of letters and styles now available in stencils. Transfer lettering systems are also now commonly used.

A scale line should be drawn on the plan relating plan length to ground length; this is not only a convenience in scaling from the plan, but also gives some indication of any shrinkage which may have taken place.

Miscellaneous Problems

It occasionally happens that a survey has to be made of a field with a pond, standing crops, or a small wood in the middle, and the normal method of chain surveying as described is not directly applicable. A typical method of solving this problem is shown in Fig. 2.21a.

Each corner must be tied as shown, and it is better if the corner triangles are checked by suitable check lines shown dotted.

It is also possible to carry out surveys of long narrow defiles, in the manner shown in Fig. 2.21b, though great accuracy is not achieved because the network of main chain lines is not closed. In these cases an angle-measuring instrument would increase the accuracy.

Obstacles. Although in general, chain lines should not be broken by obstacles, it often happens that an unhindered line is not attainable. The types of obstacle which arise may be grouped under the following

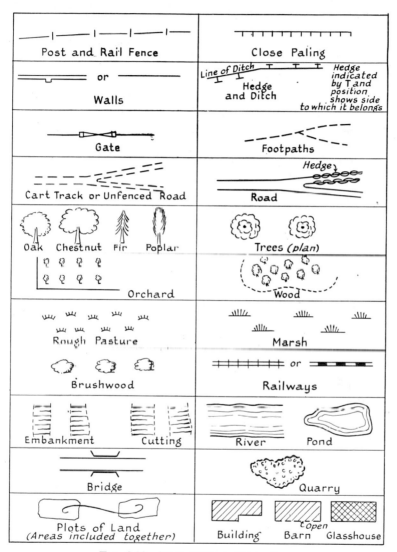

Post and Rail Fence	Close Paling
or Walls	Line of Ditch — Hedge and Ditch — Hedge indicated by T and position shows side to which it belongs
Gate	Footpaths
Cart Track or Unfenced Road	Hedge — Road
Oak Chestnut Fir Poplar	Trees (plan)
Orchard	Wood
Rough Pasture	Marsh
Brushwood	Railways
Embankment Cutting	River Pond
Bridge	Quarry
Plots of Land (Areas included together)	Building Barn (Open) Glasshouse

FIG. 2.19. SOME CONVENTIONAL SIGNS

Stencil No. U.C.2   Use Pen No. 1

ABCDEFGHIJKLMNOPQRSTUVW

Stencil No. U.L.2   Use Pen No. 1

abcdefghijklmnopqrstuvwxyz&

Stencil No. U.F.2   Use Pen No. 1

1234567890 No£/

Stencil No. U.C.3   Use Pen No. 2

ABCDEFGHIJKLMNOP

Stencil No. U.L.3   Use Pen No. 2

abcdefghijklmnopqrst

Stencil No. U.C.7   Use Pen No. 5

ABCDEF

Stencil No. U.L.7   Use Pen No. 5

abcdefghi

Stencil No. U.F.7   Use Pen No. 5

1234567

Stencil No. U.C.10   Use Pen No. 6

ABCD

Fig. 2.20

three headings: (*a*) those which obscure vision but do not prevent chaining, (*b*) those which prevent chaining but not vision, (*c*) those which prevent chaining *and* vision.

(*a*) The usual obstacle in this category is a small hill, and is dealt with by the method of *repeated alignment* (Fig. 2.22).

The surveyor and his assistant place themselves with poles at C_1 and D_1 so that each can see the other three poles in addition to his own.

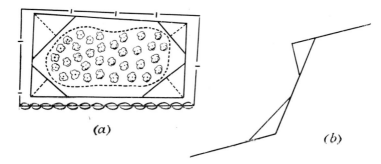

(*a*)

(*b*)

Fig. 2.21

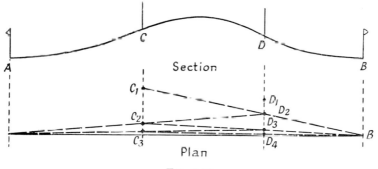

Fig. 2.22

Assuming the surveyor to be at C_1, he directs his assistant to position D_2 on the line C_1B. The assistant then ranges the surveyor to C_2 on the line D_2A, the procedure being repeated until the two poles C and D lie on the line AB. This method can be used in misty weather when poor visibility prevents ranging between the two stations.

Another method, used when one of the stations is in a depression with steep sides in the direction of the other station but with flat land to one side, is the method of *random line* (Fig. 2.23).

A random line AE is chained on the level land and a line BE is chained from B normal to AE. The length of AB can then be calculated by Pythagoras' theorem. To enable offsets to be taken from line AB,

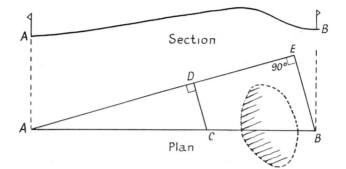

FIG. 2.23

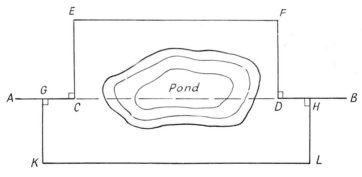

FIG. 2.24

a point C is established on this line by setting out DC perpendicular to AE such that (by similar triangles)

$$DC = EB \times \frac{AD}{AE}$$

AC can now be ranged or lined in.

(*b*) There are two types of problems in which chaining is prevented. With the first type, measurements can be made round the obstacle, which may be a pond or standing crops; with the second type, of which a river or stream of greater width than a chain length forms the obstacle, a geometrical construction is necessary.

The usual method of dealing with the former type of obstacle is illustrated in Fig. 2.24. Two equal offsets EC and FD are set out perpendicular to AB, using a tape to construct the right angles, and EF is chained to supply the missing length CD. As a check, GK and HL may be set out on the other side, if possible, and KL measured. Then

$$AB = AC + EF + DB$$
$$= AG + KL + HB$$

In the method for measuring across rivers shown in Fig. 2.25a a ranging rod is set at H on the far bank. CE is set off on the near bank perpendicular to AB, and a pole is ranged in to point F between E and

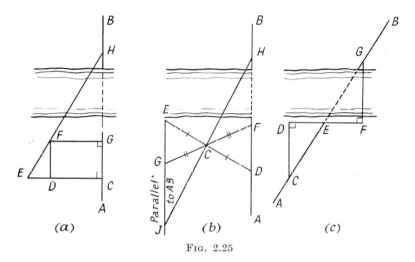

Fig. 2.25

H. The perpendicular is dropped from F on to AB, meeting at G. CE, FG, AC, CG, and HB are measured. Then from similar triangles,

$$\frac{HG}{FD} = \frac{FG}{ED}$$

∴
$$HG = FD \times \frac{FG}{ED}$$

$$= \frac{CG \times FG}{EC - FG}$$

The second method (Fig. 2.25b) gives the answer directly, and does not involve setting out any angles. A line DE is set out on the near bank and bisected at C. The line FCG is now set out such that $FC = CG$. With a pole H on the far bank and on the line AB, a pole can be set at J on the intersection of lines EG and HC produced backwards. Then $JG = FH$.

In the third method, where the line crosses the river on the skew (Fig. 2.25c), poles are placed on line AB at E and G on the near and far bank respectively. A line DF is set out along the bank, so that GF

is perpendicular to DF. A perpendicular from D is constructed to meet AB at C. Then

$$EG = \frac{CE \times EF}{ED}$$

(c) The errors involved in overcoming with chain and tape alone an obstacle which prevents both ranging and chaining are so great that a chain line passing through such an obstacle should be used only in the very last resort. Fig. 2.26 shows a way in which the problem may be dealt with. Perpendiculars of equal length are set off at C and D to

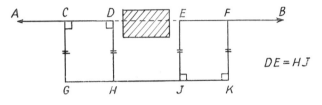

$$DE = HJ$$

FIG. 2.26

give G and H, and hence by ranging to give J and K. Further perpendiculars JE and KF are set out and EF is then produced and B fixed in a convenient position. It must be stressed, however, that without the aid of a theodolite, the chances of accurately prolonging the line past the obstacle are not high if B has been previously fixed.

It is recommended that the various methods of chaining and ranging past obstacles which are illustrated in Figs. 2.22–2.26 be practised as field exercises. By using imaginary obstacles in groups (b) and (c), the accuracy of the methods may be checked.

EXERCISES 2

1. Explain carefully the difference between mistakes and errors, giving as full a list as possible of such examples of each as occur in chain surveying.
   A line is measured with a chain of length 20·08 m, and an apparent length of 462·2 m is obtained. What is the corrected length of the line?
   Answer: 464·0 m.

2. (i) State the principles involved in choosing stations for a chain and tape survey.
   (ii) Make a sketch showing the layout of chain lines you might use in surveying a roughly L-shaped field, the legs of which are about 200 m × 100 m.
   (iii) Give an imaginary booking of one of the lines as it would appear in your field book, including measurements to a farm and some trees.

3. Derive expressions for the correction per chain length to be applied when chaining on a regular slope in terms of (a) the slope angle θ, and (b) the gradient expressed as 1 in n.
   What is the greatest slope you could ignore if the error from this source is not to exceed 1 in 1,500. Give the answer (a) as an angle, (b) as a gradient.
   Answer: (a) 2·11°; (b) 1 in 27·4.

4. Explain with the aid of sketches how you would carry out the following operations—
   (i) Pick up a boundary fence which curves rapidly towards a chain line and then away again.

(ii) Chain a line where a rise in the ground prevents you from seeing from one station to another.

(iii) Chain a line which crosses a river of about 40 m width.

(iv) Survey a field with a copse in the centre.

(v) Set out a right angle from a chain line.

5. Explain the points you would have in mind in deciding the layout of the

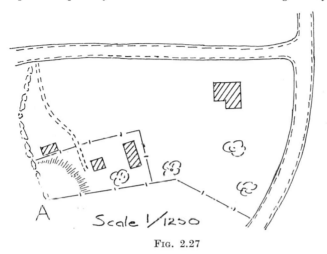

Fig. 2.27

chain lines for a chain survey and draw suitable lines on the plan attached (Fig. 2.27).

Draw up a field book approximately to size using dimensions roughly estimated from the accompanying plan and showing the booking along the last 60 m of one of the lines, near *A*, which is to include a building. (*L.U.*, *B.Sc.*)

CHAPTER 3

LEVELLING

LEVELLING is the operation required in the determination or, more strictly, the comparison, of heights of points on the surface of the earth. The qualification is necessary, since the height of one point can be given only relative to another point or plane. If a whole series of heights is given relative to a plane, this plane is called a *datum* and in topographical work the datum used is the mean level of the sea, since it makes international comparison of heights possible. In England, mean sea level was determined at Newlyn, Cornwall, from hourly observations of the sea level over a six year period from 1 May 1915.

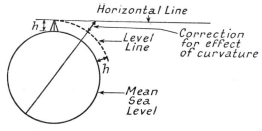

FIG. 3.1

This level is termed Ordnance Datum and is the one which will normally be used, though on small works an arbitrary datum may be chosen.

The basic equipment required in levelling is—

(*a*) a device which gives a truly horizontal line (the Level),

(*b*) a suitably graduated staff for reading vertical heights (the Levelling Staff),

(*c*) in addition, equipment is necessary to enable the points levelled to be located relative to each other on a map, plan or section; this might be, for example, chain and tape, tacheometer, or plane table, etc.

Before proceeding with the detailed description of the equipment and its use, however, some definitions are required.

A *level line* is one which is at a constant height relative to mean sea level, and since it follows the mean surface of the earth, it must be a curved line.

A *horizontal line*, however, is tangential to the level line at any particular point, since it is perpendicular to the direction of gravity at that point.

Over short distances the two lines are taken to coincide but over long distances a correction for their divergence becomes necessary.

34

Fig. 3.1 illustrates this point. In the figure, h represents the height of the instrument above mean sea level. For a distance of 100 m the correction is less than 1 mm in level.

The levelling device must be set up so that its longitudinal axis is at right angles to the direction of gravity (i.e. the line taken by a plumb bob), and the line of sight will then be horizontal, assuming the instrument to be in correct adjustment. Early levelling devices utilized the plumb bob. If a semicircular protractor, with a plumb bob attached to its centre, is held vertically, flat edge uppermost, then when the string of the plumb bob cuts the 90° graduation, the flat edge is horizontal and a sight can be taken along it. This, broadly, was the

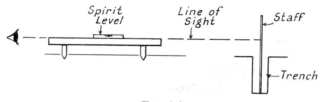

FIG. 3.2

principle of the earliest practical levels, dating to pre-Roman times, and as shown later, some modern self-levelling instruments employ a form of plumb bob as part of the self-levelling mechanism. (It is interesting to note in this context that another of the early levelling devices, the water level, was self-levelling. It consisted of a U–tube partly filled with water, and it was only necessary to sight along the two free water surfaces to obtain a horizontal sight.) Most levelling instruments do not use a plumb bob, this being replaced by a spirit level, a glass tube curved internally in longitudinal profile and partly filled with fluid, as described in the next section. The spirit level acts in effect as a very long plumb bob.

A simple type of level consists of a joiner's spirit level resting on top of a stout plank or straight-edge which is supported in such a way that the bubble of the level is central. A line of sight may then be taken on to some form of graduated staff or rule as shown in Fig. 3.2.

This method has its limitations, and is suitable only for very short lengths of sight, such as are met with on small building sites, paving work, etc.; for accurate work some better means of extending the horizontal line must be used, and in fact the extension is carried out optically by means of a telescope.

Elements of the Surveyor's Level

The general features of the conventional level are

(a) a telescope to give extended lines of sight in the horizontal plane, and

(b) a bubble tube to enable the telescope to be brought horizontal.

The Bubble Tube. This is a much more sensitive version of the builder's spirit level mentioned previously. The tubes vary in length between 50 mm and 125 mm, and are ground to a circular profile in longitudinal section. By increasing the radius to which the tube is ground, the "sensitivity" of the bublle is increased, since the distance through which the bubble is displaced by any specific tilt of the vertical

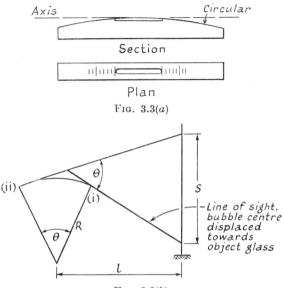

Fig. 3.3(a)

Fig. 3.3(b)

axis becomes greater. With too flat a curvature, however, the time taken for the bubble to come to rest becomes excessive.

The top surface of the tube is graduated symmetrically about its centre, as shown in Fig. 3.3(a), the two larger lines representing the positions of the ends of the bubble at normal temperature. The other graduations are necessary because in most instruments the length of the bubble varies with temperature, though manufacturers do make bubble tubes which have negligible temperature variation. For example the Hilger & Watts "constant" bubble had an approximately elliptical cross-section, and the volumes of spirit and vapour (i.e. bubble) were so designed, that the expansion of the spirit due to temperature rise was counterbalanced by the decrease of surface tension; the bubble thus maintained a constant length over a temperature range of 55°C, and it was necessary to read only one end of the bubble. Such a bubble was used in the Watts Self-adjusting Level, but apart from special instruments of this type, the growing use in modern levels of prismatic reading devices (which, as described later, enable both ends of the bubble to be read side by side) removed the need for constant length bubbles. A plane tangential to the zero or centre of the graduations contains the axis of the bubble tube (Fig. 3.3a). The bubble surface is always horizontal: thus a line tangential to a point on the internal surface of the tube, and mid-way between the ends of the bubble, will always be horizontal since it is parallel to that plane in which the ends lie. If the bubble tube is rotated in a vertical plane so that the bubble is in the centre of its run, then the axis of the bubble tube must be horizontal.

Determination of Sensitivity. To determine the sensitivity of the bubble on an instrument, sets of readings are taken on a staff held a convenient distance, l, say 30 m, from the instrument with the bubble centre displaced from the centre of its run as far as possible (i) to the object-glass end of the tube and (ii) to the eyepiece end using the instrument footscrews (Fig. 3.3b). Thus the difference in staff readings (s) is obtained and when it is divided by distance (l) the angle (θ) through which the line of sight has been rotated is determined. The bubble centre must have undergone the same movement which is deduced by noting the positions of both ends of the bubble with respect to the graduations on the tube.

If f_1 and f_2 be the readings of the object glass end of the bubble and r_1 and r_2 the readings of the rear end of the bubble respectively before and after rotation of the line of sight, the distance of the bubble centre from the centre of the graduations is $\dfrac{f_1 - r_1}{2}$ and $\dfrac{r_2 - f_2}{2}$ in each case.

Then the rotation of the line of sight

$$= \theta = \frac{s}{l} \text{ radians}$$

$$= \frac{s}{l} \frac{1}{\sin 1''} \text{ seconds}$$

$$= 206{,}265 \frac{s}{l} \text{ seconds}$$

But the bubble centre has moved a total distance of $\left(\dfrac{f_1 - r_1}{2} + \dfrac{r_2 - f_2}{2}\right)$

Then $206265 \dfrac{s}{l} - \left(\dfrac{f_1 - r_1}{2} + \dfrac{r_2 - f_2}{2}\right) = q$ say

Thus the angular value of one division

$$= 206{,}265 \frac{s}{lq} \text{ seconds}$$

If the angular value of one graduation, length z, of the tube be ϕ seconds and R be the radius of the internal curved surface, then

$$\phi = \frac{z}{206{,}265\,R} = 206{,}265 \frac{s}{lq} \text{ seconds}$$

EXAMPLE. During levelling it was noticed that the bubble had been displaced two divisions off centre when the length of sight was 100 m. If the angular value of one division of the bubble tube is 20 seconds, find the consequent error in the staff reading. What is the radius of the bubble tube if one graduation is 2 mm long?

Rotation of line of sight $= 2 \times 20$ sec

$$= 206{,}265 \frac{s}{l}$$

where s is the error in staff reading.

Therefore
$$s = \frac{100 \times 40}{206,265}$$
$$= 0\cdot019 \text{ m} = 19 \text{ mm}$$

Also
$$R = z/\phi \quad (\phi \text{ is expressed in radians})$$
$$= \frac{0\cdot002 \times 206,265}{20}$$
$$= 20\cdot627 \text{ m}$$

Optical Principles of the Surveying Telescope. These can be dealt with in the broadest outline only, since they comprise in themselves a complete branch of physics. It is hoped, however, to cover enough ground to enable the student to acquire an understanding of the

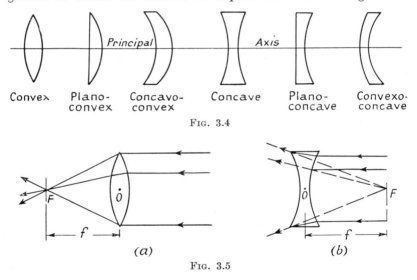

Convex   Plano-convex   Concavo-convex   Concave   Plano-concave   Convexo-concave

Fig. 3.4

(a)   *(b)*

Fig. 3.5

telescope's operation, and thus to appreciate the precise function of the telescope in a level, theodolite, or tacheometer.

LENSES. These are pieces of glass with the surfaces ground either plane or to a spherical form, the two surfaces having a common normal called the principal axis. The various types of lens are shown in Fig. 3.4.

If a beam of light composed of rays parallel to the principal axis passes through a convex lens (Fig. 3.5*a*), the rays will converge, passing through the *focal point F* (there being one on each side of the lens). The distance of this point from the optical centre of the lens *O* is known as the *focal length f*, and is a constant dependent only on the curvature of the surfaces and the type of glass used. The light actually passes through *F* so that a *real* image is formed here, i.e. one capable of being projected on to a screen.

Concave lenses are diverging lenses in that a similar beam of light would diverge as though it emanated from a focal point on the same side of the lens as the light source (Fig. 3.5*b*) and the image in this case is therefore *virtual*. In each case it will be noted that the ray of light passing through the optical centre O is not refracted or bent, and, in fact, all rays striking the centre, at any angle, pass through without refraction.

IMAGE FORMED BY A LENS. Light rays are reflected or emitted by any object which can be seen, and some of these rays, falling on a lens, are refracted to form an image whose position in relation to the lens depends on the distance of the object from the lens and on the type of

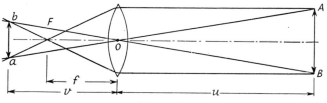

FIG. 3.6

lens. Consider the image of a distant vertical object formed by a convex lens, such as is used for the object lenses of telescopes.

One can establish the image size and position by considering particular rays from the extremities of the object, namely,

(a) rays parallel to the principal axis which pass through the focal point after refraction, and

(b) rays passing through the lens centre, and which are hence not refracted.

Note that with this lens (*see* Fig. 3.6), under these conditions the image is real, since the rays of light actually pass through it, and is inverted. The relationship between object distance u, image distance v, and focal length f, is given by

$$\frac{1}{u} + \frac{1}{v} - \frac{1}{f} \qquad . \qquad . \qquad (3.1)$$

The method of construction shown in Fig. 3.6 and equation (3.1) can be applied in the general case, i.e. for any type of lens and any position of the object, so long as in the equation the following sign convention is used—

(i) the focal length f of a convex lens is positive,

(ii) the focal length of a concave lens is negative,

(iii) the object distance u is positive if measured in the direction opposed to that of the light rays and negative if measured in the same direction as the light, starting from the lens in each case.

(iv) the image distance v is positive if measured in the direction of the light and negative if measured against it, starting from the lens.

Consider an object situated at a distance less than the focal length of a convex lens (Fig. 3.7).

Applying equation (3.1)

$$\frac{1}{u} + \frac{1}{v} = \frac{1}{f}$$

$$\therefore \qquad \frac{1}{v} = \frac{u - f}{uf}$$

since u is positive and f is positive (convex lens);

then if $\qquad u < f$

$\qquad\qquad v$ is negative

i.e. a magnified, erect, *virtual* image is formed at $a_1 b_1$ as shown.

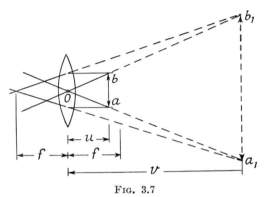

FIG. 3.7

The magnification m is shown, by considering the similar triangles Oba, $Ob_1 a_1$, to be

$$m = \frac{\text{size of image}}{\text{size of object}}$$

$$= \frac{v}{u} = \frac{\text{image distance}}{\text{object distance}}$$

Convex lenses are applied in this way in the eyepieces of simple telescopes to magnify the real image formed by the convex object lens.

DEFECTS IN LENSES. Before proceeding with the construction of a simple telescope a few words must be said about defects in lenses and their remedies, since a cursory inspection will show that, in modern surveying telescopes, single lenses are rarely used.

The two most important defects are (*a*) chromatic aberration, (*b*) spherical aberration.

So far as (*a*) is concerned, it will be recalled that when white light is refracted through a glass prism, it is split into its component colours, the red end of the spectrum being refracted less than the violet end. This phenomenon, known as *dispersion*, makes accurate focusing difficult, the image being surrounded by a rainbow-like boundary. The defect is remedied by using two lenses which are cemented together, one being a concave lens of flint glass and the other a convex lens of crown glass.

The larger lens in a surveying telescope, i.e. the object lens, is invariably so constructed. It forms its image, however, in the same way as a single convex lens.

Spherical aberration, as its name implies, arises from the use of spherical surfaces for the lenses, and again prevents accurate focusing due to rays incident on the edge of the lens being refracted more than rays incident on the centre. Such aberration may be reduced by "stopping down" the lens so that only the centre portion is used, but this also cuts down the amount of light entering the eye, so that the usual remedy is to use two lenses so arranged that the aberration of one eliminates that of the other. The pair may be a concave lens and a convex lens cemented together as objectives, or two plano-convex lenses a fixed distance apart as in eyepieces.

It is possible to obtain combinations which will eliminate both spherical and chromatic aberrations, though in practice a compromise has to be made. For example, the type of eyepiece used in English surveying telescopes is based on the Ramsden eyepiece, made up of two identical plano-convex lenses with their curved faces facing, and separated by a distance equal to two-thirds the focal length of either. Such an eyepiece is free from spherical aberration but not from chromatism.

The Simple Telescope. The Kepler type of telescope (Fig. 3.8) is the one used in surveying and this consists essentially of two convex lenses mounted so that their principal axes lie on the same line to form the optical axis of the instrument.

The object lens, i.e. the one nearest to the object, forms a real image, the rays from which pass on to the eyepiece, where they are refracted again and form a virtual image at some convenient distance in front of the eye.

To provide a positive and visible level line for use in levelling, cross-hairs are inserted in front of the eyepiece in a plane at right angles to the optical axis. The imaginary line passing through this intersection normal to the cross-hairs, and through the optical centre of the object glass, is called the *line of collimation* of the instrument, and all level readings are taken to this line.

PARALLAX. In focusing this simple telescope, the real image formed by the object lens is made to lie in the same plane as the cross-hairs. Since the eyepiece has previously been focused on the cross-hairs, the hairs and the image will be magnified by the same amount, and all readings will be made under the same conditions. If the image is not in the same plane as the cross-hairs, serious error in reading will ensue due to the phenomenon known as parallax. It is a matter of common observation, and can be readily confirmed by the student, that if, when viewing two distant objects which lie approximately along a straight line with the eye, the eye is moved to one side, then the more the distant object moves relative to the nearer one in the same direction as the eye.

This *parallax* is in fact a test to show that these two distant objects do not stand on the same spot. If no parallax occurs, then the objects lie in the same plane. Thus, if the eye is moved slightly when viewing through a telescope and no movement occurs between the image of the levelling staff and the cross-hairs, then the instrument is correctly focused.

The Surveying Telescope. Early instruments were of the external focusing type, i.e. they were of the same basic construction as the simple

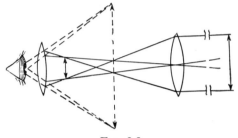

FIG. 3.8

one shown in Fig. 3.8. The Ramsden eyepiece and the compound object lens, both of which have been already described, were mounted in two tubes arranged so that one could slide inside the other, and focusing was achieved by moving one of the systems relative to the other.

This type of instrument is now superseded by the internal focusing telescope (Fig. 3.9) in which the eyepiece and object lens are mounted in a tube of fixed length, and a movable concave lens is inserted between them.

The concave lens is moved by means of a rack and pinion gearing, the pinion being connected by a spindle to the focusing screw, and the image focused on the cross-hairs as in the case of the external focusing instrument.

Although this extra lens absorbs some light, the disadvantage of this is more than offset in having a closed tube into which dust and moisture have no access. In addition, the internal focusing telescope is much more compact, while wear on the sliding surfaces is much less serious than in the external focusing type where it causes "droop" with consequent loss of alignment in the principal axes of the eyepiece and objective.

It may be noted here that with the use of "bloomed" lenses, the loss of light due to diffusion at the air-glass boundaries can be largely eliminated. This loss, which amounts to about 5 per cent per surface (10 per cent per lens) for untreated glass, is, for example, reduced to 1 per cent per surface in Zeiss telescopes due to the effect of treating each glass surface with the Zeiss T-coating. Thus modern instruments can be made much more complex, optically speaking, than formerly, without loss of image clarity.

THE DIAPHRAGM. In early instruments the diaphragm consisted of a brass ring across which were stretched two spiders' webs, but nowadays it is usually a thin glass plate on which the lines are engraved. Fig. 3.10 shows some of the arrangements of lines that are commonly met with.

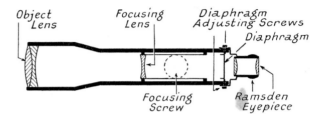

FIG. 3.9. INTERNAL FOCUSING TELESCOPE

FIG. 3.10. TYPES OF DIAPHRAGM

In addition to these, a type of diaphragm is sometimes used which employs small pointers of some non-corroding, generally platinum-based, alloy.

The diaphragm is held inside the telescope tube by four adjusting screws which enable (a) the cross-hairs to be adjusted so that the horizontal cross-hair is truly horizontal, (b) the line of collimation to be moved vertically and laterally.

EXAMPLE. Describe with the aid of a sketch the function of an internal focusing lens in a surveyor's telescope and state the advantages and disadvantages of internal focusing as compared with external focusing.

In a telescope the object glass of focal length 150 mm is located 200 mm from the diaphragm. The focusing lens is mid-way between these when a staff 25 m away is focused. Determine the focal length of the focusing lens.

(*L.U.*, *B.Sc.*, 1950)

Referring to Fig. 3.11, the focal length of the object lens is f, and the focal length of the internal lens is f_1.

Where the internal focusing lens not present, the image from the objective would be formed at P'.

$$\frac{1}{u} + \frac{1}{v'} = \frac{1}{f}$$

Therefore

$$v' = \frac{fu}{u - f}$$

Then, treating P' as a *virtual object* giving an image at P, we get

$$-\frac{1}{v'-l} + \frac{1}{v-l} = -\frac{1}{f_1}$$

$$\therefore \qquad f_1 = \frac{(v-l)(v'-l)}{v-v'} \qquad . \qquad . \quad (3.2)$$

$u = 25{,}000 \text{ mm}, \; v = 200 \text{ mm}, \; l = 100 \text{ mm}, \; f = 150 \text{ mm}$

$$\therefore \qquad v' = \frac{150 \times 25{,}000}{25{,}000 - 150} = 150{\cdot}9 \text{ mm}$$

Therefore $\qquad f_1 = \dfrac{(200 - 100)(150{\cdot}9 - 100)}{200 - 150{\cdot}9} = 103{\cdot}6 \text{ mm}$

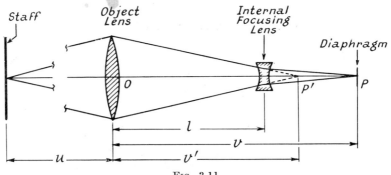

FIG. 3.11

Note the application of the sign convention. For a fuller treatment of the internal focusing lens, refer to Chapter 6, which deals with its use in tacheometry.

The Levelling Staff. There are several types of graduated staff available and only a few of the more important ones are described here.

1. TELESCOPIC STAFF. These may be made of mahogany or metal, aluminium alloy being typical in the latter case. B.S. 4484: Part 1:1969 requires lengths of either 3 m, 4 m or 5 m on extension, upon which the closed lengths naturally depend. It is possible that an extended length of 4·267 m will also be encountered since this was a typical equivalent Imperial dimension for the Sopwith staff (Fig. 3.12). Forms of graduation supplied by Messrs Holmes Bros. (Leyton) Ltd. are given in Fig. 3.13. B.S. 4484: Part 1 requires upright figuring with graduations 10 mm deep spaced at 10 mm intervals, the lower three graduations in each 100 mm interval being connected by a vertical band to form an E shape, natural or reversed. It will be seen that the 50 mm or 100 mm intervals are therefore located by these shapes. The graduations of the first metre length are coloured black on a white background, with the next metre length showing red graduations and so on alternately. Readings can be estimated to 1 mm over certain sighting distances, or

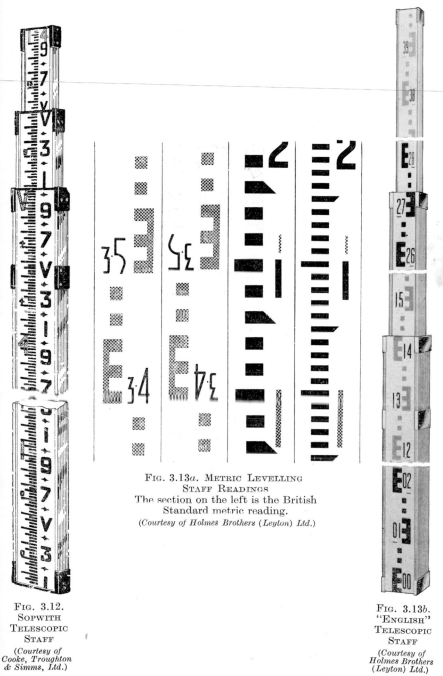

FIG. 3.12.
SOPWITH
TELESCOPIC
STAFF

(Courtesy of
Cooke, Troughton
& Simms, Ltd.)

FIG. 3.13a. METRIC LEVELLING
STAFF READINGS
The section on the left is the British
Standard metric reading.
(Courtesy of Holmes Brothers (Leyton) Ltd.)

FIG. 3.13b.
"ENGLISH"
TELESCOPIC
STAFF

(Courtesy of
Holmes Brothers
(Leyton) Ltd.)

recourse made to the parallel plate micrometer. To assist in holding the staff truly vertical a small circular spirit level and a pair of handles are sometimes incorporated, as shown in Fig. 3.14. Another form of rod which might be encountered is the Philadelphia Rod which is in two sections extending from 2·24 mm to 3·96 m. This has a target line which is adjusted to coincide with the line of sight of the level by the staff holder who then takes the reading with the aid of a vernier. This staff is not of great appeal to the civil engineer.

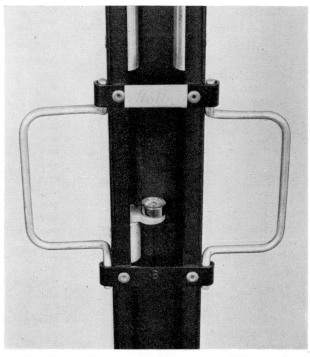

FIG. 3.14a. BUBBLE AND HANDLES OF "MUDLARK" LEVELLING STAFF
(Courtesy of Holmes Brothers (Leyton) Ltd.)

2. PRECISE LEVELLING STAFF. This staff is used for more accurate work and Fig. 3.15 shows a typical model supplied by Messrs. Hilger and Watts Ltd. An Invar steel strip carries graduations (1 mm thick) at 5 mm intervals, the left-hand scale having its zero point coincident with the underside of the steel base to the staff and being graduated from 1 to 30 (in decimetres). The right-hand scale is displaced with respect to the left-hand scale and is differently figured. The graduations are marked on an aluminium alloy body which supports the invar strip by means of a tongued groove; the strip is not constrained by the body. Another form is shown in Fig. 3.16.

FIG. 3.14b. LEVELS FOR LEVELLING STAVES
(*Upper*) Folding circular bubble. (*Lower*) Detachable level.
(*Courtesy of Holmes Brothers (Leyton) Ltd.*)

Parallel-plate Micrometer. This unit, which is normally fitted to precise
and geodetic levels, enables the interval between the horizontal cross-
hair and the nearest staff division to be read directly to 0·1 mm. The
device consists essentially of a parallel glass plate fitted in front of the
object lens and given a tilting motion by the rotation of a micrometer
head at the eye end of the telescope. Due to refraction, a ray of light
parallel to the telescope axis is displaced upwards or downwards,

FIG. 3.15. SP347 INVAR LEVELLING STAFF
3-metre, double-divided.
(Courtesy of Hilger & Watts, Ltd.)

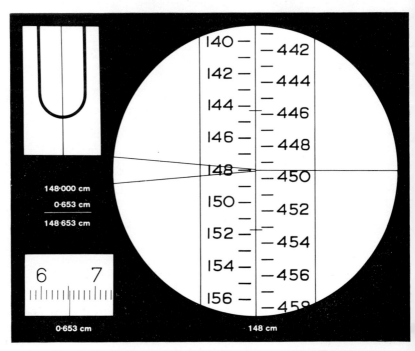

FIG. 3.16. VIEW OF A PRECISE STAFF THROUGH A GEODETIC
LEVEL TELESCOPE
(Courtesy of Wild Heerbrugg, Switzerland)

according to the direction of tilt, and by an amount varying with the angle of tilt. When the plate is vertical no displacement occurs. Fig. 3.17 shows how the readings are made, and Fig. 3.18 illustrates the theory.

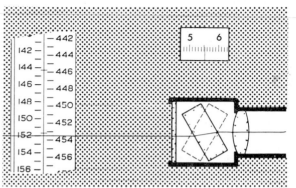

Fɪɢ. 3.17

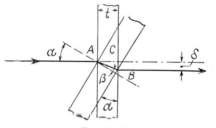

Fɪɢ. 3.18

Let μ be the refractive index for the glass used in the plate. Then in triangle ABC,

$$AB = \frac{t}{\cos \beta} \qquad BC = \delta$$

Therefore

$$\delta = \frac{t}{\cos \beta} \sin (\alpha - \beta)$$

$$= \frac{t}{\cos \beta} (\sin \alpha \cos \beta - \cos \alpha \sin \beta)$$

$$= t \left(\sin \alpha - \cos \alpha \, \frac{\sin \beta}{\cos \beta} \right)$$

$$= t \sin \alpha \left(1 - \frac{\cos \alpha}{\sin \alpha} \, \frac{\frac{1}{\mu} \sin \alpha}{\sqrt{1 - \frac{1}{\mu^2} \sin^2 \alpha}} \right)$$

Therefore
$$\delta = t \sin \alpha \left(1 - \frac{\cos \alpha}{\sin \alpha} \frac{1/\mu \sin \alpha}{\sqrt{1 - 1/\mu^2 \sin^2 \alpha}} \right)$$

$$= i \sin \alpha \left(1 - \frac{\cos \alpha}{\sqrt{\mu^2 - \sin^2 \alpha}} \right)$$

$$= t \sin \alpha \left(1 - \frac{\sqrt{1 - \sin^2 \alpha})}{\sqrt{\mu^2 - \sin^2 \alpha}} \right)$$

if α be small (in radians).

Sin $\alpha \simeq \alpha$ and $\sin^2 \alpha$ can be ignored

Therefore
$$\delta = t\alpha \left(\frac{\mu - 1}{\mu} \right) \quad . \qquad . \qquad . \qquad . \qquad . \qquad . \qquad (3.3)$$

EXAMPLE. A parallel-plate micrometer attached to a level is to show a displacement of 5 mm when rotated through 30° on either side of the vertical. Calculate the thickness of the glass when the refractive index is 1·6. State also the staff reading when the micrometer reads 2·8 in sighting the next lower reading of 1·48, the divisions running 0 to 10 with 5 for the normal position.

$$\delta = 5 \text{ mm}$$

$$= t \sin 30 \left(1 - \frac{\sqrt{1 - \sin^2 30}}{\sqrt{1\cdot6^2 - \sin^2 30}} \right)$$

$$= 23\cdot24 \text{ mm}$$

Each division corresponds to 0·001 m and therefore the reading is 1·4778 m on a staff seen inverted.

The Dumpy Level. Fig. 3.19 shows a Dumpy Level commonly in use. The telescope barrel and vertical spindle are cast in one piece,

FIG. 3.19

and the telescope is of the internal focusing type. The bubble tube is mounted on the left-hand side of the telescope, and a small circular bubble mounted on the upper plate of the levelling head enables a preliminary levelling up of the instrument to be made.

The levelling head consists essentially of two plates, the telescope being mounted on the upper plate while the lower plate screws directly on to a tripod. The two plates are held apart by three levelling screws or foot screws, and adjustments to these enable accurate levelling of the instrument to be carried out.

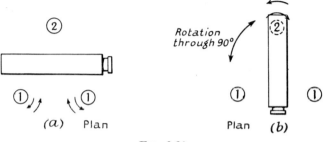

FIG. 3.20

By means of a clamping screw, the telescope may be fixed along any particular line of sight, and slight lateral deviations may be made to left or right of this line with the tangent screw.

An inclined mirror enables the bubble to be viewed from the eyepiece end of the telescope, and this when folded down protects the bubble.

SETTING UP. The following sequence of operations is required to bring the instrument ready for use—

1. Screw the lower plate of the level on the head of the tripod, whose legs have been opened and firmly fixed on the ground. The circular bubble should be nearly central; any excessive tilt on the instrument at this stage may lead to the levelling or foot screw threads running out before accurate levelling is achieved.

2. The telescope is now turned so that it is parallel to screws 1—1, which are turned so that one moves clockwise and the other anti-clockwise simultaneously (i.e. both thumbs of the surveyor move towards each other or away from each other). The bubble is thus brought to the centre of its run (Fig. 3.20).

After turning the telescope through 90° (Fig. 3.20) the bubble is again brought to the centre of its run by the third screw. If the left hand is used to operate the foot screw 2 then, in each case, the bubble will move in the direction of the surveyor's left thumb.

It is usual to repeat the operations represented by Fig. 3.20a and b, whereupon the bubble should remain central in whatever direction the telescope is pointed if the instrument is in adjustment, and the bubble is now said to traverse.

3. It is essential that parallax between the cross-hairs and the image

of the levelling staff be eliminated, for reasons already explained. There is no doubt that failure to do so is responsible for much error in levelling. To eliminate parallax: (i) Turn the telescope to the sky and focus the eyepiece so that the cross-hairs appear clear and distinct. This is usually achieved by turning the eyepiece, which is threaded into the telescope barrel. It must be realized that the eyepiece setting depends on the characteristics of the surveyor's eye, so that it will vary from one person to another. For one given operator, the setting will not vary. (ii) Now sight the levelling staff and focus its image with the focusing screw so that when the eye is moved slightly there is no relative movement between the image and the cross-hairs.

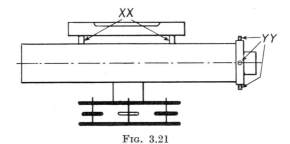

FIG. 3.21

PERMANENT ADJUSTMENTS TO THE DUMPY LEVEL. There are two conditions to be satisfied in order that a Dumpy Level shall be in perfect adjustment—

1. The bubble tube axis must be perpendicular to the vertical axis; the bubble must "traverse."
2. The line of collimation must be parallel to the bubble tube axis.

Of these two conditions, the second one must be fulfilled if accurate levelling is to be carried out. The first condition is a convenience, not a necessity, for if the second one is complied with, it is only necessary for the bubble to be brought central before taking each reading. The tests, and adjustments, are given below.

1. Bubble tube must be perpendicular to vertical axis (Fig. 3.21).
Test. Set up and level as accurately as possible, as described in the previous section. With the bubble central, turn the telescope through 180° from the position in Fig. 3.20(a), whereupon the bubble should remain central.

Adjustment. If the bubble moves out of centre, take out half the error with the foot screws and half with the bubble capstan screws XX.

2. Line of collimation must be parallel to bubble axis. There are two variations.

Test (a). Select two points A and B about 60 m apart on firm, fairly level ground and drive in pegs. With the instrument levelled in position (1) being as near A as possible (Fig. 3.22), read the staff at B

(reading Bb). Sighting back through the object glass, read staff A (reading Aa).

Without altering the focus, move the level to position (2) and repeat the process, obtaining readings Ad and Bc.

Then—

Position (1) (apparent difference in level)

$$= Bb - Aa \quad (A \text{ higher than } B)$$
$$= Be + eb - Aa$$
$$= (Be - Aa) + eb$$
$$= \text{True diff.} + eb \qquad . \qquad . \qquad . \quad (3.4)$$

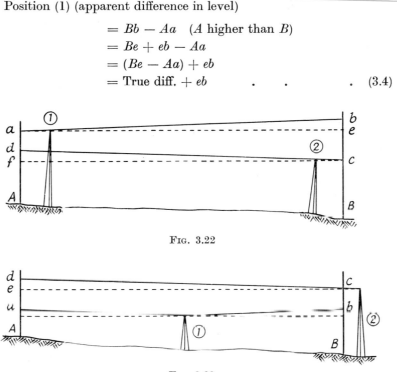

FIG. 3.22

FIG. 3.23

Position (2) (apparent difference in level)

$$= Bc - Ad$$
$$= (Bc - Af) - fd$$
$$= \text{True diff.} - fd \qquad . \qquad . \qquad . \quad (3.5)$$

but $\qquad eb = fd,$

Therefore, adding (3.4) and (3.5), we get

True difference in level between A and B

$$= D = \tfrac{1}{2} \text{ (sum of apparent differences)}$$

Adjustment. If the two apparent differences are equal, then the instrument is in adjustment. If not, with the level still in position (2), adjust the cross-hairs by means of the diaphragm adjusting screws YY until the reading Af is given, where

$$Af = Bc - D \quad (A \text{ higher than } B)$$

or $\qquad Af = Bc + D \quad (A \text{ lower than } B).$

Test (b). The Two Peg Test. With two pegs A and B as before, set up and level in position (1), exactly mid-way between the pegs, and take readings on the staff at A and B (Fig. 3.23).

Then, *true* difference in level

$$= D$$
$$= Bb - Aa \quad (A \text{ higher than } B) \text{ or}$$
$$= Aa - Bb \quad (A \text{ lower than } B)$$

since the same error occurs in each sight.

Set the instrument at (2) close to peg B, and read the staff at A and B.

If, say, A is higher than B, compute apparent difference in level which equals $(Bc - Ad)$.

Adjustment.

If this is not equal to D, then deduce $Ae = Bc - D$ and adjust cross-hairs as in test 2(a).

EXAMPLE. A Dumpy Level is set up with the eyepiece vertically over a peg A. The height from the top of A to the centre of the eyepiece is measured and found to be 1·516 m. A level staff is then held on a distant peg B and read. This reading is 0·696 m. The level is then set over B. The height of the eyepiece above B is 1·466 m and a reading on A is 2·162 m.

(1) What is the difference in level between A and B?

(2) Is the collimation of the telescope in adjustment?

(3) If out of adjustment, can the collimation be corrected without moving the level from its position at B? (*I.C.E.*)

(1) Level at A. Apparent difference in level
$$= 1\cdot516 - 0\cdot696$$
$$= 0\cdot820 \text{ m} \quad (Note: B \text{ is higher than } A)$$

Level at B. Apparent difference in level
$$= 2\cdot162 - 1\cdot466$$
$$= 0\cdot696 \text{ m}$$

$\therefore$ True difference in level
$$= \frac{0\cdot820 + 0\cdot696}{2}$$
$$= 0\cdot758 \text{ m}$$

(2) The collimation of the telescope is not in adjustment.

(3) With the level at B, the height of the eyepiece (1·466) is correct. Therefore reading of staff at A should be given by
$$X - 1\cdot466 = 0\cdot758$$
$$X = 2\cdot224 \text{ m}$$

Make the reading equal 2·224 by adjusting the cross-hairs.

EXAMPLE. A series of backsights and foresights have been taken with a Dumpy Level up a slope 500 m long to establish a temporary bench mark, the backsights being twice as long as the foresights in each case.

To check the instrument afterwards it is set up over two pegs, A and B, in turn, and the following readings taken.

Level at A—Height of eyepiece over A = 1·673 m
Reading on staff at B = 1·555 m
Level at B—Height of eyepiece over B = 1·575 m
Reading on staff at A = 1·759 m
Distance AB is 66·00 m

Determine the true R.L. of the T.B.M., supposed to be 120·960 m above datum.

Peg B is higher than peg A.

Level at A. Apparent difference in level = 1·673 − 1·555 = 0·118 m.

Level at B. Apparent difference in level = 1·759 − 1·575 = 0·184 m.

Therefore true difference in level

$$= \frac{0·118 + 0·184}{2}$$

$$= 0·151 \text{ m}$$

Thus, when the level is at B, the true staff reading on A should be 1·575 + 0·151 = 1·726 m,

∴   Level reads

$$1·759 − 1·726 = 0·033 \text{ m high in 66 m}$$

$$= 0·033 \times \frac{x}{66} \text{ m high in } x \text{ m}$$

Hence, if backsight is $2x$ m long, reading is $0·001x$ m high, and if foresight is x m long, reading is $0·0005x$ m high.

∴   Apparent rise in $3x$ m is $0·001x − 0·0005x$ m too great, i.e. error in $3x$ m is $0·0005x$ m.

∴ error in 500 m is

$$0·0005x \times \frac{500}{3x} = \frac{0·25 \text{ m}}{3} = 0·083 \text{ m}$$

Therefore true R.L. of T.B.M.

$$= 120·960 − 0·083$$

$$= 120·877 \text{ m above datum.}$$

The Tilting Level. In this level the telescope is not rigidly fixed to the vertical spindle, but is capable of a slight tilt in the vertical plane about an axis placed immediately below the telescope. This movement is governed by a fine setting screw at the eyepiece end, and the bubble is brought to the centre of its run for each reading of the level. Thus the line of collimation need not be perpendicular to the vertical axis.

Tilting levels are robust and capable of the highest accuracy, and most modern levels, whether for precise work or for ordinary levelling, incorporate the principle. Figs. 3.24 to 3.26 show some current models supplied by Hilger & Watts, Ltd., and Vickers Instruments, Ltd.

3

There are many refinements of design possible. For example, the level may be mounted on the conventional three-screw base, already described, or, where quick setting-up is important, on a ball-and-socket base where the levels are often described as "quickset" levels. Both types of base incorporate a small circular bubble to enable a preliminary levelling up to be made.

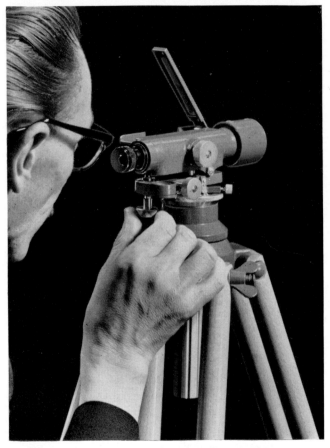

FIG. 3.24. WATTS QUICKSET LEVEL SL10
(Courtesy of Hilger & Watts, Ltd.)

PRISMATIC COINCIDENCE BUBBLE READER. Another refinement, designed to eliminate excessive movement around the instrument, is the coincidence bubble reader, and the device is now fitted to many, if not most, modern instruments. An arrangement of prisms (Fig. 3.27a illustrates the arrangement used by Wild) reflects an image of both ends of the bubble into the eyepiece. The image is split down the centre,

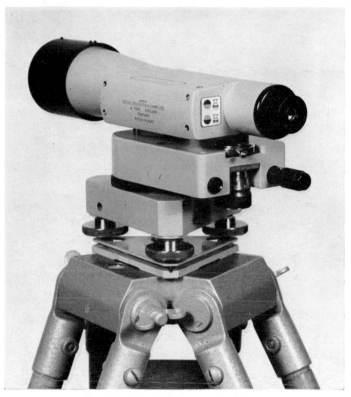

FIG. 3.25. S440 ENGINEER'S LEVEL
(Courtesy of Vickers Instruments, Ltd.)

and one half of each end is seen next to the other half in the bubble reader eyepiece, as shown in Fig. 3.27*b*.

As the telescope is tilted, the two halves appear to move in opposite directions. This magnifies the actual movement twofold, and in the Hilger & Watts model, for example, this is further magnified by the optical system, which makes very exact setting possible. Consequently, bubbles of lower sensitivity than would otherwise be necessary can be employed, and this greatly improves stability and ease of setting.

PERMANENT ADJUSTMENT OF THE TILTING LEVEL. There is only one adjustment to this type of instrument, namely, that of rendering the bubble tube axis parallel to the telescope axis, and as a consequence there is no need to touch the diaphragm. Instead, the bubble tube is brought parallel to the axis of collimation. A suitable test is the two-peg test previously described. The only deviation from the procedure for the Dumpy Level arises in the adjustment. Having deduced the correct reading on staff *A* (level near *B*), make the instrument read this

FIG. 3.26. WATTS SL80 LEVEL
(*Courtesy of Hilger & Watts, Ltd.*)

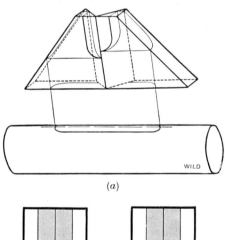

(*a*)

(*b*)

FIG. 3.27

by tilting with the tilting screw. Now make the bubble central by means of the capstan screws on the bubble.

Precise and Geodetic Levels. Many of these employ the tilting principle described in the previous section, and most of the modern types incorporate the coincidence bubble-reading principle. In addition, the telescope and main spirit level can, in some instruments, be rotated through 180° about the optical axis, thus enabling the mean of two such readings to give the true level. This method of using a reversion

Fig. 3.28. Wild NK2 Engineer's Level
(*Courtesy of Wild Heerbrugg, Switzerland*)

bubble appears to have been first suggested in 1859 by Amsler, and introduced in 1910 by Heinrich Wild in the famous Zeiss Level. It is essential that the two bubble axes (corresponding to the graduations on the two sides) are strictly parallel, and this parallelism is guaranteed by the makers to be correct to within two seconds of arc.

Fig. 3.28 shows a reversion instrument, whilst Fig. 3.29 illustrates the principle of reversion: the telescope is released for rotation by pressing a small "button" lying underneath.

If a reversion instrument be set up and levelled, a reading taken on a staff, and the telescope then rotated through 180°, relevelled by the tilting screw and a second reading taken, any error due to lack of parallelism between the collimation axis and the bubble axis will cause one reading to be too high and the other to be too low by exactly equal amounts (A and B in Fig. 3.29). The mean of the two readings will thus enable the precise level to be ascertained for a particular staff position.

Though the above results are free from instrumental errors, the instrument is readily adjusted to remove errors in the collimation adjustment by tilting the telescope until the reading corresponds to

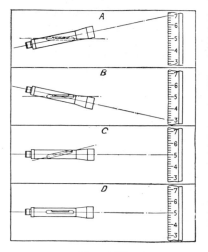

FIG. 3.29

the mean of two readings taken as above (*C* in Fig. 3.29), and then bringing the bubble central by means of the capstan screws (*D* in Fig. 3.29).

Typical modern precise levels include the Zeiss level and the Fennel High Precision level. Both the Cooke S77 Self Aligning level and the Watts Precise Autoset level 2 can be fitted with parallel-plate micrometers to become precise levels.

Fig. 3.30 shows the Wild N3 Precision level. Its three foot-screws rest on a triangular base plate which is attached to the tripod, and a spherical level is available for initial setting up. After pointing to the staff the tubular level is observed by viewing from the eyepiece end and the two half images are brought into coincidence using the small tilting screw located under the eyepiece. A clamp and tangent screw allows motion about the vertical axis of the instrument and the vertical graticule line can be set on the centre of the staff using these devices, after approximate alignment by means of the open sights on top of the instrument. The parallel plate forming the optical micrometer gives a maximum deflection of 10 mm (one gradua-tion interval on the Wild Invar staff). Its movement is registered by a scale which again can be observed at the eyepiece end of the telescope below the observation point for the tubular level. The scale readings range from 0 to 10 with 5 as the vertical position for the plate, the scale figures represent millimetres and the smaller intervals represent tenths allowing estimation to hundredths of millimetres. The movement of the parallel plate is effected by means of a knob situated on one side of the level fairly near to the telescope focusing knob.

The graticule contains two short stadia lines apart from the vertical line, and the line of sight is defined partly by a horizontal line and partly by a pair of wedge lines which can be set across a staff graduation by rotating the parallel plate (Fig. 3.16). The Wild staff contains two sets of graduations at 10 mm intervals; these can appear either erect or inverted in the field of view depending upon the type. The higher-numbered row acts as a check against errors of observation during geodetic levelling since it is displaced with respect to the other.

A typical sighting drill when two staves are available is as follows—

(*a*) Read left graduation of rear staff,

FIG. 3.30. WILD N3 PRECISION LEVEL
(Courtesy of Wild Heerbrugg, Switzerland)

(*b*) read left graduation of front staff,
(*c*) read right graduation of front staff,
(*d*) read right graduation of rear staff.

In all cases the micrometer drum reading could be added to obtain the relevant booking. This can be illustrated by reference to the following cases, using for simplicity an ordinary staff graduated to 10 mm. Let the line of sight fall between 1·03 and 1·04 when the parallel plate is vertical and the relevant scale reading is therefore 5. If the horizontal line of the graticule be set on the 1·03 reading the micrometer reading would be say 8·25: this indicates a movement of 3·25 mm upwards and the corresponding staff reading could be booked as 1·03325 m.

Let the forward staff reading then lie between 1·47 and 1·48 when the plate is vertical, and assume the micrometer gives a reading of 2·80 when the horizontal line is positioned on the 1·48 graduation. The line of sight has now been deflected downwards through 2·20 mm and the staff reading could be deduced as 1·47780 m, with a staff station level difference of 0·44455 m. If the readings were not established by considering the actual deflections, but the micrometer readings were simply added, we should book 1·03825 m and 1·48280 m respectively, resulting in the same level difference.

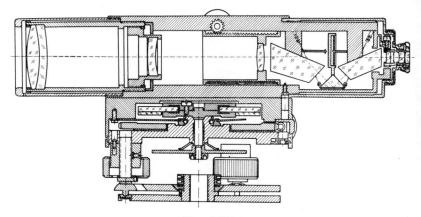

Fig. 3.31(a)

In adjusting the tubular level of this instrument a two-peg test is carried out, staff stations A and B being say 25 metres apart. After obtaining staff readings, say a and b, the level is set up on the line BA produced and distant $AB/10$ from A. Readings a_1 and b_1 are now obtained and the corrected staff reading (b_2) at B deduced, if the two differences in level do not agree, as—

$$b_2 = \frac{1}{10} \left[11(b - a) + 11\,a_1 - b_1 \right]$$

Provided that b_2 is close to b_1 the micrometer can be adjusted to give the millimetres reading by means of a special screw, and the protecting glass cover at the objective end is rotated until the relevant main staff graduation is given by the graticule lines.

The Zeiss Ni2 Self-levelling Instrument. This instrument, which is a post-war production, has the appearance of a conventional level but possesses a number of novel features. A sectional view (Fig. 3.31a) reveals some of these, which may be summarized as follows—

1. Absence of bubble tube, preliminary levelling being carried out using the conventional three-screw levelling head and a small target bubble mounted on the tribrach; this brings the line of collimation to within 10′ of the horizontal.

2. Further levelling up is unneccesary, correction for any slight tilt of the collimation axis being automatically made by a prismatic compensator fitted between the diaphragm and the focusing lens. The optical components of the compensator consist of three prisms, of which the centre prism (with its air-damped cushioning device) is suspended by two wires from the top of the telescope barrel and is thus free to swing. The two outer prisms are fixed to the barrel, and the one nearest the eyepiece is also a roof prism.

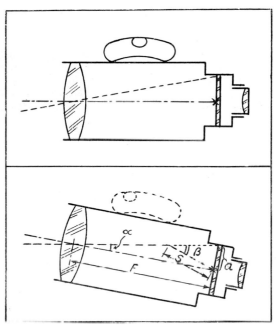

Fig 3.31(*b*)

3. Because of the three reflections in height and the lateral inversion in the roof prism, a correct, *upright image* of the staff is obtained.

4. Focusing is by a moving convex lens, the telescope containing also a fixed concave lens. With this arrangement of lenses, sufficient space is obtained between the focusing lens and the reticule to accommodate the compensator, also the instrument is almost completely anallatic (*see* Chapter 7). It can be shown* for an instrument which corrects for tilt as does the Ni2, that tilt occurs about the anallatic point of the telescope.

5. The focusing screw enables quick preliminary focusing to be made, one complete turn of the screw covering the complete focusing range of

* "Novel Surveyors Levels" by Dr. I. Drodofsky, *Zeitschrift für Vermessungs-wesen*, No. 8, 1951.

the telescope. A friction-brake operated gearing then operates automatically allowing a quarter turn of the focusing screw at a much slower (1:5) gearing, so that very fine parallax-free focusing can be made.

6. A rotatable flat cog-wheel mounted on the tribrach bears the upper part of the instrument, and the tangent screw worm gear, mounted to the upper part, mates into this cog. The friction between the cog bearing and the tribrach is made greater than that between the upper-part bearing and the cog, so that when the tangent screw is turned, the upper part will rotate relative to the cog wheel, thus giving an endless tangential slow-motion device without any clamping. For rough hand setting, of course, the telescope is turned in the usual manner and the worm gear pulls the cog wheel round with it.

OPERATION OF THE COMPENSATOR. Fig. 3.31*b* shows schematically the operation of the instrument. Although the collimation axis is inclined at a small angle to the horizontal, if the horizontal rays of light passing through the optical centre of the objective can be deflected so as to strike the horizontal cross-hair of the diaphragm, as shown, then the correct staff reading will be given despite tilt. From the figure

$$a = F\alpha = s\beta$$

$$\therefore \quad \frac{F}{s} = \frac{\beta}{\alpha}$$

When the telescope barrel tilts through α, the centre prism of the compensator takes up a new position in which the vertical line through its centre of gravity passes through the intersection point of the two supporting wires. The geometry of the quadrilateral formed by the prism base, the two wires and the top of the telescope is such that the prism base is now inclined at 3α to the horizontal. By the laws of reflection (the effects of other reflections and refractions being compensating)

$$\beta = 6\alpha \text{ and } s = \frac{F}{6}$$

The swinging prism weighs only 5 gm so that inertia effects are absent, and the motion is frictionless and free from wear. The damping device prevents prolonged oscillation and enables readings to be made within one-half a second after disturbances.

ACCURACY AND USES OF THE Ni2. The accuracy of the Ni2 is of the same order as that of a precise level and since it may be used with a parallel-plate micrometer, it follows that precise levelling may be carried out with it. The main attribute of the instrument, however, is that it enables accurate levelling to be carried out quickly; linear levelling using one level and two staffmen is approximately twice as rapid with the Ni2 compared with a conventional bubble level. The instrument will be economical therefore for the following operations—

1. Levelling in mines and tunnels; check levels on gantry crane stanchions; similar cases where only brief periods are available.

2. Checking of bench-marks; levelling for irrigation or railway projects; subsidence investigations; similar cases involving large amounts of accurate levelling.

Watts Autoset Level 2. Cooke S 700 Level. These British levels possess a compensator which consists of two swinging prisms and one fixed prism. The action of this device, which may also be seen in Fig. 14.1, is indicated in Fig. 3.32 after the level has been slightly displaced. The compensator acts as a pendulum, ensuring that the

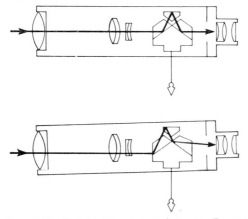

Fig. 3.32. Principle of the Autoset Level

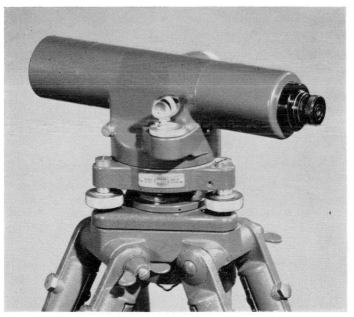

Fig. 3.33. Precise Autoset Level 2
(Courtesy of Hilger & Watts, Ltd.)

horizontal line is defined by the diaphragm. An air damping piston ensures that the oscillations of the suspended prisms are damped, restricting the total swing to ± 20 min. The manufacturers claim that the compensating device is unaffected by the sun's rays, whereas conventional precise levels are normally shielded from direct rays during use; a parallel-plate micrometer can be attached to these levels to enable them to act as precise levels.

Fig. 3.33 shows the Autoset Level which gives a magnification of $\times 32$ and produces an erect image; the shortest focusing distance is 1·8 m.

Procedure in Levelling

The basic operation is the determination of the difference in level between two points. Consider two points A and B as shown in Fig. 3.34. Set up the level, assumed to be in perfect adjustment, so that readings may be made on a staff held vertically on A and B in turn.

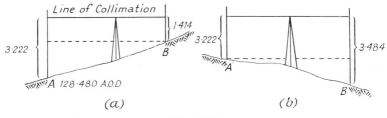

FIG. 3.34

If the readings on A and B are 3·222 m and 1·414 m respectively (Fig. 3.34a), then the difference in level between A and B is equal to AC, i.e. $3·222 - 1·414 = 1·808$ m, and this represents a *rise* in the height of the land at B relative to A. If the reading at B is greater than that at A (Fig. 3.34b), say 3·484 m, then the difference in level would be $3·484 - 3·222 = 0·262$ m, and this would represent a *fall* in the height of the land at B relative to A. Thus we have that in any two successive staff readings—

2nd reading less than 1st represents a *rise*,
2nd reading greater than 1st represents a *fall*.

If the actual level of one of the two points is known, the level of the other may be found by either adding the rise or subtracting the fall, e.g. if the level at A is 128·480 m above Ordnance datum (A.O.D.), then

(a) Level at B = Level at A + Rise
                 = $128·480 + 1·808$
                 = $130·288$ m above datum

(b) Level at B = Level at A − Fall
                 = $128·480 - 0·262$
                 = $128·218$ m above datum.

The levels at A and B are known as *reduced levels* (R.L.) as they give the level of the land at these points "reduced" or referred to a datum level (in this case, Ordnance datum, which is the mean sea height at Newlyn), and this method of reducing the staff readings gives a system of booking known as the *Rise and Fall Method*.

A second method, known as the *Height of Collimation Method*, also exists, and since the two methods are in common use they must both be known. In the second method, the height of the line of collimation above the datum is found by adding the staff reading obtained with the staff on a point of known level to the R.L. of that point. Thus in Fig. 3.34 the height of collimation is $128 \cdot 480 + 3 \cdot 222 = 131 \cdot 702$ m A.O.D., and this will remain constant until the level is moved to another position. The levels of points such as B are determined by deducting the staff reading at these points from the height of collimation—

(a) Level at B = Height of Coll. — Reading at B

$\qquad\qquad\qquad = 131 \cdot 702 - 1 \cdot 414$

$\qquad\qquad\qquad = 130 \cdot 288$ m A.O.D.

(b) Level at B = Height of Coll. — Reading at B

$\qquad\qquad\qquad = 131 \cdot 702 - 3 \cdot 484$

$\qquad\qquad\qquad = 128 \cdot 218$ A.O.D.

General Procedure. This is best dealt with by means of an example, and we will consider the line of levels down the centre line of the road as shown in the plan of Fig. 3.35. The object of such a line of levels is the production of a *longitudinal section* and this is shown schematically in Fig. 3.35 with the level readings marked thereon.

The instrument is set up at a convenient position P such that a *bench mark* (B.M.) may be observed. Bench marks are points of known

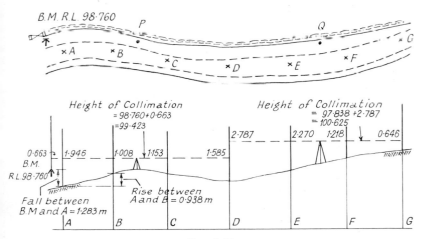

Fig. 3.35

elevation above Ordnance datum which have been established by surveyors of the Ordnance Survey. The commonest types are chiselled in the form of a broad arrow on permanent features such as bridge parapets, church plinths, etc., and the centre of the bar across the arrow gives the level (to 0·01 m on the 1/1,250 and 1/2,500 scale maps) at which the toe of the staff should be held, as shown in Fig. 3.36.

This first reading, made with the staff on a point of known reduced level (which need not, of course, be a bench mark), is known as a *backsight* (B.S.), and this term will now be used to denote that reading taken

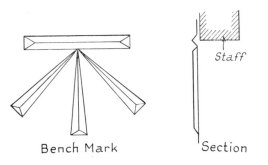

Bench Mark   Section

FIG. 3.36

immediately after setting up the instrument, with the staff on a point of known level. The staff is now held at points *A*, *B*, and *C* (*see* Fig. 3.35) in turn, and readings, which are known as *intermediate sights*, are taken. It is found that no readings after *D* are possible, due either to change in level of the ground surface or some obstruction to the line of sight, and it is necessary therefore to change the position of the instrument. The last reading on *D* is then known as a *foresight* (F.S.), and is the final reading taken before moving the instrument. The point *D* itself is known as a *change point*, because it is the staff position during which the position of the level is being changed.

The instrument is moved to *Q*, set up and levelled, and the reading, a backsight, taken on the staff at the change point *D*, followed by intermediate sights with the staff on points at which levels are required, until a further change becomes necessary, resulting in a foresight on point *G*. This procedure is repeated until all the required levels have been obtained.

Booking. (1) RISE AND FALL METHOD. The readings are booked in a *level book* which is specially printed for the purpose, as shown below. The reduction of these readings is carried out in the same book, so that two types of ruling are available, corresponding to the two methods of booking. The booking of the readings given in Fig. 3.35 on the Rise and Fall system would be—

Back-sight	Inter-sight	Fore-sight	Rise	Fall	Reduced Level	Distance m	Remarks
0·663					98·760		B.M. on gate, 98·76 O.D.
	1·946			1·283	97·477	0	Staff station A
	1·008		0·938		98·415	20	B
	1·153			0·145	98·270	40	C
2·787		1·585		0·432	97·838	60	D (change pt.)
	2·270		0·517		98·355	80	E
	1·218		1·052		99·407	100	F
		0·646	0·572		99·979	120	G

3·459		2·231	3·079	1·860	99·979		
2·231			1·060		98·760		
1·219		1·219	1·219		1·219		

It will be noted that estimations have been made to 1 mm when making the staff readings. The levels referred to in Figs. 3.24 to 3.26 would easily allow this; alternatively the parallel-plate micrometer could be used in certain cases. In view of the precision quoted for the bench-mark levels it is doubtful if there is any merit in working to 1 mm in some of the more usual civil engineering projects; the reduced levels could of course be rounded off to 0·01 m later to fit in with the precision of the bench-mark levels. A compromise of reading to 0·005 m might be considered sometimes, as will be seen in later examples, but the engineer will have to decide in certain cases, such as say sewer construction at flat gradients, to what accuracy to work. However, it may often be possible to base the levellings on one bench mark only, whose assigned value can be expressed to the third place of decimals as in the levelling book examples in the above table, rather than to two places (in this case 98·76 m) as on the Ordnance Sheets. Staff readings taken to 1 mm will then be in accordance with the assigned value.

Note that each reading is entered on a different line in the applicable column, except at change points, where a foresight and a backsight occupy the same line. The reason for this is made quite clear by referring to the remarks column—the change point occurs at staff station *D*, and since the staff is not moved only one reduced level is involved, requiring the one line. The R.L. is obtained by applying the rise or fall shown by the foresight compared with the previous intersight. The backsight is taken with the staff still on this point of known level, and the next rise or fall is obtained by comparing this backsight with the next intersight.

As a check on the arithmetic involved in reducing the levels, the backsights and foresights and the rises and falls must be summed up. The checks are then—

$$\Sigma(\text{Backsights}) - \Sigma(\text{Foresights})$$
$$= \Sigma(\text{Rises}) - \Sigma(\text{Falls})$$
$$= \text{Last R.L.} - \text{First R.L.}$$

The application of the checks is shown in the example, and the student is advised to prove the rules for the general case, i.e. by using letters instead of numbers. It must be stressed that these checks concern only the accuracy of the reductions, and have no effect on the accuracy of the readings themselves.

(2) HEIGHT OF COLLIMATION METHOD

Back-sight	Inter-sight	Fore-sight	Height of collimation	Reduced level	Distance	Remarks
0·663			99·423	98·760		B.M. on gate, 98·76 O.D.
	1·946			97·477	0	Staff station A
	1·008			98·415	20	B
	1·153			98·270	40	C
2·787		1·585	100·625	97·838	60	D (change pt.)
	2·270			98·355	80	E
	1·218			99·407	100	F
		0·646		99·979	120	G
3·45*0* 2·2*3*1	2·*2*31			99·97*0* 98·7*6*0		
1·219				*1*·219		

The height of collimation is obtained by adding the staff reading, which must be a backsight to the known R.L. of the point on which the staff stands. All other readings are deducted from the height of collimation, until the instrument setting is changed, whereupon the new height of collimation is determined by adding the backsight to the R.L. at the change point.

The arithmetical checks to be applied to this system of booking are

$$\Sigma(\text{B.S.}) - \Sigma(\text{F.S.}) = \text{Last R.L.} - \text{First R.L.} \qquad (3.6)$$
$$\Sigma(\text{all R.L.'s except the first})$$
$$= \Sigma(\text{each instrument height}) \times (\text{no. of Inter. Sights and}$$
$$\text{F.S.'s deduced from it}) - \Sigma(\text{F.S.} + \text{I.S.}) \qquad . \qquad . \qquad (3.7)$$

This second check is cumbersome and is often ignored, so that as a consequence, the intermediate R.L.'s are unchecked. In this case, errors could go unchecked (compared with the Rise and Fall method where errors in all R.L.'s are detected).

Reduction is easier with the height of collimation method (or height of instrument method, as it is sometimes called) when levelling for earthworks, and large numbers of intersights are taken from each position of the instrument.

In lengthy levelling operations where pages of readings are involved, each page should be checked separately. Book the last reading on each page as a foresight, and repeat it as a backsight at the top of the next page.

EXAMPLE. The following figures were extracted from a level field book, some of the entries being illegible owing to exposure to rain. Insert the missing figures and check your results. Re-book all the figures by the Rise and Fall method, and state the advantage of this method of booking.

B.S.	I.S.	F.S.	H. of I.	R.L.	Remarks
?			279·08	277·65	O.B.M
	2·01			?	
	?			278·07	
3·37		0·40	?	278·68	
	2·98			?	
	1·41			280·64	
		?		281·38	T.B.M.
					(*L.U., B.Sc Exam*)

Each line in a level book can be regarded as an equation which can be drawn up by applying the principles just outlined.

Thus, line 1,   H. of I. = R.L. + B S.
∴   B.S. = H. of I. − R.L.
                   = 279·08 − 277·65
                   = 1·43

line 2,   R.L. = H. of I. − I.S.
                   = 279·08 − 2·01
                   = 277·07

In this way the field book is completed as follows—

B.S.	I.S.	F.S.	H.I.	R.L.	Remarks
1·43			279·08	277·65	O.B.M.
	2·01			277·07	
	1·01			278·07	
3·37		0·40	282·05	278·68	
	2·98			279·07	
	1·41			280·64	
		0·68		281·37	T.B.M.
4·80		1·08		281·37	
1·08				277·65	
3·72				3·72	

B.S.	I.S.	F.S.	Rise	Fall	R.L.	Remarks
1·43					277·65	O.B.M.
	2·01			0·58	277·07	
	1·01		1·00		278·07	
3·73		0·40	0·61		278·68	
	2·98		0·39		279·07	
	1·41		1·57		280·64	
		0·68	0·73		281·37	T.B.M.
4·89	1·98	4·39	4·39	0·58	281·37	
1·08		0·58	0·58		277·65	
3·72		3·72			3·72	

This is a purely academic exercise, and in practice, under such circumstances the whole of the work would be relevelled.

Uses of Levelling

Apart from the general problem of determining the difference in level between two points, which has already been fully dealt with, the main uses of levelling are—

(1) the taking of longitudinal sections, (2) cross-sections, (3) contouring, (4) setting out levels.

Longitudinal Sections. An example of such a section has been given in Fig. 3.35, from which it will be seen that the object is to reproduce on paper the existing ground profile along a particular line—often, though not invariably, the centre line of existing or proposed work, e.g. the centre line of a railway, road or canal, or along the proposed line of a sewer, water main, etc. Staff readings to 0·01 m should be generally adequate for this purpose.

The accuracy with which the ground profile is represented on the section is dependent on the distance between staff stations, and this in turn depends on the scale of the section. As a general basis, however, levels should be taken at—

(1) every 100 links,
(2) points at which the gradient changes, e.g. top and bottom of banks,
(3) edges of natural features such as ditches, ponds, etc.,
(4) in sections which cross roads, at the back of the footpath, on the kerb, in the gutter, and on the centre of the road.

The sections are usually plotted to a distorted scale, a common one for road work being 1/500 scale horizontal and 1/100 vertical.

The following points should be borne in mind during the actual levelling, particularly when levelling long sections, to avoid the build up of error.

(*a*) Start the work from a bench mark if possible, and make use of any nearby bench marks which lie within the length being levelled.

(*b*) Try to keep backsights and foresights equal in length to minimize errors which will occur if the line of collimation is not parallel to bubble tube axis.

(*c*) Make all changes on firm ground, preferably on identifiable features on which check levels can be taken if required.

(*d*) Take the final foresight on a bench mark or, better, close back on the starting point by a series of "flying levels," i.e. a series of equal foresights and backsights, each about 80 m long.

(*e*) Do not work with the staff extended in high wind.

Cross Sections. Works of narrow width such as sewer and pipe lines require only one line of levels along the centre line of the proposed trench, since there will generally be little change of the ground surface level over the proposed width. Wider works, however, such as roads, railways, embankments, large tanks, etc., will necessitate the use of ground on either side of the centre line, and information regarding relative ground levels is obtained by taking *cross sections* at right angles to the centre line. The width of these must be sufficient to cover the proposed works, e.g. 15 m either side of the centre line for a normal road. The longitudinal spacing of the sections depends on the nature of the ground, but should be constant if earthworks are to be computed. A spacing of 100 links, i.e. 20 m, is common.

The centre line is first set out, pegs being placed at points where cross sections are required, and the cross sections themselves may then also be set out—using an optical square or similar instrument where precision is required in setting out the right angle—with white arrows or ranging rods marking the points where levels are required. The choice of points will be governed by the same principles as govern the taking of longitudinal sections, and the aim is again to reproduce the ground profile accurately.

In the actual levelling, cross sections may be completed one at a time, setting up the instrument as many times as is necessary.

This method facilitates booking but may be tedious in that where the ground has steep cross gradients an excessive number of change points will be required. In this case it is often quicker to take staff readings on other cross sections as the ground allows, though it is obvious that great care must be exercised in booking such readings, that none are forgotten, and that they are identified with the correct longitudinal and lateral measurements.

It is common to plot cross sections to a natural, i.e. undistorted, scale, and since only the ground profile and a limited depth are required, the plots can be kept compact by judicious choice of datum or base height. Fig. 3.37 gives examples of plotting.

Contouring. A *contour* is a line joining points of equal altitude. Contour lines are shown on plans as dotted lines, often in distinctive colour, overlaying the detail. The vertical distance between successive contours is known as the *vertical interval*, and the value of this depends

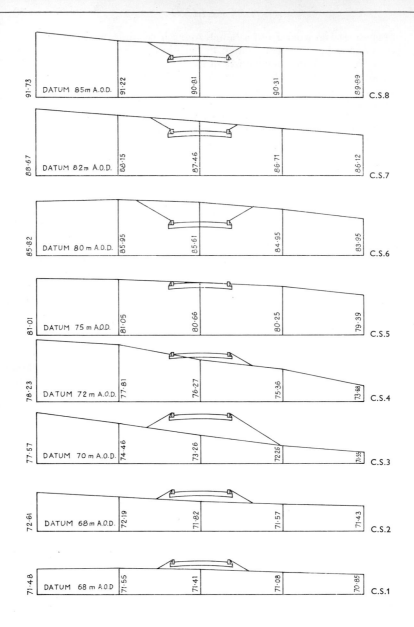

DATUM 85m A.O.D.   91·73   91·22   90·81   90·31   89·89   C.S.8

DATUM 82m A.O.D.   88·67   88·15   87·46   86·71   86·12   C.S.7

DATUM 80 m A.O.D.   85·82   85·95   85·61   84·95   83·95   C.S.6

DATUM 75 m A.O.D.   81·01   81·05   80·66   80·25   79·39   C.S.5

DATUM 72 m A.O.D.   78·23   77·81   76·27   75·36   73·68   C.S.4

DATUM 70 m A.O.D.   77·57   74·46   73·26   72·26   71·55   C.S.3

DATUM 68m A.O.D.   72·61   72·19   71·82   71·57   71·43   C.S.2

DATUM 68 m A.O.D   71·48   71·55   71·41   71·08   70·85   C.S.1

Fig. 3.37a

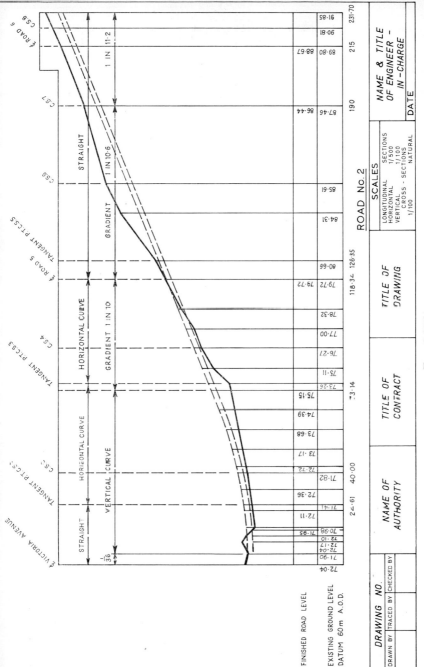

Fig. 3.37(b)

on (*a*) the scale of the plan, (*b*) the use to which the plan is to be put. For example, a 1/5,000 plan prepared by photogrammetric methods (Chapter 11) for the planning of a highway project may have contours at 5 m intervals.

As regards the interpretation of contours, when they are close together, steep gradients exist, and as they open, the gradients flatten. Two contour lines of different value cannot intersect, and a single contour cannot split into two lines having the same value as itself. A contour line must make a closed circuit even though not within the area covered by the plan. The main value of a contour plan, therefore,

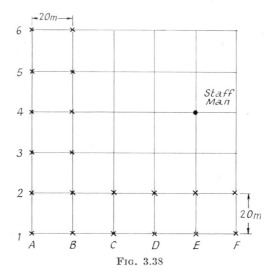

FIG. 3.38

is that it enables an assessment to be made of the topography; such plans will be commonly prepared when layouts of large projects such as housing estates are under consideration. Accurate contour plans are invariably prepared when reservoir projects are being designed, but for general civil engineering work, a vertical interval of 1 m, say, may be required, and the preparation of such a contour plan is a lengthy and tedious job. The following methods are probably the most commonly used ones for contouring.

"GRIDDING." This method is ideal on relatively flat land, especially on comparatively small sites. Squares of 10 m to 20 m side are set out (according to the accuracy required) in the form of a grid, and levels are taken at the corners. To save setting out all the squares, two sets of lines may be established using ranging rods as shown in Fig. 3.38, and to locate any particular square corner the staff man lines himself in, using pairs of ranging rods. For booking the reading, the staff man is at *E*4 in the example shown and this would be noted in "Remarks." The reduced levels are then plotted on the plan, which has been gridded in the same manner, and by any suitable means of interpolation (e.g. radial dividers) the required contours are plotted.

RADIATING LINES. Rays are set out on the ground from a central point, the directions being known. Levels are taken along these lines at measured distances from the centre (*see* Fig. 3.39). Again, interpolation is used to give the contour lines; this method is particularly suitable for contouring small hills or knolls.

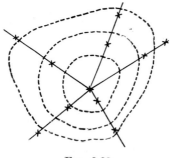

FIG. 3.39

DIRECT CONTOURING. This method may be applied in any case, but is used to best advantage in hilly terrain. The actual contour is located on the ground and marked by coloured laths or other convenient means. The levelling technique is shown in Fig. 3.40; the level is set up and levelled at some convenient position, and the height of collimation established by a backsight on to some point of known level. In Fig.

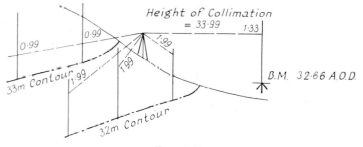

FIG. 3.40

3.40 this gives a height of collimation of 33·99 A.O.D. Thus, any intermediate sight of 0·99 means that the staff is on a point 103·99 — 0·99 = 33 m A.O.D. The surveyor directs the staff man up or down the hill until the staff reads 0·99, and on being given a signal, the staff man sticks a lath into the ground (one colour being used for one particular contour line), and moves to another point on the contour. A series of staff readings of 33·99 − 32 = 1·99 would enable the 32 m contour to be set out, and so on. Speed of setting out depends largely on the skill of the staff man; Staff readings to 0·01 m are quite adequate for this purpose.

The positions of the laths are later surveyed by some suitable method, e.g. chain survey, traverse survey, plane tabling, tacheometry, etc.

Setting Out Levels—Boning-in. An example of the basic operation involved in setting out levels has already been given in the previous section, and it is proposed to amplify this only to cover the simple but highly important task of setting up sight rails. These are set up to enable excavator operators to cut out earth to an even gradient, and to enable the pipe-layer to lay his pipes to a given gradient.

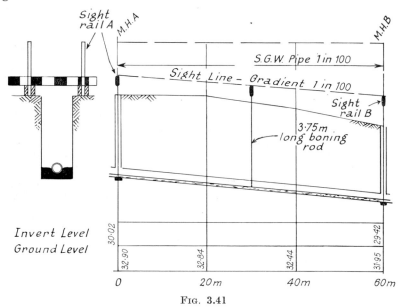

F ɪ ɢ. 3.41

As an example, consider a length of sewer being laid from manhole *A*, with an invert level of 30·02 m, to manhole *B*, 60 m away, the gradient from *A* to *B* being 1 in 100 and falling from *A* to *B*. Fig. 3.41 shows the proposed sewer, and it will be seen that the general depth of the sewer is below 3 m. Thus if two rails are fixed over stations *A* and *B* about 1 m above ground level, and each a fixed height above invert level, then an eye sighting from rail *A* to rail *B* will be sighting down a gradient equal to that of the proposed sewer. In the example, a convenient height above invert would be 3·75 m, so that a boning rod (a plate with a sight bar across one end, and looking like a T-square) of this length, held vertically so that its sight bar just touched the line of sight between sight rails *A* and *B*, would give at its lower end a point on the sewer invert line. (*Note:* the invert of a sewer is the lowest point on the inside surface.) Staff readings might well be required to 0·001 m when establishing sight rails even though the nearby bench mark heights are known only to 0·01 m.

To fix these sight rails for use with a 3·75 m-long boning rod, therefore, we drive two posts on either side of the manholes and nail the rails between these at the following levels—

$$\text{Sight rail } A, \text{ R.L.} = 30\cdot02 + 3\cdot75$$
$$= 33\cdot77 \text{ m A.O.D.}$$
$$\text{Distance } AB = 60 \text{ m}$$

Therefore
$$\text{Fall} = \frac{60}{100} = 0\cdot60 \text{ m}$$

∴
$$\text{Invert level } B = 30\cdot02 - 0\cdot60$$
$$= 29\cdot42 \text{ m A.O.D.}$$

and
$$\text{sight rail B, R.L.} = 29\cdot42 \mid 3\cdot75$$
$$= 33\cdot17 \text{ m A.O.D.}$$

If a level set up nearby has a height of collimation of, say 34·845 m A.O.D., then the staff is moved up and down the posts at M.H.A. until a reading of $34\cdot845 - 33\cdot770 = 1\cdot075$ m is obtained. Pencil marks are made on each post and the black and white sight rail is nailed in position as shown in Fig. 3.41. For rail B, the staff reading would, of course, be

$$34\cdot845 - 33\cdot170 = 1\cdot675 \text{ m}$$

Frequent checking of sight rails is required, as they are liable to be disturbed by excavators, dumpers, lorries, etc.

HEADROOM OF BRIDGES. Reduced levels of bridge soffits and similar measurements are made by using the staff in an inverted position (Fig. 3.42), taking care when using a telescopic staff that the catches are properly engaged.

The headroom of the bridge in Fig. 3.42 is

$$1\cdot555 + 2\cdot535 = 4\cdot09 \text{ m}$$

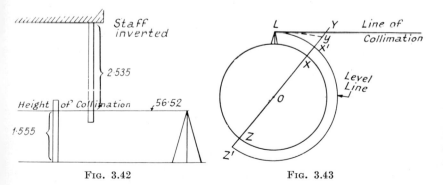

FIG. 3.42   FIG. 3.43

The inverted staff reading is booked with a negative sign, so that when reducing this reading we get

$$\text{R.L. of soffit} = 56 \cdot 52 - (-2 \cdot 535)$$
$$= 59 \cdot 055 \text{ m A.O.D.}$$

This value could be rounded off to $0 \cdot 01$ m, upwards or downwards, without much loss of accuracy here.

Curvature and Refraction. Referring again to the definitions given at the beginning of this chapter, it will be remembered that the line of collimation is not a level line but is tangential to the level line. As a consequence corrections must be applied when the sights are long, i.e. when the deviation of the tangent from the circle becomes appreciable.

Taking a level line as being the circumference of a circle whose centre is the earth's centre (an assumption which is accurate enough for this purpose), the required correction can be deducted as follows—

Let L be the position of the instrument, which is directed towards a staff held vertically at X—i.e. held along the extension of the radius OX (*see* Fig. 3.43).

Now $\qquad (LY)^2 = X'Y \cdot Z'Y$
$$= X'Y(X'Y + X'Z')$$
and $\qquad X'Z' = XZ + 2XX'$
where $\qquad XZ \backsimeq 12{,}734$ km
and $\qquad XX' \backsimeq 1 \cdot 5$ m so that $X'Y \ngtr 2 \cdot 5$ m if a 4 m staff be used
Therefore $\quad (LY)^2 = X'Y \cdot XZ$, neglecting $(X'Y)^2$ as a second order term
i.e. $\qquad (LY)^2 = 2R \cdot X'Y$
and $\qquad X'Y = $ correction for *curvature*
$$= \frac{(LY)^2}{12{,}734} \text{ km}$$
$$= 0 \cdot 081 \ (LY)^2 \text{ m}$$

where LY is expressed in kilometres.

Owing to the gradual reduction in air density with altitude, however, light rays from a higher altitude are refracted downwards as they pass through progressively denser air. Thus the line of collimation actually cuts the staff at y, where

$$Yy = \frac{X'Y}{7}$$
$$= 0 \cdot 012 \ (LY)^2 \text{ m}$$

Therefore total correction for refraction and curvature
$$= 0 \cdot 069 \ (LY)^2$$
$$= 0 \cdot 069 \cdot D^2 \text{ m where } D = LY \text{ km} \qquad . \qquad (3.8)$$

EXAMPLE. A staff is held at a distance of 200 m from a level and a reading of 2·758 m obtained. Establish the reading corrected empirically for curvature and refraction.

$$\text{Correction} = 0\cdot067 \; (0\cdot2)^2$$
$$= 2\cdot8 \text{ mm}$$
$$\text{Corrected reading} = 2\cdot758 - 0\cdot003$$
$$= 2\cdot755 \text{ m}$$

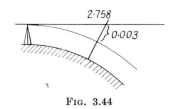

FIG. 3.44

Fig. 3.44 indicates why the correction is applied in this manner.

RECIPROCAL LEVELLING. By means of reciprocal levelling, the need for applying these corrections may be avoided. If taking levels across a wide ravine, for example, an instrument would be set up near a staff held at X and sights taken on this staff and on another staff held on the far bank Y, as in Fig. 3.45a. Immediately afterwards similar readings

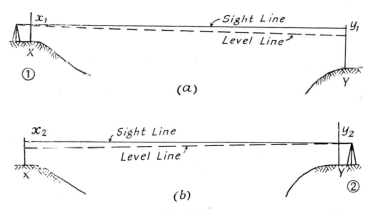

FIG. 3.45

are made with a second instrument set up on the opposite bank, as in Fig. 3.45b.

Note that with this method *two* similar instruments in correct adjustment are required.

Level (1). Apparent difference in level between X and Y

$$= Xx_1 - Yy_1$$

True difference in level will be the apparent difference less the total correction for curvature and refraction

$$= \alpha \text{ (say)}$$

$$\therefore \qquad \text{True difference} = Xx_1 - [Yy_1 - \alpha]$$
$$= Xx_1 - Yy_1 + \alpha$$
$$= \text{Apparent difference}_{(1)} + \alpha \quad . \quad (3.9)$$

Level (2). Apparent difference in level
$$= Xx_2 - Yy_2$$
$$\text{True difference} = [Xx_2 - \alpha] - Yy_2$$
$$= [Xx_2 - Yy_2] - \alpha$$
$$= \text{Apparent difference}_{(2)} - \alpha \quad . \quad (3.10)$$

Add equations (3.9) and (3.10),

True difference in level $= \frac{1}{2}$[sum of apparent differences].

EXAMPLE. In levelling across a river, reciprocal levelling observations gave the following results for staffs held vertically at X and Y from level stations A and B on each bank respectively—

Staff reading of X from $A = 1 \cdot 753$ m
Staff reading of X from $B = 2 \cdot 080$ m
Staff reading of Y from $A = 2 \cdot 550$ m
Staff reading of Y from $B = 2 \cdot 895$ m

If the R.L. of X is 90·37 A.O.D., obtain that of Y. (*R.T.C.S.*)

It will be noted from the staff readings that Y is lower than X.

Instrument at A. Apparent difference $= 2 \cdot 550 - 1 \cdot 753 = 0 \cdot 797$ m
Instrument at B. Apparent difference $= 2 \cdot 895 - 2 \cdot 080 = 0 \cdot 815$ m

$$\therefore \qquad \text{True difference} = \frac{0 \cdot 797 + 0 \cdot 815}{2}$$

$$= 0 \cdot 806 \text{ m}$$

Therefore $\qquad$ R.L. of $Y = 90 \cdot 37 - 0 \cdot 81$
$$= 89 \cdot 56 \text{ A.O.D.}$$

Note that where only *one* instrument is used, this method corrects not only for curvature and refraction but also for any lack of adjustment in the collimation axis (cf. Permanent Adjustment of Level, Test 2*a*). In these circumstances, however, it must be appreciated that temperature change, and hence change in refraction effects, may take place while the level is being moved. Where two levels not in perfect adjustment are available reciprocal levelling is carried out, after which the instruments are interchanged and another set of readings taken.

It has been assumed in this section that the line of sight is concave towards the earth, as in Fig. 3.43. Geisler,* in a paper given to the

* *Refraction near the Ground and its Influence on Precise Levelling*, M. Geisler. 12th Congress F.I.G., London.

12th Congress of F.I.G. in London, has discussed the influence of topography and ground cover upon the air layers near to the ground. Light
rays from staff to level pass through such layers, and climatic conditions
in this zone differ from those in the overlying atmosphere. The refraction correction of one-seventh the curvature correction is in fact based
upon the latter consideration.

Geisler shows that the line of sight can be convex towards the earth,
depending upon the temperature gradient. If the ground is warmer
than the adjacent air layers the temperature gradient (dT/dh) is negative and then the line of sight curves upwards (Fig. 3.46) rather than
downwards as shown in Fig. 3.43, which is in fact the case when dT/dh
is positive and which can occur mainly after sunset.

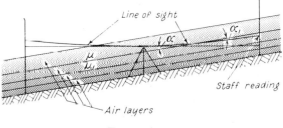

Fɪɢ. 3.46

In Fig. 3.46 we have $\mu \cos \alpha = \mu_1 \cos \alpha_1 = a$ constant, and on
differentation we obtain $d\mu = \mu \tan \alpha \, d\alpha$. The relationship between
μ and μ_0 (the refractive index at 0°C and 760 mm of mercury) is

$$\mu - 1 = \frac{\mu_0 - 1}{1 + kt} \cdot \frac{p}{760}$$

where $k = 1/273$

Also we have

$$d\mu = \frac{\mu}{1 + kt} \cdot 1 \, (-k)dt$$

and

$$d\mu = (\mu - 1) \frac{dp}{p}$$

It is known that change in pressure is of much less influence than
change in temperature and accordingly one can write

$$d\mu = - (\mu - 1) \frac{dT}{T}$$

where $T = 273 + t$ (i.e. absolute temperature)
Thus

$$\mu \tan \alpha \, . \, d\alpha = - (\mu - 1) \frac{dT}{T}$$

$$\therefore \qquad d\alpha = -\frac{\mu - 1}{\mu T} \cdot \frac{dT}{dh} \cdot dL$$

where $\tan \alpha$ is written as dh/dL, dh being the rise over a length dL before refraction through α. The expression can also be written as

$$d\alpha = \frac{1}{R_1} \cdot dL$$

where
$$\frac{1}{R_1} = \left(\frac{\mu - 1}{\mu T}\right)\left(\frac{-dT}{dh}\right)$$

At 20°C, $\mu = 1.00027$ and $T = 293°$

$$\frac{1}{R_1} = -\frac{0.00027}{1.00027 \times 293} \cdot \frac{dT}{dh}$$

$$= -10^6 \frac{dT}{dh}$$

It will be seen that if $dT/dh = 0.16°$ per metre then R_1 is virtually equal to the earth's radius.

The line of sight to staff at the higher station in Fig. 3.46 deviates more than that to the lower station since it passes through layers in which dT/dh is of greater magnitude, and accordingly the apparent difference in level is reduced; this would be true even if the two sighting lengths were the same, nor would the error be eliminated by also levelling in the other direction. If a sighting length of 50 m were used and dT/dh was $-0.16°$ per metre "upwards" and $-0.04°$ per metre "downwards" then using the expression $L^2/2R_1$ as the effect of curvature on distance, deduced as previously, the error in levelling is

$$\tfrac{1}{2} \times 10^{-6} \times 50^2 \times (0.16 - 0.04)$$
$$= \tfrac{1}{2} \times 10^{-6} \times 2,500 \times 0.12$$
$$= 0.15 \text{ mm per 100 m}$$

Only if the temperature gradients were the same for each direction would refraction have had no influence.

Geisler carried out experimental work using two points 100 m apart on an asphalt surface and having a height difference of some 2·3 m, the levels being of the Zeiss Ni2 type, and similar, with parallel plate micrometers attached. Amongst his conclusions were (i) for 50 m sighting distances with upper staff reading of about 0·4 m, refraction may reduce height differences by, say, 2·5 mm per kilometre and (ii) for best results, short sights with upper staff readings greater than 0·5 m should be observed during morning hours.

Accuracy in Levelling. The main errors affecting accuracy in levelling, apart from the above, are—

1. Errors in reading the staff.
2. Errors due to the bubble not being central.
3. Errors due to the instrument not being in adjustment.

4. Errors due to differential settlement of the tripod.

5. Errors due to tilting and settlement of the staff.

ORDINARY LEVELLING. Assuming a tripod on firm ground, and equal backsights and foresights, we can ignore the effects of (4) and also of (3) since the main systematic error in ordinary levelling will be due to any lack of parallelism between the bubble tube and the collimation axis. The effect of staff tilt and settlement can be reduced to the second order by use of a staff with a target bubble on firm ground.

1. *Reading Errors.* These in turn depend on the magnification and image clarity afforded by the telescope and on the manner in which the staff is marked. At a distance of 80 to 100 m estimation can be made to 0·001 m, and the probable error can be taken as ± 0·5 mm.

With equal backsights and foresights, we have 10 readings per kilometre of levelling, so that the p.e. due to reading per kilometre (M_r) is ± 0·5$\sqrt{10}$

$$\therefore \qquad M_r = \pm 1\cdot6 \text{ mm}$$

2. *Bubble Displacement.* The accuracy of bubble centring depends on the method used to view the bubble and W. Schneider* gives the following figures—

Type of viewer	p.e.
Viewed directly	$\pm \dfrac{\phi''}{7}$
Viewed by hinged mirror	$\pm \dfrac{\phi''}{10}$
Prismatic viewer	$\pm \dfrac{\phi''}{25}$
Zeiss compensator	$\pm 0\cdot5''$

where ϕ is the bubble sensitivity in seconds per 2 mm displacement. These would appear to be limiting figures and for ordinary levelling, viewing directly, a value of ± $\phi/4''$ is probably more reasonable.† For a bubble sensitivity of 20", the p.e. is thus ± 5". Over a sight of 100 m this represents an error of $\dfrac{5}{3,600} \times \dfrac{\pi}{180} \times 100$

$$= \pm 0\cdot0024 \text{ m} = \pm 2\cdot4 \text{ mm}$$

Therefore p.e. per km of levelling due to displacement (M_b)

$$= \pm 2\cdot4\sqrt{10}$$
$$M_b = \pm 7\cdot5 \text{ mm}$$

Combined Effects

$$\text{total p.e.} = M_t = \pm \sqrt{M_r^2 + M_b^2} = \pm \sqrt{1\cdot6^2 + 7\cdot5^2}$$
$$\simeq \pm 7\cdot7 \text{ mm per km}$$

This computation, however, ignores the other effects which tend to be systematic (and hence more directly proportional to the distance

* "The Evolution of Modern Levelling Instruments," by Dr. I. W. Schneider, *Vermessungstechnische Rundschau*, No. 8/9, 1953.

† "The Economics of Levelling," by Dr. J. G. Förstner. *Allgemeine Vermessungs-Nachrichten*, No. 7, 1953.

levelled) so that for third-order accuracy of levelling a maximum error of $\pm\,12\sqrt{K}$ mm is suggested, although a value of $\pm\,25\sqrt{K}$ mm for ordinary levelling is not uncommon, where K is the length of the circuit covered in km.

There is no doubt, however, that by carefully following the rules given earlier in this chapter (Longitudinal Sections) and using a modern tilting level with coincidence bubble reader, greater accuracy than this is possible. The errors arise mainly from field technique, since for example the Watts Microptic SL80 Level is said to have an accuracy of $\pm\,2\cdot5$ mm per kilometre of single levelling. In a round of levels, taken in N. Lancashire in connexion with an engineering project, a total length of some 16 km closed within 40 mm, giving an error of $\pm\,10\sqrt{K}$.

In the above, since

$$40 = C\sqrt{16}$$

$\therefore$

$$C = 10 \text{ mm}$$

Great care was taken to keep backsights and foresights equal and approximately 75 m in length, while only firm features were used as change points.

The above figure lies within the limits of second- and third-order accuracy suggested by Prior in a paper "Accuracy of Highway Surveys" delivered to the 11th Congress of the International Society for Photogrammetry, i.e. $8\cdot4\sqrt{K}$ mm and $12\sqrt{K}$ mm respectively. First-order accuracy was quoted as $4\sqrt{K}$ mm and fourth-order accuracy as $120\sqrt{K}$ mm, i.e. less than 10^{-4} relatively.

Precise Levelling. By further refinement of field technique and by using precise staffs and levels of the types described earlier, much greater accuracy is attainable and the term *precise levelling* is then applied. Standards for "levelling of high precision" were fixed by the International Geodetic Association in 1912 along lines suggested by Lallemand, and these are—

Probable accidental error not $> \pm\,1\sqrt{K}$ mm, K in kilometres

Probable systematic error not $> 0\cdot2\sqrt{K}$ mm

These standards represent a very high degree of precision and will be applicable to work carried out by such bodies as the Ordnance Survey. In the Second Geodetic Levelling of England, which was carried out between 1912 and 1920, the errors were within the above limits. Geisler (*op. cit., see* p. 82), in fact suggests a relative accuracy of $\frac{1}{2} \times 10^{-6}$, i.e., $\frac{1}{2}$ mm per kilometre of levelling in precise surveying, and points out that such high accuracy involves great expense: in fact there is an exponential relationship between the two. Typical rules governing field technique in precise levelling are given below, and these can be usefully applied to levelling where medium accuracy is required.

1. Backsights and foresights to be equal in length or a correction applied for curvature and refraction: readings above 0·5 m level.

2. All lines of levels to be run twice in opposite directions, each run to be made on different days, with different change points.

3. Levelling to be carried out only during favourable conditions of light and temperature, and not during high wind or in the rain.

4. All change points to be made on a special metal footplate.

5. If the standards laid down regarding allowable error (in the form $\pm C\sqrt{K}$ mm) are not complied with, the work shall be repeated.

EXERCISES 3

1. Describe with the aid of a sketch the function of an internal focusing lens in a surveyor's telescope, and state the advantages and disadvantages of internal focusing as compared with external focusing.

In a telescope, the object glass of focal length 178 mm is located 230 mm from the diaphragm. The focusing lens is mid-way between these when a staff 20 m away is focused. Determine the focal length of the focusing lens. (*L.U.*, *B.Sc.*)

Answer: 148 mm.

2. A level set up in a position 30 m from peg *A* and 60 m from peg *B* reads 1·914 on a staff held on *A* and 2·237 on a staff held on *B*, the bubble having been carefully brought to the centre of its run before each reading. It is known that the reduced levels of the tops of the pegs *A* and *B* are 87·575 m and 87·280 m O.D. respectively.

Find (*a*) the collimation error, and (*b*) the readings that would have been obtained had there been no collimation error. (*L.U.*, *B.Sc.*)

Answer: (*a*) 1 mm in 1·15 m, (*b*) 1·888 m; 2·183 m.

3. A sewer is to be laid at a uniform gradient of 1 in 200, between two points *X* and *Y*, 240 m apart. The reduced level of the invert at the outfall *X* is 150·82. In order to fix sight rails at *X* and *Y*, readings are taken with a level in the following order—

Reading	Staff Station
B.S. 0·81	T.B.M. (near *X*), R.L. 153·81
I.S. "*a*"	Top of sight rail at *X*
I.S. 1·07	Peg at *X*
F.S. 0·55	C.P. between *X* and *Y*
B.S. 2·15	C.P. between *X* and *Y*
I.S. "*b*"	Top of sight rail at *Y*
F.S. 1·88	Peg at *Y*

(i) Draw up a level book and find the reduced levels of the pegs.

(ii) If a boning rod of length 3 m is to be used, find the level readings *a* and *b*.

(iii) Find the height of the sight rails above the pegs at *X* and *Y*. (*L.U.*, *B.Sc.*)

Answer: (iii) 0·27 m, 0·68m.

4. In order to find the rail levels of an existing railway, a point *A* was marked on the rail, then points at distances in multiples of 20 m from *A*, and the following readings were taken—

Backsight 3·39 on O.B.M. 23·10.

Intermediate sights on *A*, *A* + 20, and *A* + 40, 2·81, 2·51, and 2·22 respectively.

A + 60: change point: foresight 1·88, backsight 2·61.

Intermediate sights on A + 80 and A + 100, 2·32 and 1·92 respectively; and finally a foresight of 1·54 on *A* + 120, all being in metres.

Tabulate the above readings on the Collimation system; then assuming the levels at *A* and *A* + 120 were correct, calculate the amounts by which the rail would have to be lifted at the intermediate points to give a uniform gradient throughout.

Repeat the tabulation on the Rise and Fall system, and apply what checks are possible in each case. (*L.U.*, *B.Sc.*)

Answer: 0; 0·03; 0·08; 0·07; 0·11; 0·05; 0 m.

5. Describe the parallel-plate micrometer, and show how it is used in precise work when attached to a level.

If an attachment of this type is to give a difference of 5 mm for a rotation of 20°, calculate the required thickness of glass when the refractive index is 1·6. Describe how the instrument may be graduated to read to 0·0001 m for displacements of 5 mm above and below the mean. (*L.U.*, *B.Sc.*)

Answer: 39 mm.

6. Complete the levelling table given below. If an even gradient of 1 vertically in every 70 horizontal starts 1 m above peg 0, what is the height of the gradient above, or its depth below, peg 7 ?

Station	Distance	Back-sight	Inter-sight	Fore-sight	Rise	Fall	R.L.
B.M.		3·10					193·62
0	0		2·56				
1	20		1·07				
2	40	1·92		3·96			
3	60	1·20		0·67			
4	80		4·24				
5	100	0·22		1·87			
6	120		3·03				
7	140			1·41			

Answer: 0·01 m above. (*I.C.E.*)

7. Describe with diagrams the methods of setting out inverts of sewers to their proper gradients with the aid of sight rails and boning rods.

Assume in your description that the gradient is to be 1 in 250 and the distance between rails at points *A* and *B* is 75 m. The lower end of the sewer is to be at a depth of 3 m below ground level at *A* and a fixed boning rod of 4 m length is to be used. The ground level at *B* is 0·55 m below that at *A*. (*I.C.E.*)

8. A line of levels at 100 links centres, run from a point at the top of a quarry face, around the edge of the excavation, and then across it to a point immediately below the starting point, yielded the bookings given in the table below. Complete the level book by the rise and fall method, and apply the usual checks.

Direct measurement down the face indicated the level difference to be 0·43 m greater than that indicated by the levelling. Subsequent investigation revealed that the level had been tampered with, and was out of adjustment, and that the levelling party had consistently read the staff near the top and bottom at change points, and had set up the instrument on the line of travel with unequal backsight and foresight distances. If the gradient is regular enough at any instrument setting for one to assume reading distance proportional to the difference between the staff reading and the average instrument height of 1·37 m, estimate the instrument error, and the true levels at *A* and *L*.

Station	Dist. (links)	B.S.	I.S.	F.S.	Rise	Fall	Red. lev.	Remarks
A	0	0·44						Top of face
B	100		1·64					
C	200		2·94					
D	300	0·17		3·98				
E	400		2·12					
F	500	0·32		4·11				
G	600	0·54		3·88				
H	700	0·11		4·08				
I	800	4·00		3·96			149·73	B.M. at commencement of cut
J	900		2·44					
K	1,000	3·86		0·11				
L	1,100			0·59				Bottom of face below A

(*L.U.*, *B.Sc.*)

Answer: 0·19 m in 100 links; level of *L* = 157·12 m.

CHAPTER 4

THE THEODOLITE AND TRAVERSE SURVEYING

THE THEODOLITE is used to measure horizontal and vertical angles. It is without doubt the most important instrument for exact survey work, and many types are available to meet varying requirements of accuracy and precision, ranging from say the Wild T0 Compass Theodolite, with a horizontal circle reading to 1 min, to precision instruments which read direct to 0·5 sec and 0·1 sec. There is thus a wide selection from which to choose to satisfy the surveyor's needs.

The inventor of the name "theodolite" was Leonard Digges, and his description* of the instrument under the title *The Construction of an Instrument Topographicall Serving most Commodiously for all Manner of Mensurations* was published in the sixteenth century by his son, Thomas. From this time, English surveyors concentrated their attention on the development of the instrument, and, about 1785, Ramsden produced his famous telescopic theodolite which was used by Roy in 1787 for the first tie-up between the English and French triangulation systems. The horizontal circle was three feet in diameter, and by means of micrometers, readings could be made to single seconds.† Vertical angles could not be measured as the instrument had no vertical circle, though less accurate instruments, equipped with these, were made early in the nineteenth century, culminating in the so-called "transit" instruments which are described below.

In this chapter, types met in modern engineering practice in this country will be discussed, and the reader should turn to more comprehensive works for reference to such instruments as the Y-theodolite.

Basic Components and Principles of the Transit Vernier Theodolite

As will be seen in the illustrations (Figs. 4.1 and 4.2), one of the components of the instrument is a telescope which may be revolved through 360° about its transverse horizontal axis (an operation known as transitting). The name *transit theodolite* is then applied to the instrument.

The telescope is provided with an object glass, diaphragm and eyepiece, as described in the previous chapter, and in modern instruments is focused internally. When elevated or depressed it rotates about its transverse horizontal axis (*trunnion axis*), which is placed at right angles to the line of collimation, and the vertical circle, which is connected to the telescope, rotates with it. The trunnion axis is supported at its ends on the standards which are carried by the horizontal

---

* *A Geometrical Practise named Pantometria*, by Leonard Digges, published by his son, Thomas Digges (London, 1571).

† *Maps and Map-making*, by E. A. Reeves (Royal Geographical Society, London, 1910).

FIG. 4.1. VERNIER THEODOLITE ST2
Sexagesimal reading.
(*Courtesy of Hilger & Watts, Ltd.*)

vernier plate (*upper plate*). Vertical angles are measured on the graduated vertical circle by means of a pair of diametrically-opposite stationary verniers which are carried independently of the circle and telescope, but centred on the trunnion axis, on a vertical frame shaped roughly in the form of a letter T. A bubble tube is often found on this frame, which can be moved by the clip screws. A clamp is provided for fixing the telescope in any position in the vertical plane, and a slow-motion or tangent screw then allows some small angular movement

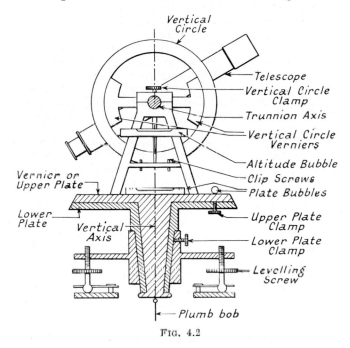

Fig. 4.2

for final coincidence of the image of the point observed with the cross-hairs.

The vernier plate carries one or two small bubble tubes (*plate bubbles*) and also the verniers which are used in the measurement of horizontal angles. This plate is carried on an inner bearing from which a plumb bob may be suspended so as to allow the placing (*centring*) of the instrument over a survey station. The plumb bob should lie on the vertical axis of the instrument, and this in turn should be at right angles to the trunnion axis. The vertical plane in which the telescope revolves should coincide with the vertical axis of the instrument as should the centre line of the inner axis.

The inner bearing rotates within an outer sleeve bearing which carries the horizontal circle (*lower plate*) on which are marked the graduations by which horizontal angles are read. By means of the lower clamp the outer bearings can be prevented from rotating, and similarly the upper clamp fixes the upper and lower plates together. Provision for small angular movements is made by tangent screws for both clamps. If the lower clamp be tightened, the inner bearing may rotate inside the outer bearing so long as the upper clamp is free: if the upper be tightened and the lower clamp is freed, both bearings move together as one: with both clamps tightened, no movement will occur. The standards, and hence the telescope, will rotate about the vertical axis of the instrument so long as the upper plate is free to turn. This rotatable portion of the theodolite is known as the alidade. A thin film of oil is required between the bearing surfaces to prevent stiff movement.

The instrument must be set up with the vertical axis truly vertical when angular measurements are made, and this is effected by the levelling head which has a footplate and three footscrews (in old patterns four footscrews may be found). The ball ends of the footscrews fit into recesses in the footplate and the footscrews operate on bushes fitted on the levelling head. For ease in centring the instrument over a station, a centring device is usually provided whereby the instrument and plumb bob can be moved independently of the levelling head when the footplate has been screwed on to the tripod: this device may be often seen just above the footscrews. It is important that there should be no movement of the footscrew feet on the footplate when this has been screwed to the tripod and the instrument levelled.

Setting Up and Levelling the Theodolite. The instrument must be correctly levelled (thereby making the vertical axis truly vertical) over the station, and in setting up, the footplate should be kept approximately horizontal to prevent excessive movement of the footscrews. The tripod legs can be moved inwards or outwards and sideways, and if a centring device is fitted to the theodolite these legs are moved so as approximately to centre the instrument over the station as indicated by the plumb bob; final centring is carried out using that plumb bob, and the device, which is then clamped. The tripod legs must be firmly pressed into the ground so that no movement can occur in the instrument as the surveyor moves round or when traffic moves nearby, and the wing nuts clamping the legs must be tight. If a centring device is not fitted then more patience will be required in the actual positioning of the tripod legs. Small movements of legs, or pairs, at a time must be made to secure final centring of the plumb bob over the station.

When the instrument has been centred, it must be levelled. Assuming three footscrews only, and using the bubble on the horizontal vernier plate, the procedure is as follows—

(*a*) Rotate the inner axis so that the bubble tube is parallel to two of the footscrews. Turning those footscrews, the bubble is brought to the centre of its run. The footscrews are turned simultaneously with the thumbs moving towards each other or away from each other. The left thumb movement gives the direction of the consequent movement of the bubble.

(*b*) Rotate the inner axis so that the bubble tube is at right angles to its former position, when it should be parallel to a line joining the third footscrew to the mid-point of the line joining the other two. Bring the bubble to the centre of its run using the third screw only.

A correctly-adjusted instrument will now be levelled, and as the vernier plate rotates and takes the bubble tube round, then the bubble should remain at the centre of its run. In practice, the above procedure is carried out at least twice, the telescope being wheeled successively through 90° back to position (*a*) and then, after checking with the two screws, to position (*b*).

The instrument is now set up ready for the measurement of horizontal angles.

The Vernier. Fig. 4.3 shows a direct vernier, together with part of the main graduated circle of a theodolite, which allows the observer

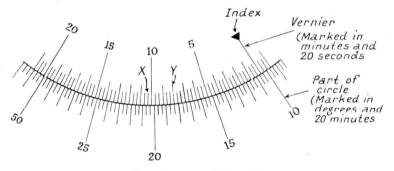

Vernier Reading :- 10° 28' 00"
59 small divisions on circle are equivalent to
60 small divisions on vernier

FIG. 4.3

to read the value of an angle to 20 sec. The main circle is graduated in degrees and third parts of a degree (i.e. 20 min). If the index only were present, any fractional part of 20 min would have to be estimated by the observer, and the lowest value he would obtain would depend upon his skill in reading. He may be able to estimate to say half a graduation, i.e. 10 min, which would mean that his value of the angle for the position of the index mark in Fig. 4.3 would be 10° 30'. By means of the vernier, smaller fractional parts of the smallest subdivision of the main circle are obtainable and, in the particular case shown, the vernier allows readings to 20 sec, this being termed the least count of the instrument.

Let α be the angular value of the smallest subdivision of the vernier, β the angular value of the smallest subdivision of the main scale, and n be the number of subdivisions of the vernier. In a direct vernier, n subdivisions of the vernier have the same total length as $(n-1)$ subdivisions of the main scale. Thus $n\alpha = (n-1)\beta$, and so $\alpha = \dfrac{n-1}{n}\beta$,

$$\therefore \qquad \beta - \alpha = \frac{1}{n}\beta$$

In the case illustrated, $\beta = 20$ min, $\alpha = 20$ sec, $n = 60$, and $(\beta - \alpha) = \dfrac{20'}{60} = 20$ sec, which is the *least count* of the instrument, and is the difference in length of the graduations on the vernier and main circle.

To read the vernier the observer first notes the position of the index mark, and Fig. 4.3 shows it to point between $10°\ 20'$ and $10°\ 40'$. The observer now looks along the vernier scale until he finds a graduation on that scale which coincides with a graduation on the main circle. Inspection of the figure will show that this occurs at the eighth division Y (or the twenty-fourth subdivision) on the vernier. This reading, which is 8 min (or 24×20 sec) is added on to the reading $10°\ 20'$, which is the *lower* of the two graduations straddling the index mark.

To obtain a reading of, say, $10°\ 30'\ 20''$, the index mark would be moved into the region of just over $10°\ 20'$ and the upper clamp then fixed (the lower clamp must be tight throughout). Then, using the slow-motion screw, the index mark is made to travel alongside the main scale so that the graduation X in Fig. 4.3 coincides with a graduation on the main circle. Note that in both reading and setting out angles, after noting the next lowest main circle reading nearest the index mark, the remainder of the reading is obtained on the vernier, the value of the coincident main circle graduation being of no account whatsoever.

To assist in reading the vernier, small eyepieces are provided on the instrument. Even so, sometimes two graduations appear to be in coincidence with main circle graduations, and their mean should be taken. It will be seen in Fig. 4.3 that vernier graduations to the right of the coincident division are left of the main circle graduations, while those vernier graduations to the left of the coincident divisions are to the right of the main circle graduations, and this fact will assist in locating the coincident graduation.

Measurement of Horizontal Angles. To measure angle ABC, the instrument is set up over station B in the manner described, and carefully levelled by means of the footscrews. The *face* of the instrument must be checked at this stage. Most telescopes have sights similar to those on a gun, fitted on top of the barrel, to assist in sighting the target. With these sights on top and the telescope pointing to the target, the vertical circle, which is known as the *face* of the instrument, will be left or right of the telescope. Suppose it is to the left; the theodolite is said to be in the *face left* position. By rotating the telescope through $180°$ in the vertical plane (i.e. about the trunnion axis), and then through $180°$ in the horizontal plane, the telescope will again be pointing at the signal, but with the gunsights on the underside of the barrel, and the vertical circle to the right—i.e. the theodolite is in the *face right* position. Starting with all clamps tightened, then—

(*a*) The plates are unclamped and the vernier zero index brought to the zero of the horizontal circle, coincidence being obtained by the upper tangent screw after clamping the plates by means of the upper clamp. Actually there is no real necessity to bring the vernier index to a zero reading, and if so desired any arbitrary setting of the index in the region of the zero may be used, the exact value being read from the verniers.

(*b*) The lower plate will now be unclamped, and the telescope directed so that *A* appears in the field of view; turning the telescope moves the scales and vernier plates together so that the setting of the verniers is the same as in (*a*). Exact coincidence of the vertical cross-hair upon *A* is obtained by means of the lower tangent screw. The vernier readings may now be taken, or checked to be at zero.

(*c*) With the lower clamp fixed, the upper clamp will be freed, and the telescope directed towards *C*, a rough setting being obtained by hand, the upper clamp then applied, and coincidence on the vertical hair obtained by means of the upper tangent screw.

(*d*) The vernier readings are again noted and the angle value found as in the examples below. In theodolite traverse surveying other readings will be taken to increase the accuracy of the measurement. The face of the instrument may be changed and new vernier zero settings chosen so as to obtain several values for the same angle and a mean then computed. Taking the mean of face left and face right readings will eliminate the errors caused if permanent adjustments 2 and 3 have not been carried out correctly.

EXAMPLES OF BOOKING AND REDUCTION

Example 1.

	Vernier A	*Vernier B*
Instrument directed to *C* (*BC*)	59° 31′ 40″	239 31′ 40″
Instrument directed to *A* (*BA*)	00° 01′ 20″	180° 01′ 20″
	59° 30′ 20″	59° 30′ 20″

Mean 59° 30′ 20″

Example 2.

Station B	*Vernier A*	*Vernier B*
BC	93° 34′ 40″	273° 34′ 40″
BA	01° 15′ 20″	181° 15′ 40″
	92° 19′ 20″	92° 19′ 00″

Mean 92° 19′ 10″

It may be found, as in the second example, that the vernier readings give values varying by the least count of the instrument.

It is advisable to sight the intersection of the cross-hairs as near as possible to the bottom of the observed signal to reduce to a minimum any effects due to that signal, perhaps a ranging rod, not being vertical. Ensure that the lower clamp or tangent screw is not disturbed after setting the instrument in position (*b*), otherwise the scale plate may be moved and a false reading obtained.

Accurate bisection of the mark by the vertical hair is essential, and accordingly the mark (or signal) must not be so wide that bisection errors can occur, but should be of such a size that a fine target is presented, although at the same time it should be distinctly visible. As mentioned previously, a ranging rod may be used, but for short sights this can prove too wide for bisection unless the lower end is in the line of sight. Two more suitable marks are: (i) a chaining arrow at the station, or (ii) a plumb bob suspended from a rod so that it hangs over the station, the plumb line being protected from the wind. The instrument is directed on to this line, not on to the plumb bob.

METHOD OF REPETITION. (*a*) Set up and level the instrument exactly over station *B*. By means of the upper tangent screw the vernier is made to read zero on the scale plate, and the instrument is then directed to *A* with the two plates clamped together. Coincidence and bisection by the vertical hair on *A* is obtained by means of the lower tangent screw. Check and note both vernier readings.

(*b*) With the lower plate clamped, the telescope is directed towards *C*, final coincidence and bisection of the signal by the vertical hair being obtained by the upper tangent screw. The vernier readings would now give the value of angle *ABC*, which is noted as a check on the final result.

(*c*) The lower plate is now unclamped (with the verniers still retaining the reading obtained in (*b*)), and the telescope rotated clockwise back to *A*, obtaining a bisection in the normal way by the lower tangent screw.

(*d*) After clamping the lower plate, the upper clamp is freed and the telescope redirected to *C* in the normal way and the verniers should now read double the angle obtained in (*b*).

(*e*) This procedure could be repeated, say, once more, and then the face of the instrument changed and the whole repeated a further three times, always swinging in a clockwise direction. The mean total angle is now found from the vernier readings, and divided by the number of movements carried out, and a value obtained which should be lower than the least count of the instrument, i.e. if the instrument reads to 20 sec this method may give a value to say 5 sec. Improved accuracy may be obtained for reading the external angle *ABC* in a similar manner.

Measurement of Vertical Angles. The angles of elevation (+) or depression (−) are measured with respect to the horizontal plane containing the trunnion axis of the instrument. Assuming, as in the previous section, that the permanent adjustments have been checked, then the instrument will be set up and levelled, over the station, using the plate bubble. The altitude bubble on the vertical circle should now be nearly, if not quite, central. Level up this bubble using the levelling screws, and wheel through 180° to see whether the bubble "traverses" (*see* tests on the Dumpy level). If it does not, take out half the bubble displacement on the *clip screws* and the other half on the levelling screws. Repeat until the bubble traverses. If only a few vertical angles are required, it is probably quicker to level the altitude bubble accurately before each reading, as shown on pages 97 and 98.

The telescope is now directed to one of the signals and exact coincidence on the mark obtained, using both horizontal and vertical slow-motion devices. The mean reading of the vertical circle verniers will now give the angle (+ or −) subtended by the signal at the instrument relative to the horizontal plane. If the telescope is directed to the other signal, obtaining coincidence as before, the mean vernier readings give a second angle subtended, and applying one mean reading to the other will give that vertical angle which the two signals subtend at the instrument. These verniers may be figured in both directions and the observer must use those figures which count in the same direction as those on the main scale. It is immaterial whether the signals are in the same vertical plane or not, so long as the instrument is in adjustment.

If there be some error in adjustment, then a true value of the vertical angle will not have been obtained. In Fig. 4.4, the line joining the zeros on the verniers is inclined at β to the horizontal when the altitude bubble is brought to the centre of its run in each case by means of the clip screw. Also in this example an error of α is present, this being

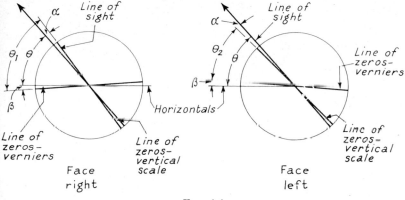

Fig. 4.4

the angle between the line joining the zeros on the vertical circle and the line of collimation.

The true vertical angle in each case is θ and the observed values are θ_1 and θ_2 respectively. Change of face will reverse the conditions insofar as angles α and β are concerned.

$$\begin{array}{l}\textit{Face Right} \quad \theta_1 = \theta + \alpha + \beta \\ \textit{Face Left} \quad \theta_2 = \theta - \alpha - \beta\end{array}, \text{ whence } \frac{\theta_1 + \theta_2}{2} = \theta$$

Thus, provided the altitude bubble is brought central by the clip screw, the mean of two readings, one taken face left and the other face right, will give the required value of the vertical angle.

If the altitude bubble is not central, the line connecting the vernier zeros would be inclined to that line given when the altitude bubble is central, and any movement of the altitude bubble would cause an equivalent rotation of the line of vernier zeros in the same direction. Let

O_R be reading of objective end of bubble, instrument FR,
O_L be reading of objective end of bubble, instrument FL,
E_R be reading of eyepiece end of bubble, instrument FR, and
E_L be reading of eyepiece end of bubble, instrument FL,
γ be the angular value of one division of the bubble tube.

Refer to Fig. 4.4; angle β would be reduced by $\dfrac{O_R - E_R}{2}\gamma$ when the instrument is face right, and would be increased by $\dfrac{O_L - E_L}{2}\gamma$ in the face left case.

Thus
$$\theta_1 = \theta + \alpha + \left(\beta - \frac{O_R - E_R}{2}\gamma\right)$$

and
$$\theta_2 = \theta - \alpha - \left(\beta + \frac{O_L - E_L}{2}\gamma\right)$$

where θ_1 and θ_2 are the observed readings in each case. Adding the equations gives

$$\theta_1 + \theta_2 = 2\theta - \frac{\gamma}{2}[O_L + O_R - E_L - E_R]$$

$$\therefore \qquad \theta = \frac{\theta_1 + \theta_2}{2} + \frac{\gamma}{4}[\Sigma O - \Sigma E]$$

and the correct angle is given by applying to the mean of the observed altitudes for both faces, the factor $\dfrac{\gamma}{4}(\Sigma O - \Sigma E)$. This factor may be written in general as the product of the angular value of one division and the sum of objective end readings, less the sum of the eyepiece end readings, all divided by the *number of ends* read.

Setting Out a Horizontal Angle ABC

(a) To the Least Count of the Instrument. In this case it will be assumed that the angle to be set out with, say a 20-sec instrument, is to a value which is not expressed any lower than 20 sec, say 1° 40′ 20″.

The instrument will be set up and accurately levelled over B and by means of the upper tangent screw a vernier is brought to give a zero reading. The lower plate is permitted to rotate and the telescope wheeled to sight A (the left-hand mark), bisection on this point being obtained by means of the lower tangent screw after clamping. The upper clamp is released and the vernier plate rotated with the lower plate fixed until the above vernier has a reading which gives the angle to be set out, the final setting being obtained by means of the upper tangent screw. Any suitable signal (e.g. an arrow) may be moved to and fro until an accurate bisection by the stationary vertical hair is made.

(b) To a Value below the Least Count. In this case an angle of the order of say 51° 49′ 5″ may have to be set out using the 20-sec instrument.

The same approach as in method (a) is adopted initially, setting out an angle ABC at B which would apparently be 51° 49′ 0″ by the vernier readings.

This angle may now be more accurately measured using the method of repetition and a more reliable value so found, and the difference (θ) between the required value and that measured then computed (*see*

Fig. 4.5). Now measure the length BC and this will lead to the displacement required to shift the trial position of C to the required position. Since $CC' = BC \tan \theta$, and θ and BC being known, the calculated value CC' can be set out at right angles to CB. A check on $A\hat{B}C'$ should now be possible by means of the method of repetition.

Permanent Adjustments of the Theodolite

The following adjustments may be required–

(1) To set the vertical axis of the instrument truly vertical and to adjust the plate bubble.

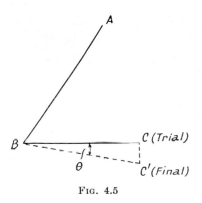

FIG. 4.5

(2) To set the telescope sighting line at right angles to the horizontal or trunnion axis of the instrument.

(3) To set the horizontal axis at right angles to the vertical axis.

(4) To adjust the altitude bubble and vertical circle zero.

An analysis of the errors caused by failure to make these adjustments correctly may be found in Jameson's *Advanced Surveying*.*

1. ADJUSTMENT OF PLATE BUBBLE—TO SET VERTICAL AXIS TRULY VERTICAL. This bubble, set on the upper plate, must be central when the plate is horizontal; and the vertical axis must then be vertical, since the plate and vertical axis are set at right angles by the maker. If this adjustment is correct then the plate bubble will remain in the centre of its run throughout a complete rotation of the telescope in azimuth.

The instrument is set up and levelled as accurately as possible with the aid of the tripod legs. The plate bubble tube is placed over a pair of foot screws and by means of these the bubble is brought to the centre of its run (Fig. 4.6a). The instrument is rotated through 180° in azimuth (Fig. 4.6c) and any deviation of the bubble noted. Half this deviation is corrected by the same pair of footscrews and the remainder by means of the plate bubble capstan nuts. The bubble tube is now arranged over the other footscrew, and half any deviation is taken out by that

* *Advanced Surveying* by Jameson (Pitman).

screw. This procedure is repeated until the bubble remains central for
any position.

In Fig. 4.6c, the plate has been turned through 180° in azimuth;
the bubble tube is shown with ends 1 and 2 reversed in position, and
the angle 90° − ε between the vertical axis and the bubble is as
indicated. This means that an angle of 2ε will be given by the intersection of the horizontal plane and the plane through the bubble axis.
Using the footscrews, the movement of the bubble through an angle ε
will evidently move the vertical axis into its true position and the
capstan nuts will take out the remainder and place the line 1.2 parallel

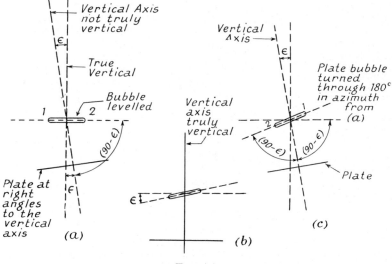

Fig. 4.6

to the plate (since inspection of Fig. 4.6b shows the angle between the
bubble axis and the plate then to be ε).

2. SETTING THE TELESCOPE SIGHTING LINE AT RIGHT ANGLES TO
THE HORIZONTAL AXIS. The sighting line joining the vertical cross-hair
to the optical centre of the object glass must be at right angles to the
horizontal axis. If the vertical hair is displaced to one side or the other
of its true position then the line of sight will no longer be along the telescope and at right angles to the horizontal axis (Fig. 4.7). The line will
trace a flat cone and not a plane if this adjustment is not carried out.

The instrument is set up and levelled on reasonably level ground in
such a position that a field of view up to one hundred metres is available
on both sides of the instrument.

(a) A small sharp object (e.g. a chaining arrow) is placed at A about
one hundred metres from the instrument and the telescope, say face left,
is directed towards it, obtaining exact coincidence on the cross-hairs
using the slow-motion devices.

(*b*) The telescope is transitted and a second arrow *B* set at about the same distance away from the instrument as *A* on the line of sight. *A* and *B* should be at the same level if possible so that any error due to the next adjustment is avoided. The instrument is now face right (Fig. 4.8).

(*c*) The clamping arrangement is freed and the telescope rotated

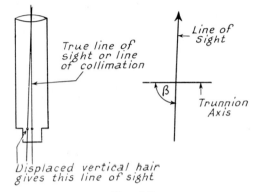

True line of sight or line of collimation

Line of Sight

β   Trunnion Axis

Displaced vertical hair gives this line of sight

Fig. 4.7

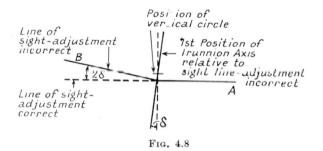

Line of sight-adjustment incorrect

Position of vertical circle

1st Position of Trunnion Axis relative to sight line-adjustment incorrect

B

2δ

A

Line of sight-adjustment correct

¼δ

Fig. 4.8

through 180° in azimuth and exact coincidence obtained at *A* with the slow-motion screw after clamping. The instrument is still face right.

(*d*) The telescope is again transitted towards the position of *B*. The vertical hair position is again noted. If the line of sight bisects *B*, then the adjustment is correct; if not, a second arrow B_1 is placed at the same distance from the instrument as *B* and on the line of sight. The instrument is now face left. The line of sight is directed to some point *C* positioned such that $B_1C = \frac{1}{4}BB_1$ (i.e. the line of sight is turned through an angle δ relative to the horizontal axis, Fig. 4.9). This redirection is effected by means of the two side diaphragm screws. Repeat the adjustment as necessary until correct. It is as well to check that the vertical hair is truly vertical at this stage. If the telescope is directed on to a very small sharply-defined nearby point, this point should appear to move along the vertical hair as the telescope is moved in a vertical plane. The diaphragm adjusting screws enable the cross-hairs to be turned until the vertical hair is truly vertical. (Otherwise, a sight can be taken on to a plumb-bob line.)

3. To Set the Horizontal Axis at Right Angles to the Vertical Axis. In the previous adjustment the line of sight has been arranged to trace out a plane and to be at right angles to the horizontal axis. The horizontal axis is now to be parallel to the scale plate and the telescope will transit in a true vertical plane.

(*a*) The instrument is set up and correctly levelled in such a position that the telescope can be directed, with the line of sight inclined at 45°

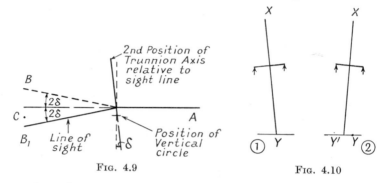

FIG. 4.9   FIG. 4.10

approximately, on to a small sharply-defined mark *X* (Fig. 4.10) such as the tip of a lightning conductor. After clamping the plates, coincidence is obtained with the slow-motion screws in both the horizontal and vertical movements.

(*b*) Depress the telescope and take a reading *Y* on a scale at the level of the instrument axis (alternatively, make a fine mark).

(*c*) Transit the telescope and again obtain an accurate bisection on *X*.

(*d*) Depress the instrument, the face having been changed since the previous depression, and again note the scale reading. If this is the same as in (*b*) (or the line of sight coincides with the mark), then the adjustment is correct. If not, the line of sight must be moved to give a reading on the scale mid-way between the other two. Direct the telescope here, and then on elevation it will not intersect at *X*. By means of an adjustment screw on one of the trunnion axis bearings, the vertical hair is sighted on *X*.

The adjusting screw lifts or lowers the end of the axis shown in Fig. 4.10 to bring the line of sight into the vertical plane.

4. Adjustment of Altitude Bubble and Vertical Circle Zero. The instrument is set up, levelled, and the telescope directed towards a levelling staff, held vertically, about 50 to 60 m away. The altitude bubble is now brought to the centre of its run using the clip screw and the telescope is placed (with suitable manipulation of the vertical circle tangent screw) so that the verniers read zero on the vertical scale (Fig. 4.11*a*).

(*a*) Take a reading on the levelling staff using the middle hair if three stadia hairs are provided.

(*b*) Transit the telescope and redirect it on the staff.

(*c*) Bring the altitude bubble back to the centre of its run using the clip screw and set the vertical circle to read zero again by moving the telescope as required (Fig. 4.11*b*).

(*d*) Note the staff reading, and if the two separate readings agree

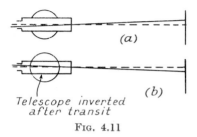

*Telescope inverted
after transit*

Fig. 4.11

then the adjustment is correct. If not, the mean of the two readings, and the trunnion axis, will be in the same horizontal plane. Then—

1. Set the telescope to this mean reading by means of the vertical circle tangent screw.

2. Arrange the verniers to read exactly zero by means of the clip screw.

3. Since (2) now moves the altitude bubble, return this to the centre of its run by means of its capstan screw.

The test is repeated to remove all discrepancies as far as possible.

It is possible that this bubble will not stay exactly central for all vertical angles, and this is caused by eccentricity of the telescope axis.

Note that though the adjustments are carried out separately, they are not completely independent of one another. It may be necessary therefore to repeat the cycle of tests before complete adjustment is achieved, and now the line of sight, the horizontal or trunnion axis and the vertical axis are mutually at right angles.

Traverse Surveys

A theodolite traverse survey consists of the measurement of (*a*) angles between successive lines (or bearings of each line), and (*b*) the length of each line.

A number of points (*stations*) will be chosen to fulfil the demands of the survey, the lines joining these stations being the *treavrse lines*. If the figure formed by these lines closes at a station, i.e. if they form a polygon, then a *closed traverse* (Fig. 4.12) has been obtained. In the special case where two of the stations have been previously located on the ground by triangulation (*see* Chapter 9), and the traverse lines are based on these points, although the line joining them need not be measured the survey will still be considered to be a closed traverse.

On the other hand a traverse starting at say station A (not a triangulation station) and ending at E is called an *unclosed traverse* (*see* Fig. 4.12). Each type has its particular uses, but the closed traverse is the more satisfactory figure since it is the easiest one to which to apply corrections for the errors which invariably occur. The unclosed traverse survey can be carried out when the survey is comparatively long and narrow, such as that required for a trunk sewer, pipe line, main trunk

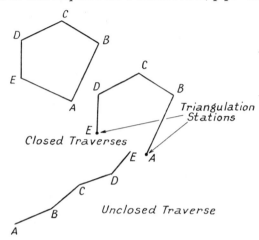

Fig. 4.12

road or rail construction (though where the length is great, consideration should always be given to the possibility of tying in to Ordnance trigonometrical stations).

A closed traverse survey may be used for the framework of surveys for housing or factory sites, and determination of the perimeters of lakes, etc. They may also have to be undertaken when setting out shafts to tunnels which are being driven under built-up areas: in such cases, of course, it may be impossible to set out a surface line directly above the tunnel.

Choice of Stations and Procedure. On the surveys which form a fair part of the duties of an engineering assistant in this country, the party consists of the surveyor himself and two chainmen or helpers. The angles will be read by the surveyor and booked by him, while the helpers will often be stationed near the signals at the adjacent stations to make any adjustments to their verticality (and sometimes, particularly in traffic, to protect them). On surveys of great extent, several parties under the control of one chief surveyor may function simultaneously. The chief surveyor will carry out the reconnaissance for positioning the stations, and will generally co-ordinate the work.

The stations should be chosen with the requirements of the survey in mind. When surveying land for a housing site, for instance, the traverse lines will be used for picking up much of the detail to be plotted, so that they will follow the perimeter of the site. Cross lines, sited to serve as checks, will also be arranged, if possible, to enable the collection of other detail. For setting out purposes, the nature of the district will fix the stations. This is often the case in unclosed traverses, where bends in roads, etc., dictate the precise pattern. When there is no such restriction and the framework is for control purposes only, then long sights are preferable, and to avoid refraction the lines of sight should be well clear of the ground. Stations when chosen should be placed in such a way that there will be no displacement, since some time may elapse between angular and linear measurements, plotting the survey, and actual construction work, if any. The stations may serve for the survey, or for control of levelling or contouring operations over the site, and also for setting out. On roads, etc., short heavy nails may be driven in and located by ties to nearby permanent features, while in fields, stout wooden pegs with small-headed nails driven in their tops are often favoured. It is preferable to set these in concrete, if possible, and to locate them by ties from nearby features. Alternatively it is possible to purchase proprietary forms of plastic or metallic marking systems for purposes of monumentation. Areas liable to flooding and settlement are always suspect, and it must be remembered that although a station position may be very acceptable insofar as the lines of sight from adjacent stations are acceptable, the station itself will have to be occupied and should therefore be suitable for (a) setting up the instrument, and (b) the surveyor to be able to read the angle(s) subtended there.

The Linear Measurements will require three persons: two actually carrying out the work, and the surveyor directing operations and booking the data. On small traverse surveys the chain may be used and detail collected as the measurement along the traverse lines proceeds. On the other hand—and particularly on large surveys—it is preferable to split this work so that the lines are measured first and the detail collected by chaining later. The equipment will now include a steel band, together with a spring balance, Abney Level, and thermometer to improve the accuracy, since the errors in traverse surveying are more likely to arise from linear than from angular measurement. The steel band and other equipment should always be used for control surveys and for those traverses devised for setting out works.

When the chain is being used, the procedure for measurement will be as described in Chapter 1. With the steel band, alignment may be either by ranging rod or by a theodolite directed down the line. The purpose of the spring balance is to ensure that the band is tensioned up to the value at which it was calibrated, i.e. if the band is 30 m long at 20°C under a 5 kg pull on the flat, then a tension of 5 kg should be applied to eliminate any correction for pull. The balance is usually attached to a ranging rod or arrow, and thence to one end of the band, by means of a short cord. The far end of the band will then be attached to a second rod, and if these rods be set firmly on or in the ground and levered backwards, the tension applied to the tape can be regulated to any value. The procedure is then the same as for the chain. The spring balance is at the leader's end, and he is lined in when the follower has the band correctly positioned on his mark: an arrow is inserted by the leader at his end of the band after tensioning. The band is stretched on the ground and the temperature of the tape is found and booked. It is important that the thermometer bulb be shaded from the sun's rays. Finally, the slope of the ground is measured by means of the Abney Level and booked.

Corrections are now made to the length as taped. If the temperature of the tape be $T°C$ and the standard temperature $t°C$, then the correction for temperature is

$$L\alpha(T - t)$$

where L = length of the band and α = coefficient of linear expansion of the band material. The correction may be positive or negative depending upon whether T is greater or less than t. The slope correction is applied to reduce the sloping length to the horizontal length required for plotting, as outlined in Chapter 1. Thus, if α is the angle measured, then the correction applied for slope is

$$- L(1 - \cos \alpha)$$

If the ground slope varies appreciably within the length of the band, say 20 m on a 3° slope and 10 m on the flat, then L will equal 20 in the above expression.

When the lines have been accurately measured by the band, the chain is run out and detail collected in the usual manner, the final chained lengths serving as checks on the values given by the band.

Angular Measurements. If internal angles are being read, it is usual to proceed from station to station round the traverse in an anti-clockwise direction. Starting at A, Fig. 4.13, the instrument will be directed to F and then wheeled to B as previously described. The next station to be occupied will be B, where the telescope is directed first on A and then on C. E would perhaps be observed here as a check. It is advisable to change face and zeros at each station. The angles may be booked in the field book on separate pages, or probably, at most, two sets to the page. For example, one might book at stations A and B,

Station A	Face Left		Face Right	
	Vernier A	*Vernier B*	*Vernier A*	*Vernier B*
AB	93° 03′ 20″	273° 03′ 00″	133° 06′ 20″	313° 07′ 00″
AF	00° 01′ 40″	180° 01′ 40″	40° 05′ 00″	220° 05′ 20″
	93° 01′ 40″	93° 01′ 20″	93° 01′ 20″	93° 01′ 40″

$$F\widehat{A}B = 93° 01′ 30″$$

Station B	Face Left		Face Right	
	Vernier A	*Vernier B*	*Vernier A*	*Vernier B*
BC	163° 43′ 20″	343° 43′ 20″	222° 55′ 40″	42° 55′ 40″
BE	84° 56′ 40″	264° 56′ 20″	144° 08′ 40″	324 09′ 00″
BA	01° 02′ 20″	181° 02′ 20″	60° 14′ 40″	240° 14′ 40″

$$A\widehat{B}E = 83° 54′ 10″ \qquad A\widehat{B}C = 162° 41′ 00″$$

Note that (i) the final reading BC (Vernier B) is $42° 55' 40''$ which means that the vernier zero has crossed the zero on the graduated circle, (ii) the instrument has been wheeled clockwise, reading from the backward station to the forward one.

When all the internal angles have been measured, it is possible to make a check, since the sum of the angles should equal $(2n - 4)$ right angles, where n is the number of sides.

Should there be a large error in the sum of the observed angles when related to $(2n - 4)$ right angles, then the angles must be rechecked. The value of the check sight BE is now apparent, since the error mentioned above may be isolated into one of the two quadrilaterals

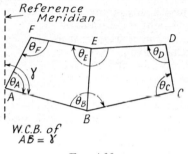

Fig. 4.13

by balancing their values as separate units. A small error may be apportioned equally among the angles.

Bearings. It will be noticed that in Fig. 4.13, the direction of line BC can be related to line BA when angle ABC is measured, so that the *bearing* of BC is given relative to BA, which thus serves as a fixed line or *meridian*. In practice, however, the meridian would be chosen from one of the following—

(a) *The true meridian* at a station. This is the trace which a plane containing the north and south poles and the station itself makes on the earth's surface. It may be located by astronomical observations.

(b) *Magnetic meridian* at a station. This is determined by means of a floating compass needle (*N.B.* there must be no local attraction due to magnetic materials, and the needle must be unrestricted and balanced). True and magnetic meridians do not coincide at most stations, and the magnetic meridian may lie east or west of the true meridian (*see* Chapter 5).

The theodolite may have a compass needle above the vernier plate and between the standards, or alternatively a tubular (trough) compass fitted to the standards. The north-south line of the compass should be parallel to the line of collimation of the telescope when the instrument is set up and levelled at the station. A vernier is set to read zero on the scale plate and the plates clamped. The lower clamp is then freed and the telescope moved round until the compass needle (which must be floating) is lying north–south. Final adjustment is made by the lower plate tangent screw. If the lower plate is now clamped, and the upper clamp freed, then by sighting through the telescope the bearing of any point may be observed relative to the magnetic meridian at the station. It will be read directly from that vernier which read zero initially.

(c) *Any chosen meridian.* It is not essential for either (a) or (b) to be adopted, although on the completed plan the north point should be shown and this would usually entail fixing the direction of the chosen meridian relative to the true meridian. Any permanent and sharply-defined point (say a church spire or weather cock) could be accepted, and the bearing of a line AB (*see* Fig. 4.14) at a station A relative to this meridian would be determined by first sighting the point and then swinging to sight on station B.

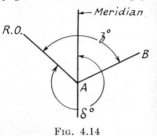

Fig. 4.14

In the case of the true meridian, it is possible to obtain the bearing of traverse line AB astronomically, or, alternatively, by determining the true bearing of some reference object which is not a survey station, from A, and then relating AB to that bearing by measuring the angle subtended by the reference object $R.O.$ and station B, at A.

In Fig. 4.14,

$$\text{Bearing of } AB \text{ (relative to meridian)} = (\delta + z)^c - 360°$$

TYPES OF BEARING. Bearings may be of two types: (i) Whole circle bearings (W.C.B.), and (ii) Reduced (quadrantal) bearings (R.B.).

Whole circle bearings are measured clockwise from north, from 0° to 360°. (For ease in computation the chosen point in (c) above could be taken to be north.) Reduced bearings are measured from 0° to 90° only, and a line can thus lie in any of the four quadrants. The bearing gives the direction of the line—whether north or south, followed by its angular deviation, either east or west, from this direction.

Fig. 4.15 indicates the connexion between reduced and whole circle bearings; the former may be deduced from known values of the latter.

Referring to Fig. 4.13, the mean internal angles are found to be θ_A, θ_B, etc., while the whole circle bearing of AB has been determined as γ. The whole circle bearings of the remaining traverse lines are required. Conditions at B (Fig. 4.13) are reproduced in Fig. 4.16, the dotted line through B being the north-south meridian NBS.

The whole circle bearing required is $N\hat{B}C = \gamma_1$. The whole circle bearing of BB_1, which equals that of $AB = \gamma$

$$N\hat{B}B_1 = A\hat{B}S \text{ (vertically opposite)}$$
$$= \gamma$$

Therefore

$$\gamma_1 = S\hat{B}A + A\hat{B}C - N\hat{B}S$$
$$= \gamma + \theta_B - 180°$$

i.e. the whole circle bearing of *BC* is given by the sum of the whole circle bearing of *AB* and the internal angle at *B* *minus* 180°.

Inspection of *C* shows that the whole circle bearing of *CD*, which

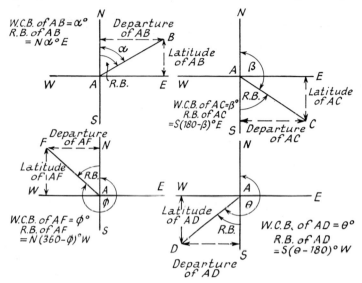

FIG. 4.15

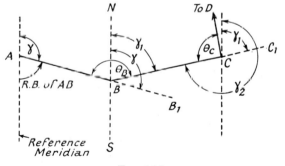

FIG. 4.16

equals γ_2, is given by the sum of the whole circle bearing of $BC(\gamma_1)$ and the internal angle at $C(\theta_2)$ *plus* 180°.

To summarize, then, for the general case: to determine the whole circle bearing of a line at a station—

(i) Add the included angle at the station to the whole circle bearing of the previous line.

(ii) If the sum obtained is below 180°, then add 180° to it (i.e. as for line *CD*).

(iii) If the sum exceeds 180°, then deduct 180° from it (i.e. as for line BC).

In Fig. 4.13, $\gamma = 115°\ 20'\ 00''$

$$\theta_A = \ \ 99°\ 49'\ 30'' \qquad \theta_D = \ \ 93°\ 40'\ 20''$$
$$\theta_B = 149°\ 23'\ 00'' \qquad \theta_E = 182°\ 16'\ 50''$$
$$\theta_C = \ \ 91°\ 17'\ 30'' \qquad \theta_F = 103°\ 32'\ 50''$$

W.C.B. of $AB = 115°\ 20'\ 00''$
$\theta_B = 149°\ 23'\ 00''$

$264°\ 43'\ 00''$
Ddt.　$180°\ 00'\ 00''$

∴　W.C.B. of $BC = \ \ 84°\ 43'\ 00''$
$\theta_C = \ \ 91°\ 17'\ 30''$

$176°\ 00'\ 30''$
Add　$180°\ 00'\ 00''$

∴　W.C.B. of $CD = 356°\ 00'\ 30''$ 　　*Check*
$\theta_D = \ \ 93°\ 40'\ 20''$ 　　W.C.B. of FA 　$195°\ 30'\ 30''$
　　　　　　　　　　　　　　　　　　　θ_A 　　　　$99°\ 49'\ 30''$
$449°\ 40'\ 50''$
Ddt. $= 180°\ 00'\ 00''$ 　　　　　　$295°\ 20'\ 00''$
　　　　　　　　　　　　　　Ddt.　$180°\ 00'\ 00''$

∴　W.C.B. of $DE = 269°\ 40'\ 50''$ 　　W.C.B. of $AB = \underline{\underline{115°\ 20'\ 00''}}$
$\theta_E = 182°\ 16'\ 50''$

$451°\ 57'\ 40''$
Ddt.　$180°\ 00'\ 00''$

∴　W.C.B. of $EF = 271°\ 57'\ 40''$
$\theta_F = 103°\ 32'\ 50''$

$375°\ 30'\ 30''$
Ddt. $= 180°\ 00'\ 00''$

Therefore W.C.B. of $FA = 195°\ 30'\ 30''$

The reduced bearings may now be deduced as follows—

AB	$180°-115°\ 20'\ 00''$	S $\ 64°\ 40'\ 00''$ E
BC	$84°\ 43'\ 00''$	N $\ 84°\ 43'\ 00''$ E
CD	$360°-356°\ 00'\ 30''$	N $\ 03°\ 59'\ 30''$ W
DE	$269°\ 40'\ 50''-180°$	S $\ 89°\ 40'\ 50''$ W
EF	$360°-271°\ 57'\ 40''$	N $\ 88°\ 02'\ 20''$ W
FA	$195°\ 30'\ 30''-180°$	S $\ 15°\ 30'\ 30''$ W

The whole circle bearing values indicate in which of the four quadrants the lines lie.

Latitudes and Departures. In the position now reached the lengths of the lines are known, the internal angles have been measured and adjusted, and bearings have been calculated. It is possible at this stage to plot the survey using a scale and protractor, and in fact this may be done in order (*a*) to determine a suitable scale for plotting (if this has not already been decided by, for example, statutory requirement), or (*b*) to obtain a suitable layout scheme on several sheets, if the survey is a large one.

For plotting the survey proper, however, it is preferable to use *co-ordinates* for fixing the positions of the traverse stations. This is the most satisfactory method because

(1) it enables errors to be assessed and adjusted,
(2) each station is plotted independently,
(3) it is not dependent on any angle measuring device.

The co-ordinates are calculated from the projections of the lines on the meridian and at right angles to the meridian. Referring to Fig. 4.15, it will be observed that the projections of each line on the co-ordinate axes are termed latitudes or departures. The *latitude* of a line is the projection on the axis representing the reference meridian, and the *departure* is the projection on the axis at right angles to the meridian.

Line	*Latitude*	*Departure*
AB	$+ AB \cos \alpha$	$+ AB \sin \alpha$
AC	$- AC \cos (180 - \beta)$	$+ AC \sin (180 - \beta)$
AD	$- AD \cos (\theta - 180)$	$- AD \sin (\theta - 180)$
AF	$+ AF \cos (360 - \phi)$	$- AF \sin (360 - \phi)$

It is apparent that the latitude is the product of the length of line and the cosine of its reduced bearing, while the departure of that line is the product of length and the sine of the reduced bearing.

Signs are given to latitudes and departures to indicate the direction which the line takes, the convention used being as follows—

Positive latitude denotes a line pointing northwards,
negative latitude denotes a line pointing southwards,
positive departure denotes a line pointing eastwards,
negative departure denotes a line pointing westwards.

These signs are obtained from the reduced bearings.

Latitude and departures will give the co-ordinates of one station at the end of a line with respect to the station at the other end of the line.

The latitude of AB in Fig. 4.17 is negative (B is south of A), but that of BA would be positive (A north of B), since AB has a reduced bearing of S α° E, whereas BA has a reduced bearing of N α° W. As mentioned previously, both latitudes and departures of lines from stations must be referred to the same axes. It can be seen from Fig. 4.17 that for the closed traverse starting and ending at the same point, the sum of the

positive latitudes should equal the sum of the negative latitudes, and
similarly the sum of the positive departures should equal the sum of the
negative departures.

The surveyor is now in a position to calculate the latitude and departure of each line and check the above statement for his closed traverse
survey. Generally, he will find that in fact the sums of positive and
negative latitudes, and the sums of positive and negative departures,
are not equal, with a result that a closing error is present, which must
be adjusted. Note that plotting by length and angles set off on a protractor would leave the surveyor with a closing error which he could

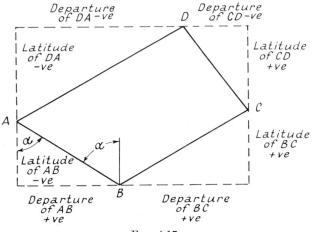

Fig. 4.17

not balance out logically. Consider as an example the closed traverse
ABCEGJLA, field data being given in Table I and the computations
in Table II.

(i) In Table I, *AB* indicates not only the line in question but also
the included angle at *A*; similarly *BC* gives the included angle at *B*,
and so on.

TABLE I

Line		Mean Included Angle	Length (m)
AB	$(\hat{A})$	94° 10′ 00″	103·40
BC	$(\hat{B})$	178° 19′ 00″	157·25
CE	$(\hat{C})$	118° 21′ 45″	143·36
EG	$(\hat{E})$	94° 42′ 25″	169·08
GJ	$(\hat{G})$	158° 07′ 30″	176·74
JL	$(\hat{J})$	89° 03′ 55″	110·60
LA	$(\hat{L})$	167° 15′ 50″	140·83

The whole circle bearing of *AB* is 187° 22′ 20″

(ii) The sum of the internal angles of the traverse should be $(2n - 4) = (2 \times 7 - 4) = 10$ right angles, and there is thus an overall error of $+ 25''$ on the sum of the angles. All the angles were measured with the same accuracy and under the same conditions. It is considered that they are all equally liable to error and so the overall error is removed by applying a correction of $- 3''$ or $- 4''$ to alternate angles.

(iii) Knowing the whole circle bearing of AB, the whole circle bearings and reduced bearings of all lines are now computed using the corrected included angles.

i.e. W.C.B. of $AB = 187°\ 22'\ 20''$

$$\hat{B} = 178°\ 18'\ 57''$$

$$\begin{array}{c} \hline 365°\ 41'\ 17'' \\ \text{Ddt.} \quad 180°\ 00'\ 00'' \\ \hline \end{array}$$

W.C.B. of $BC = 185°\ 41'\ 17''$ $\therefore$ R.B. of $BC = $ S $05°\ 41'\ 17''$ W

$$\hat{C} = 118°\ 21'\ 41''$$

$$\begin{array}{c} \hline 304°\ 02'\ 58'' \\ \text{Ddt.} \quad 180°\ 00'\ 00'' \\ \hline \end{array}$$

W.C.B. of CE $124°\ 02'\ 58''$ $\therefore$ R.B. of $CE = $ S $55°\ 57'\ 02''$ E

etc. etc.

(iv) Latitudes and departures are calculated and entered on Table II after applying the sign convention.

Lat. of $AB = 103 \cdot 40 \cos 7°\ 22'\ 20'' = 102 \cdot 55$ m

and is negative since AB is running southwards.

Dep. of $AB = 103 \cdot 40 \sin 7°\ 22'\ 20'' = 13 \cdot 27$ m

and is negative since AB is also running westwards, i.e. B is $102 \cdot 55$ m south and $13 \cdot 27$ m west of A.

(v) From the four columns the differences of the positive and negative latitude and departure totals are found. If the latitude and departure of each line be plotted, a point A_1 would result, near A, but $0 \cdot 07$ m south and $0 \cdot 13$ m east of A. $A_1 A$ is the closing error of the survey.

Closing error $= \sqrt{0 \cdot 07^2 + 0 \cdot 13^2} = 0 \cdot 15$ m

at a bearing of $\tan^{-1} \dfrac{0 \cdot 13}{0 \cdot 07} \left(\text{or, } \tan^{-1} \dfrac{\text{Departure Error}}{\text{Latitude Error}} \right)$, which is approximately S $61°\ 42'$E.

Accuracy. The closing error of $0 \cdot 15$ m deduced from the measurements obtained during the traverse implies an error of, say, 1 in 6,700 in horizontal distance. This lies between the limits of second and third accuracy suggested by Prior (*op. cit., see* p. 86) as follows—

TABLE II

Line	Length	Included Angle	Corrected Included Angle	W.C.B.	R.B.	Latitude +	Latitude −	Departure +	Departure −
AB	103·40	94° 10′ 00″	94° 9′ 56″	187° 22′ 20″	S 7° 22′ 20″ W		102·55		13·27
BC	157·25	178° 19′ 00″	178° 18′ 57″	185° 41′ 17″	S 5° 41′ 17″ W		156·48		15·59
CE	143·36	118° 21′ 45″	118° 21′ 41″	124° 2′ 58″	S 55° 57′ 2″ E		80·27	118·78	
EG	169·08	94° 42′ 25″	94° 42′ 22″	38° 45′ 20″	N 38° 45′ 20″ E	131·85		105·85	
GJ	176·74	158° 7′ 30″	158° 7′ 26″	16° 52′ 46″	N 16° 52′ 46″ E	169·12		51·32	
JL	110·60	89° 3′ 55″	89° 3′ 52″	285° 56′ 38″	N 74° 3′ 22″ W	30·38			106·35
LA	140·83	167° 15′ 50″	167° 15′ 46″	273° 12′ 24″	N 86° 47′ 36″ W	7·88			140·61
	1,001·26	900° 00′ 25″	900° 00′ 00″			339·23	339·30	275·95	275·82
							339·23	275·82	
							Error − 0·07	+ 0·13	

Order	Unadjusted Horizontal Distance	Unadjusted Horizontal Angles
First	1 in 25,000	$2\sqrt{N}$ sec
Second	1 in 10,000	$10\sqrt{N}$ sec
Third	1 in 5,000	$30\sqrt{N}$ sec
Fourth	1 in 2,000	$60\sqrt{N}$ sec

It will be noted that the angles were obtained to, or satisfy, second-order accuracy.

Adjustment of Errors. The co-ordinates of the traverse stations can be determined after the closing error is eliminated. There are various methods for correcting the latitudes and departures so as to give no closing error.

(*a*) BOWDITCH'S METHOD. This method assumes that the probable error in the bearing of a line due to some inaccuracy in angular measurement gives a displacement of one end of that line, relative to the other end, equal and at right angles to that displacement, due to the probable error in measurement of its length. The probable error in length (l) is taken to be proportional to $\sqrt{l}$. Thus in Fig. 4.18 the displacement of B due to the probable error in linear measurement is BC (and is proportional to $\sqrt{l}$), and the displacement BD due to an error in bearing

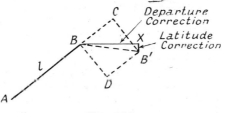

FIG. 4.18

is equal to BC, with a result that B has a final position of B'. The total probable error BB' thus equals $\sqrt{BC^2 + BD^2}$.

The method of Least Squares states that the sum of the weighted errors squared will be a minimum, and after weighting the lines inversely as the square of the probable error of each line (i.e. $\sqrt{(l)^2} = l$), then $\Sigma\left(\dfrac{\sqrt{BC^2 + BD^2}}{l}\right)$ is to be a minimum.

This leads (refer to Jameson's *Advanced Surveying*) to the following conclusions—

(1) Correction to a latitude

$$= \text{total latitude correction} \times \frac{\text{length of the side}}{\text{perimeter of the traverse}}$$

$$= \text{LC} \times \frac{l_1}{\Sigma l}$$

(2) Correction to a departure

$$= \text{total departure correction} \times \frac{\text{length of the side}}{\text{perimeter of the traverse}}$$

$$= \text{DC} \times \frac{l_1}{\Sigma l}$$

i.e. point B moves to B' over a distance of

$$\sqrt{\left(\frac{l_1}{\Sigma l}\right)^2 (\text{LC})^2 + \left(\frac{l_1}{\Sigma l}\right)^2 (\text{DC})^2} = \frac{l_1}{\Sigma l}\sqrt{(\text{LC})^2 + (\text{DC})^2}$$

The bearing of the movement is $\tan^{-1} \dfrac{BX}{B'X}$

where

$$\tan^{-1} \frac{BX}{B'X} = \frac{\dfrac{l_1}{\Sigma l}\,\text{DC}}{\dfrac{l_1}{\Sigma l}\,\text{LC}} = \frac{\text{DC}}{\text{LC}}$$

and this is the bearing of the closing error as seen in Table II.

Thus, generally speaking, the bearing of each line of the traverse will be altered after applying Bowditch's Method of Correction. When using a theodolite, however, the probable error in angular measurement should be smaller than that due to linear measurement, and it would appear then that this method is not a truly sound approach to the adjustment of a theodolite traverse. It is, in fact, most suitable for use in correcting a compass traverse. since the probable angular errors in this case will be much greater than for the theodolite. This method is nevertheless popular in practice for the average type of engineering survey, since (i) it is easy to apply, and (ii) the corrections, although affecting the bearings of the lines, do not affect the plotting to a noticeable extent. Table III illustrates typical corrections to the traverse $ABCEGJLA$ (Table II).

Notes on Table III

(i) Correction to latitude

$$AB = \frac{103 \cdot 40}{1{,}001 \cdot 26} \times (+\ 0 \cdot 07) = +\ 0 \cdot 01 \text{ m}$$

Correction to departure

$$AB = \frac{103 \cdot 40}{1{,}001 \cdot 26} \times (-\ 0 \cdot 013) = -\ 0 \cdot 01 \text{ m}$$

and similarly for other lines.

The total corrections equal the total errors for latitude and departure respectively, but are opposite in sign. Thus, in the example given, corrections to latitudes will be positive and to departures will be negative, and are then applied with those signs, i.e.

Corrected latitude $AB\ \ = -\ 102 \cdot 55 + 0 \cdot 01 = -\ 102 \cdot 54$ m

Corrected departure $AB = -\ \ \ 13 \cdot 27 - 0 \cdot 01 = -\ \ \ 13 \cdot 28$ m

TABLE III

Line	Correction to Latitude	Corrected Latitude	Correction to Departure	Corrected Departure	Co-ordinates		
AB	+ 0·01	− 102·54	− 0·01	− 13·28	A	2,000·00 N	1,000·00 E
BC	+ 0·01	− 156·47	− 0·02	− 15·61	B	1,897·46 N	986·72 E
CE	+ 0·01	− 80·26	− 0·02	+ 118·76	C	1,740·99 N	971·11 E
EG	+ 0·01	+ 131·86	− 0·02	+ 105·83	E	1,660·73 N	1,089·87 E
GJ	+ 0·01	+ 169·13	− 0·02	+ 51·30	G	1,792·59 N	1,195·70 E
JL	+ 0·01	+ 30·39	− 0·02	− 106·37	J	1,961·72 N	1,247·00 E
LA	+ 0·01	+ 7·89	− 0·02	− 140·63	L	1,992·11 N	1,140·63 E

+ 0·07
− 0·13

(ii) It was assumed that the co-ordinates of A were 2,000·00 N, 1,000·00 E. Then, co-ordinates of B are

$$2,000·00 - 102·54 = 1,897·46 \text{ N}$$
and
$$1,000·00 - 13·28 = 986·72 \text{ E}$$

co-ordinates of C are

$$1,897·46 - 156·47 = 1,740·99 \text{ N}$$
and
$$986·72 - 15·61 = 971·11 \text{ E}$$

As a check, the co-ordinates of A should be derived from those of L, proceeding in a cyclic manner round the traverse.

Co-ordinates of L	1,992·11 N		1,140·63 E
apply corrected lat.	+ 7·89	corrected dep.	− 140·63

$\therefore$ Co-ordinates of A + 2,000·00 N   + 1,000·00 E

which are those from which all the others were deduced.

(*b*) Transit Rule. In the previous method all lines will have some correction made in both latitude and departure, as called for in the totals of latitudes and departures (Table II). The Transit Rule or method has no mathematical background, and lengths of lines do not enter into the calculations. The rule is—

$$\frac{\text{Correction of Latitude}}{\text{of a line}} = \text{LC} \times \frac{\text{Lat. of line}}{\text{Sum of Latitudes}}$$

$$\frac{\text{Correction of Departure}}{\text{of a line}} = \text{DC} \times \frac{\text{Dep. of line}}{\text{Sum of Departures}}$$

and Table IV indicates corrections by this method for the traverse in Table I.

TABLE IV

Line	AB	BC	CE	EG	GJ	JL	LA
Correction to Lat. .	+ 0·01	+ 0·02	+ 0·01	+ 0·01	+ 0·02	0·00	0·00
Correction to Dep. .	0·00	0·00	− 0·03	− 0·03	− 0·01	− 0·03	− 0·03

Hence if a line has no latitude, then it will not have a latitude correction, and similarly if it has no departure then there will be no departure correction. So long as the line is running parallel to one of the co-ordinate axes, then the corrections applied will not alter the bearing of the line (compare with Bowditch's method), but when the line is inclined to the axes corrections will be made which do affect the bearings.

Comparison of Tables III and IV indicates the effects of the two methods. Note that line LA by Bowditch's Rule will have some alteration in bearing, whereas by the Transit Rule the alteration is virtually zero.

(c) UNALTERED BEARINGS. A further method which may be employed is one which leaves the bearings of lines unaltered after the corrections have been made. The proof and an example can be seen at the end of Chapter 13, but briefly, corrections are applied to each line, such that

$$\frac{\text{Correction to latitude}}{\text{Correction to departure}} = \frac{\text{latitude}}{\text{departure}}$$

Much more calculation is involved in this method, and it is likely that the work involved would not be justified in the usual type of traverse, particularly when the lines have been measured by chain. The main value of this method lies in its application to the correction of precise traverses.

EXAMPLE. During a theodolite survey the following details were noted.

Line	AB	BC	CD	DA
Chainage (m)	114·17	162·76	99·21	?
W.C.B.	115° 00′	17° 00′	219° 10′	?

Obtain the length and bearing of *DA* and draw up a table giving co-ordinates taking those of *A* as (100, N; 0, E).

Procedure: (1) Calculate latitudes and departures of the lines from the data.

Line	Length	W.C.B.	R.B.	Latitude +	Latitude −	Departure +	Departure −
AB	144·17	115° 00′	S 65° 00′ E		60·93	130·66	
BC	162·76	17° 00′	N 17° 00′E	155·65		47·59	
CD	99·21	219° 10′	S 39° 10′W		76·92		62·66
DA	?	?					
				155·65	137·85	178·25	62·66
				137·85		62·66	
				17·80		115·59	

(2) For closure the latitude of *DA* must be −17·80 m and its departure −115·59, from which it will be noted that *DA* is in the third quadrant with a reduced bearing of S tan^{-1} $\dfrac{115·59}{17·80}$ W

$$= \text{S } 81° \, 15' \text{ W (W.C.B. } 261° \, 15').$$

The length of the line *DA*

$$= \sqrt{17·80^2 + 115·59^2}$$
$$= 116·95 \text{ m}$$

5

(3) Co-ordinates

Point	Co-ordinates	
A	100·00 N	00·00 E
B	39·07 N	130·66 E
C	194·72 N	178·25 E
D	17·80 N	115·59 E

DETECTION OF ERROR IN LINEAR MEASUREMENT. The size of the closing error can indicate the presence of some error which is not admissible for balancing out by Bowditch's, or other, method.

Should there be a large error in linear measurement, i.e. failure to book one chain or tape length, then so long as only one such error has been made it will be found that the bearing of the closing error will be similar to the bearing of the line in error.

EXAMPLE. The following table purports to give uncorrected lengths and bearings of the legs of a closed traverse, but it contains an error in transcription of one of the values of length.

Leg	AB	BC	CD	DE	EA
Length (m)	210·67	433·67	126·00	294·33	223·00
Bearing	20° 31′ 30″	357° 16′ 00″	120° 04′ 00″	188° 27′ 30″	213° 31′ 00″

Find the error.

(*B.Sc., Int.*, London)

Leg	Length	R.B.	Latitude +	Latitude −	Departure +	Departure −
AB	210·67	N 21° 31′ 30″ E	197·29		73·86	
BC	433·67	N 02° 44′ 00″ W	433·19			20·69
CD	126·00	S 59° 56′ 00″ E		63·13	108·98	
DE	294·33	S 08° 28′ 30″ W		291·12		43·38
EA	223·00	S 33° 31′ 00″ W		185·91		123·12
			630·48	540·16	182·84	187·19
			540·16		182·84	182·84
			90·32			4·35

$$\tan \theta = 4·35/90·32$$
$$\theta = 02° 45′ \text{ (refer to Fig. 4.19)},$$

i.e. line *BC* is in error and should be reduced in length so as to diminish the excessive latitude.

For closure,

the latitude of $BC = + (540·16 - 197·29)$
$$= + 342·87 \text{ m}$$
departure $\quad = - (182·84 - 166·50)$
$$= - 16·34 \text{ m}$$

Thus the length of *BC* entered in the table should have been 343·67 m which results in a closing error of the order of Lat. $+ 0·42$ m, and Dep. $- 0·05$ m.

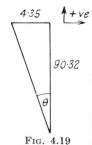

FIG. 4.19

Unclosed Traverses. These types cannot be balanced by the three methods mentioned for the closed traverse. On long traverses checks may be applied by means of astronomical observations for azimuth of lines and for latitude and longitude at various stations. Some account may have to be taken of the convergence of meridians over long distances and the student should refer to Chapter 12 for information on this subject.

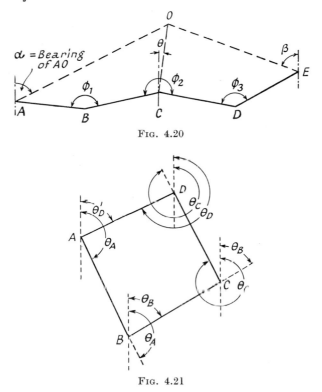

FIG. 4.20

FIG. 4.21

It may be possible to obtain bearings, from occasional stations, on to a salient point O, as shown in Fig. 4.20.

The co-ordinates of O may be determined from the observations within $ABCO$. From the observations in $OCDE$, the co-ordinates of E may be deduced and checked with those obtained directly from the angular and linear measurements along $ABCDE$.

Direct Measurement of Bearings. In this method the whole circle bearings of the lines with respect to the chosen meridian are obtained directly on the horizontal circle of the instrument. The instrument is set up and levelled at A (Fig. 4.21) with vernier A reading zero, and then the bearing of $AB(\theta_A)$ with respect to the meridian is obtained as previously described (the bearing θ'_D may also be found). With the horizontal plates clamped together, the instrument is taken to B, and set up and levelled. The telescope is directed to A and coincidence on the signal obtained by means of the lower clamp and tangent screw, the telescope then being transitted. The horizontal circle reading is still θ_A, but the instrument is directed along AB produced, and is correctly oriented, since if the telescope were rotated back through θ_A it would be directed along a line parallel to the meridian at A. Direct the telescope to C and read θ_B, which is the whole circle bearing of BC, on vernier A. This procedure is now repeated at C and D. The bearing θ_D of DA should compare with the back bearing θ'_D as found at A, since in this case $\theta_D = 180 + \theta'_D$.

It will be realized that the face of the instrument will be changed at each station, since the telescope is transitted each time, and if the line of collimation is not at right angles to the trunnion axis then alternate bearings will be in error.

Connexion of Surface and Underground Lines. It can occur that a sewer or pipe line has to be laid in a tunnel and naturally its alignment must be established accurately. Plumb lines are used to transfer lines below ground and two possible cases may be met when setting out, namely the case in which the underground line lies directly below the surface traverse line, and secondly the case in which it does not. This second case means that the proposed underground line must be initially established from the overground traverse and then the plumb lines put down. Typical calculations are as follows—

EXAMPLE. A length of sewer RQS is to be constructed in heading, the straights QR and QS having reduced bearings of S 22° 46′ W and N 20° 14′ E respectively, whilst manhole Q has co-ordinates of 448·62 m N, 127·05 m E. If the co-ordinates of a nearby station A on a street traverse are 300·00 m N, 60·00 m E and the reduced bearing of a traverse line AB is N 21° 33′ E, obtain data for setting out the two straight lengths of sewer.

Assuming ground conditions allow a line PQ to be set out, P can be located along AB, so that PQ may be set off at right angles to AB and Q then located by direct measurement (Fig. 4.22a).

Bearing $AQ = \tan^{-1} (127·05 - 60·00)/(448·62 - 300·00)$

$\qquad\qquad = 67·05/148·62$

$\qquad\qquad = $ N 24° 17′ E

$AQ = \sqrt{(67·05^2 + 148·62^2)} = 163·08$ m

$AP = 163·08 \cos (24° 17′ - 21° 33′) = 162·90$ m

$PQ = 163·08 \tan (24° 17′ - 21° 33′) = 7·79$ m

Thus Q can be positioned on the surface

The bearing of PQ is S 68° 27′ E and thus angles, $P\hat{Q}R$ and $P\hat{Q}S$, equal to 88° 47′ and 88° 41′ respectively, can be set out on the ground. If the sewer were not to be laid at great depth plumb lines could be established on QR or QS after a shaft had been sunk at Q (Fig. 4.22b). A theodolite would be set up as near as possible to Q and placed on the line RQ, say, using the centring device if fitted. Fine marks could be scribed on timbers, or metal plates thereon, across the top of the shaft and plumb lines or wires, R' and Q', suspended on the line RQ. Oscillations could be damped by immersing the plumb bobs in a liquid of suitable viscosity. Visual alignment will allow the heading to be commenced along QR at the shaft bottom, but at the earliest oppor-

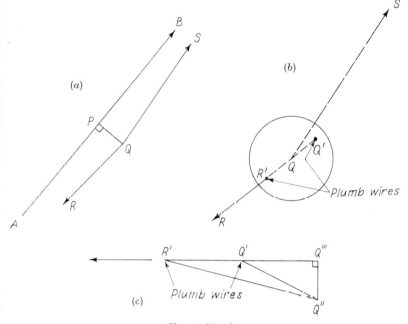

FIG. 4.22a, b, c

tunity the theodolite is taken below and again lined in on the two wires. Scales can be established behind the wires if these be oscillating appreciably, and the mean positions of swing derived for the proper establishment of the theodolite. The co-ordinates of R' and Q' could be found from the surface survey.

Alternatively the *Weisbach Triangle* (Fig. 4.22c) can be adopted since it is difficult to align the theodolite on two plumb wires, particularly when the headings are at great depth. Extreme scale readings must now be obtained by following the path of the swinging plumb bob to obtain a satisfactory mean. The theodolite, reading to one second, is established just off the line $R'Q'$ given by the plumb wires with $R'Q'$ nearly equal to $Q'Q''$. Both these lengths are measured, along with angle $R'Q''Q'$, the instrument being at Q''. This allows $Q'\hat{R}'Q''$ to be estimated, since by the sine rule

$$\frac{R'Q'}{\sin R'\hat{Q}''Q'} = \frac{Q'Q''}{\sin Q'\hat{R}'Q''}$$

For small angles we can put $\sin R'\hat{Q}''Q' = \sin 1'' \times (R'\hat{Q}''Q')$

and $\sin Q'\hat{R}'Q'' = \sin 1'' \times (Q'\hat{R}'Q'')$

where $(R'\hat{Q}''Q')$ and $(Q'\hat{R}'Q'')$ are expressed in seconds of arc. Thus

$$\frac{R'Q'}{(R'\hat{Q}''Q')} = \frac{Q'Q''}{(Q'\hat{R}'Q'')}$$

$Q''Q'''$ can also be estimated now and Q''' placed on the tunnel centre line. Alternatively other stations off that line can be established from Q'' to give one or more reference lines below ground, which in turn can be

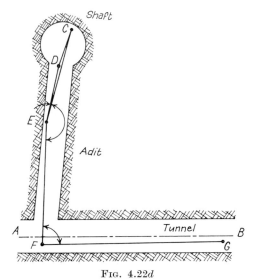

Fig. 4.22d

$CD = 3\cdot64$ m; $DE = 4\cdot46$ m; $EF = 13\cdot12$ m; $FG = 57\cdot50$ m.
Angle $DEC = 38''$; angle $CEF = 167°\ 10'\ 20''$; angle $EFG = 87°\ 23'\ 41''$.
The drawing is not to scale.

used to determine the tunnel centre line from the theoretical co-ordinate data. It will be appreciated that the Weisbach Triangle can also be used in the ground survey to establish the bearing of $R'Q'$ from ground stations.

The measurement of depth is mentioned in a later chapter.

EXAMPLE. The centre-line of the tunnel AB shown in Fig. 4.22d is to be set out to a given bearing. A short section of the main tunnel has been constructed along the approximate line and access is gained to it by means of an adit connected to a shaft. Two wires, C and D, are plumbed down the shaft, and readings are taken onto them by a theodolite set up at station E slightly off the line CD produced. A point F is located in the tunnel, and a sighting is taken onto this from station E. Finally a further point G is located in the tunnel and the angle EFG measured.

From the surface survey initially carried out, the co-ordinates of C and D have been calculated and found to be N 1,119·32, E 375·78 m for C, and N 1,115·70, E 375·37 m for D.

Calculate the co-ordinates of stations F and G. Without making any further calculations describe how the required centre-line could then be set out. (*I.C.E.*)

In the Weisbach Triangle CDE (Fig. 4.22*d*)

$$D\hat{C}E = \frac{4\cdot46}{3\cdot64} \times 38'' = 46\cdot6'',$$

or external angle $C\hat{D}E = \dfrac{4\cdot46 + 3\cdot64}{3\cdot64} \times 38''$

$$= 84\cdot6'' = 01'\ 24\cdot6''.$$

The bearing of DC	$= \text{N tan}^{-1}\dfrac{0\cdot41}{3\cdot62}\ \text{E}$
	$= \text{N } 06°\ 27'\ 42\cdot4''\ \text{E}$
$\therefore$ bearing of ED	$= \text{N } 06°\ 27'\ 42\cdot4''\ \text{E} - 01'\ 24\cdot6''$
	$= \text{N } 06°\ 26'\ 17\cdot8''\ \text{E}$
i.e.	$= \text{N } 06°\ 26'\ 18''\ \text{E}$

The reader can now check that

   (i) bearing of EF $= \text{S } 06°\ 22'\ 44''\ \text{E};$

  (ii) bearing of FG $= \text{N } 81°\ 00'\ 57''\ \text{E};$

 (iii) co-ordinates are

          F 1,098·23 N, 376·33 E

          G 1,107·22 N, 433·12 E

Modern Instruments

This section gives a brief outline of some instruments now being manufactured which are somewhat different from the ordinary theodolite described earlier in the chapter.

THE CENTESIMAL SYSTEM. In this chapter so far angular measurements have been referred to circles with major graduations from 0° to 360° with secondary graduations which subdivide each degree into 10 minute or 20 minute intervals (Fig. 4.3). Vernier or micrometer subdivision then gives the reading down to seconds, and since there are sixty minutes in a degree and sixty seconds in a minute, the system is known as the *sexagesimal* system (Latin: *sexaginta*, sixty). It is possible, however, to obtain instruments graduated in 400 major parts from 0^g to 400^g (read as 400 grade). The grade is subdivided into five intervals each of 20 minutes and since there are 100 minutes to the grade on this system it is known as the *centesimal* system (Latin: *cent*, hundred). Angles can be expressed as decimals on this system.

GLASS CIRCLE THEODOLITES. The instruments now to be described differ greatly from the vernier instrument previously described in that the metal scale plates read by vernier are replaced by glass circles which are read by means of internal optical systems. The circles provided in Messrs. Hilger & Watt's theodolites are photographic copies of glass master circles which in turn have been graduated by means of an automatic dividing engine.

Another feature of modern theodolites worthy of note is that most telescopes are provided with focusing rings or sleeves on the telescope barrel near to the eyepiece. These replace the knurled focusing screw previously fitted at trunnion axis level outside one of the standards.

A discussion of accuracies attainable in the field, by several glass arc theodolites, can be found in an article by Bannister and Schofield in *The Surveyor*, 23 Dec., 1967.

Fig. 4.23(*a*). The DKM 2 Triangulation Theodolite

This instrument has been set up on a tripod having a "ball and socket" head

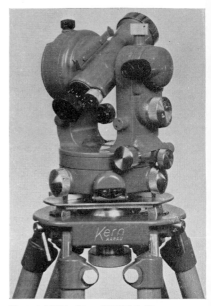

Fig. 4.23(*b*). The DKM 1 Triangulation Theodolite

(4.23(*a*) and 4.23(*b*) by courtesy of Kern & Co., Ltd., Arrau)

Kern DK Theodolites. These instruments have a different vertical axis system from that of the described theodolite, in which a pivot spindle and cylindrical bearings were used. Instead, a high precision ball-bearing movement acts as the main bearing. There are two flat bearing surfaces, precisely machined, with perfectly true balls and ball races, the whole being situated between the upper plate and the tribrach. The footscrews have been eliminated and in their place are three screws each having a longitudinal axis and helical thread. Rough levelling of the instrument is allowed by a tilting plate on the tripod, and the fine levelling is effected by the three horizontal screws. Two different types of tripod head may be seen in Fig. 4.23.

In Fig. 4.23*b* the instrument rests on the plate which is suspended by gimbals in the frame at the head of the tripod: this plate will thus tilt in any required direction and may be locked by three screws. Two small level tubes at right angles to each other are provided for ensuring the rough levelling of the plate. A red dot is marked on the tripod head, and the bubble pointing towards it is centred first, the locking screw nearby then being tightened. By means of the second bubble the plate is finally levelled and then fixed by locking the remaining two screws situated on either side of this level tube. The rough levelling is now complete and the instrument may be centred over the station by means of an optical plummet or plumb bob, and then finely levelled.

The instrument shown in Fig. 4.23a is fitted with a centring tripod and the fluted grip of the centring rod can be clearly seen. This rod replaces the plumb bob, its tip being placed over the station and the tripod head being moved until a circular bubble carried on the rod is centred. The grip is then tightened and the support plate, on which the theodolite will be placed, is now approximately level and is centred within 1 mm.

The DKM 1 small triangulation theodolite has a horizontal circle 50 mm in diameter, and reads direct to ten seconds with a possible estimation to one second, while the DKM 2 type reads direct to one second on a circle 75 mm in diameter: these instruments weigh 1·8 kg and 3·6 kg, and measure 100 mm and 160 mm from base plate to trunnion axis respectively. The vertical and horizontal circle readings are observed in a reading microscope, in the field of view of which the surveyor sees double line graduations.

Light from a lighting mirror is directed by an optical system through the circles to the reading microscope. Each circle has two graduations, one being numbered, and the optical system is such that the double graduations seen in the field of view are formed from diametrically-opposite single graduations. These double lines are moved in the image plane by an optical micrometer until a fixed vertical index line bisects a double graduation. Three apertures are visible in the field of view, the uppermost giving the vertical circle readings, and the middle one giving the horizontal circle reading, both having the double line graduations. The lowest aperture shows the micrometer drum reading. In the DKM 1 instrument, the double lines give readings at 20′ intervals and the micrometer gives readings of single minutes subdivided into six parts, while in the DKM 2 type the double lines are at 10′ intervals, the micrometer giving minute and ten second readings, subdivided into ten equal parts. Thus the former instrument reads direct to 10 sec, and the latter direct to 1 sec. Each reading is the arithmetical mean of two diametrically opposed points of a circle, the micrometer drum serving both vertical and horizontal circle.

Readings of the Kern DKM3 instrument are obtained by bisecting the double-line graduations of one circle with diametrically opposite single-line graduations of the other circle using the micrometer as shown in Fig. 4.24. This particular theodolite reads directly to 0·5 second.

It is claimed for these instruments that (i) the accuracy of centring is increased, (ii) sideplay of footscrews is eliminated and (iii) they are easier to level and set up.

Wild T2 Universal Theodolite (Fig. 4.25). The normal range of this instrument is 0 to 20 km; the vertical and horizontal circle diameters are 70 mm and 90 mm respectively, and the theodolite weighs 5·6 kg. The upper end of the main bearing has a conical opening in which ball bearings are placed, the underside of the shoulder of the hollow vertical

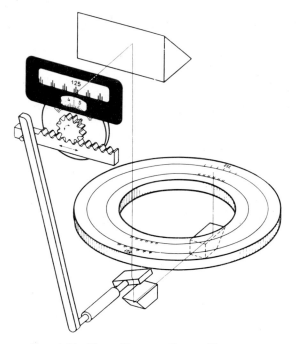

FIG. 4.24. KERN DOUBLE CIRCLE PRINCIPLE
(Courtesy of Survey & General Instrument Co., Ltd.)

axis rests on these bearings allowing fluent movement in azimuth. The horizontal circle can be seen near tribrach level, whilst the vertical circle is conventionally positioned; the tribrach itself is detachable from the theodolite.

Light rays enter by means of a mirror, and are deflected by prisms to diametrically opposite points of the horizontal circle. The rays then return through one of the prisms and are turned through 90° to pass through a lens, up the hollow vertical axis, into a further prism which diverts them through the parallel plate micrometers. Finally, by means of other prisms, the images of the two points of the circle reach the reading microscope. By means of light from another mirror at vertical circle level, the readings of that circle are also obtained,

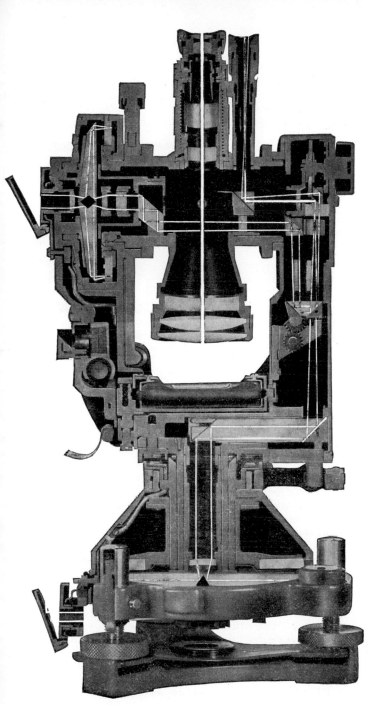

FIG. 4.25. CROSS-SECTION THROUGH UNIVERSAL THEODOLITE
(*Courtesy of Wild Heerbrugg, Switzerland*)

the light rays passing through the same parallel plate micrometers which served the horizontal circle. When the micrometer knob is turned, the parallel plate micrometers are inclined through equal but opposite amounts. The rays are refracted, and the knob is turned until the images of adjacent graduations of the diametrically opposite

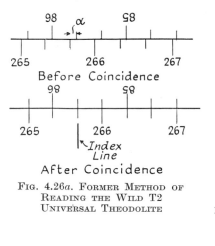

FIG. 4.26*a*. FORMER METHOD OF READING THE WILD T2 UNIVERSAL THEODOLITE

FIG. 4.26*b*. NEW READING FOR THE WILD T2 UNIVERSAL THEODOLITE

Reading example 360° model or vertical circle:

Degree number	094
Tens of minutes	1
Micrometer scale	2′ 44″
Booking	94° 12′ 44″

(*Courtesy of Wild Heerbrugg, Switzerland*)

points of the circle are exactly coincident. The range of the micrometer is 10′ and its gives readings to minutes and seconds within that range. Whole numbers of degrees and tens of minutes are obtained from the relevant circle readings. In Fig. 4.26*a* it will be seen that the index line reads 265° 40′, but in fact this reading should be deduced from the numbers of divisions between main graduations 180° apart, and it will be observed that there are four such between 265° and 85°. This indicates a reading of 265° plus 4 × 10′ and the micrometer reading is added afterwards. It will be noted that the interval between graduations is 20 minutes on the circles: the two sets move in opposite directions in the field of view when the telescope is rotated, thereby giving coincidence every ten minutes but the parallel plates ensure that coincidence is possible within that interval.

In Fig. 4.26*a* it will be seen that the parallel plate micrometers move the images through a distance of α for coincidence, i.e. each scale image moves through α/2; the drum reading, not shown in the diagram, appears below the scale reading, as in Fig. 4.26*b* which illustrates the current readout of the circles of the T2 theodolite. Coincidence is made as before (upper window) but now degrees and tens of minutes are indicated directly in the middle window so that the circle reading as illustrated is 94° 10′ + 02′ 44″, i.e. 94° 12′ 44″. The optical micrometer is shown in Fig. 4.27.

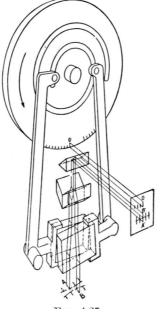

Fig. 4.27

Watts Microptic Theodolites. Hilger & Watts, Ltd. manufacture the following models of Microptic theodolites—

(*a*) No. 1, reading direct to 20 sec with estimation to 5 sec,

(*b*) No. 2, reading direct to 1 sec, and

(*c*) No. 3, reading direct to $\frac{1}{5}$ sec.

In this section the second (*see* Fig. 4.28) will be described, but the third is mentioned in Chapter 9.

ST 200 Microptic Theodolite No. 2 Mk II. This instrument (Fig. 4.28) is a redesign of the well-known Microptic No. 2 Mark 1 which first appeared about 1950. The horizontal and vertical circles are 98 mm and 76 mm respectively in diameter and are graduated at 10 minute intervals. Reading is direct to 1 second. The instrument is in two main types, one of which is for use in three-tripod surveying discussed in the following section. Each type is obtainable with sexagesimal and centesimal systems of graduation. A built-in optical plummet, which rotates with the instrument, enables accurate centring to be made over the stations. Observations are taken from diametrically opposed positions and the mean taken (if there be any discrepancy) to give the correct centring.

The images of either the vertical or horizontal circles are brought into the field of view of the reading eyepiece (outside the right-hand upright) by means of a right-angled prism actuated by a circle selector knob. The letters H or V appear in the field of view and help to avoid possible reading mistakes. The large central aperture gives the reading of one side of the selected circle; the upper aperture shows two sets of three lines each, and the bottom aperture gives the relevant micrometer reading. The sets of three lines are images of graduation lines from

Fig. 4.28. The ST 200 Microptic Theodolite No. 2 Mk.II

KEY

1. Telescope clamp
2. Circle reading eyepiece
3. Foresight
4. Circle selector knob
5. Horizontal circle micrometer knob
6. Plate bubble
7. Azimuth clamp
8. Case support leg
9. Cover to circle repetition knob
10. Azimuth slow motion screw
11. Optical plummet eyepiece
12. Cover screw for altitude bubble adjustment
13. Telescope eyepiece
14. Telescope focusing sleeve

(*Courtesy of Hilger & Watts, Ltd.*)

diametrically opposite positions on the circles, the middle one being from one position and the two outer ones from the position 180° away (*see* Fig. 4.29). Although the circle is graduated at 10 minute intervals, alternate lines are lengthened inwards and short parallel lines are marked at their ends. The two outer lines in the upper aperture are the images of these lines, which are at 20 minute intervals. When the

FIG. 4.28 (*contd.*) THE ST 200 MICROPTIC THEODOLITE No. 2 MK.II

KEY

15. Vertical circle illumination mirror	20. Lighting socket
16. Telescope slow motion screw	21. Horizontal circle illumination mirror
17. Altitude bubble slow motion screw	22. Horizontal circle lamp socket
18. Wear adjusting screw for levelling screw	23. Altitude bubble
	24. Prism reader for altitude bubble
19. Levelling screw spring plate retaining bolts	25. Vertical circle micrometer knob
	26. Vertical circle lamp socket

(*Courtesy of Hilger & Watts, Ltd.*)

circles rotate the single lines move in one direction and the double lines in the other.

Light is admitted by circle illumination mirrors and by means of

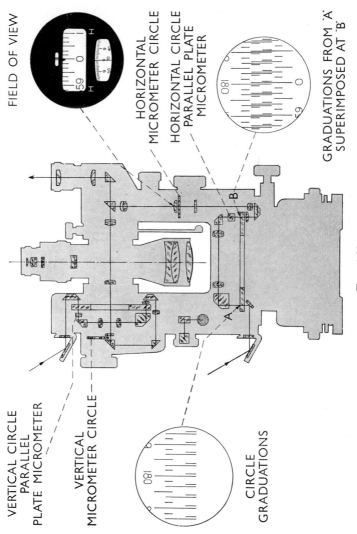

FIELD OF VIEW

HORIZONTAL
MICROMETER CIRCLE

HORIZONTAL
CIRCLE
PARALLEL PLATE
MICROMETER

GRADUATIONS FROM 'A'
SUPERIMPOSED AT 'B'

VERTICAL CIRCLE
PARALLEL
PLATE MICROMETER

VERTICAL
MICROMETER CIRCLE

CIRCLE
GRADUATIONS

FIG. 4.29

separate systems of lenses and prisms it passes through diametrically opposite points on the H or V circle (as selected) to the right-angled prism previously mentioned and thence to the reading eyepiece. There are two micrometer knobs, one for each circle (set not only on different uprights but at different levels), which each actuate a parallel-sided glass block. This enables the double lines in the field of view to be placed symmetrically astride the single line, which does not move since the main circle is stationary while this operation is carried out. The image of the reading of the relevant small micrometer circle then appears in the bottom aperture, and is added to the reading in the central aperture to give the required value of the angle, this being the mean of the diametrically opposed points (*see* Fig. 4.30).

HORIZONTAL CIRCLE

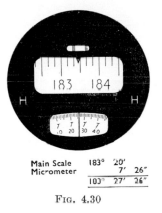

Main Scale	183°	20'	
Micrometer		7'	26″
	103°	27'	26″

FIG. 4.30

The internal optical system of the Mark I instrument has been rearranged in this instrument and is said to be simpler both for adjustment and overhaul. The reading eyepiece has been moved from the position it occupied alongside the telescope in the Mark I pattern. The third main difference is that the altitude bubble is now totally enclosed within the left-hand upright at a position just above the horizontal circle, and like the plate bubble is kinematically mounted. Both ends of the altitude bubble are seen side by side in one field of view by means of a coincidence setting prism.

The horizontal circle can be set to zero by means of the circle repetition knob. Similar devices are found on the Kern and Wild instruments. The vertical circle is graduated from 0° to 360° (or 0 to 400 g), the reading being zero when the telescope is horizontal and face left. When the telescope is elevated, the vertical circle reading in the sexagesimal system will be between 0° and 90° (face left), or 90° and 180° (face right). On the other hand, if depressed, the reading will be between 180° and 270° (face right), or 270° and 360° (face left). The value of the vertical angle depends upon the quadrant in which the circle reading lies.

If readings of 334° 24′ 36″ and 205° 35′ 26″ were obtained, the value of the angle of depression would be the mean of 25° 35′ 24″ and 25° 35′ 26″.

Cooke Theodolites. The Cooke Tavistock theodolite is another well proved major British instrument, manufactured by Vickers Instruments Ltd., which reads directly to one second, the coincidence method

Fig. 4.31*a.* Cooke V22 Theodolite
The levelling base is detachable; the instrument
body can be inserted in the base shown in
Fig. 4.31*b.*
(*Courtesy of Vickers Instruments*)

being adopted. A single optical micrometer serves both circles, and a control knob is used to select the circle to be viewed. Coincidence occurs every 10′ as with the Wild T2, but a triangular index mark is used to read the circles to 10′ rather than counting intervals of 10′ between whole degree readings; the micrometer reading appears alongside the circle reading and is added directly. The diameters of the horizontal and vertical circles are 85 and 70 mm respectively, and the vertical circle reads zero when the telescope points vertically upwards in the face-right position.

The Cooke V22 Theodolite (Fig. 4.31a) uses a scale, and not a micrometer, to read directly to 20″ with estimation to 10″. Both circle readings are seen in the field of view of the reading eyepiece and, as shown in Fig. 4.32, the horizontal circle can be graduated increasing clockwise (as is conventional and normal) or increasing anti-clockwise if so wished: this can be advantageous in particular when setting out

FIG. 4.31b. V221900 CENTRING AND LEVELLING BASE
(*Courtesy of Vickers Instruments*)

curves. It will be noted in Fig. 4.31 that the lower clamp found on the vernier theodolite has been replaced by a repetition clamp and that no lower tangent screw is fitted. When this clamp is on the vertical position the horizontal circle is fixed with respect to the tribrach so that on releasing the one horizontal clamp the telescope can be pointed from one station to another, and the angle between them obtained from the circle readings. If the clamp is horizontal the circle readings do not change as the telescope rotates in azimuth, and so when obtaining an angle by the repetition method the clamp is raised and depressed alternately for successive pointings but only one clamp and tangent combination is in use.

A centring rod can be provided with this instrument for setting up, in lieu of a plumb bob or optical plummet.

Three-tripod Traverse (Forced Centring) Equipment is provided by many firms for use with optical micrometer instruments. Basic components are a theodolite, optical plummet, and targets, all of which are interchangeable on tripod heads. After measuring the angle at a traverse station, instead of moving both the instrument and its tripod to the next station only the instrument is moved, being interchanged with the target at the next station. This target will be placed on the tripod vacated by the theodolite, and the other target and tripod involved in the previous angular measurement will be carried forward to the next station, so that a further angle can be read; although a

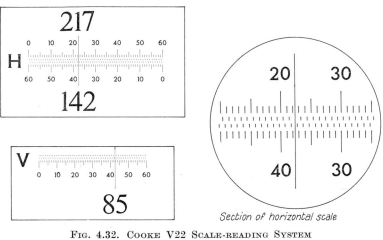

Section of horizontal scale

FIG. 4.32. COOKE V22 SCALE-READING SYSTEM
With dual-reading horizontal circle.
Horizontal reading 217° 22′ 20″ or 142° 37′ 40″.
Vertical reading 85° 42′ 40″.
(*Courtesy of Vickers Instruments*)

tripod can be set in advance if more than three are available. The purpose of the plummet is to centre the tripods over the stations so that targets and theodolite will occupy the same position over the stations. The main advantage of this system of surveying is that the centring errors are much reduced, especially on short sights. Also, an increased rate of progress is possible.

The accuracy of the whole traverse is influenced by the exactness with which the vertical axis of the theodolite can be placed in the lines previously occupied by the vertical axes of the targets. Centring errors must be of low magnitude, and alternative methods of centring have been developed, one frequently encountered being the stub axis system. Such axes, on theodolites or targets, fit closely into recesses in the tribrachs carried on the tripod heads; locking devices are provided in the tribrachs. German theodolites (Chapter 14) tend to use stub axes whose diameters are less than their lengths, but Watts and Vickers instruments have shallow stubs of much larger diameter. These are located by two studs in the tribrach and locked by a third Fig. 4.31*b*.

A mean square error of $\pm$ 0·1 mm has been mentioned by some writers in respect of centring errors, but a German specification demands a value of $\pm$ 0·03 mm for the stub axis system. Over a sighting distance of 10 m, which may not be excessive in certain circumstances underground, the former value suggests an error of $\pm$ 2 sec. The lateral shift in an unclosed traverse is given by the expression

$$\pm \frac{s_{\bar{x}} L \sqrt{\dfrac{n}{3}}}{206{,}265}$$

where L is the length of traverse, n the number of angles read, and $s_{\bar{x}}$ the mean square error of angular observation, including centring effects. The importance of restricting centring errors to low magnitude can be appreciated if the case of a survey of length 2 km and sighting distances 100 m with $s_{\bar{x}} = \pm 5''$ be considered: a lateral shift of $\pm$ 0·124 m can occur. The Hungarian Optical Works manufacture a forced centring device enclosed in a rotating sleeve. This allows the stub axis system to be rotated through 180° in order that eccentricity errors can be eliminated by taking the means of pairs of observations.

EXERCISES 4

1. During a theodolite survey the following details were noted—

Line	AB	BC	CD	DA
Chainage (m)	55	120		105
W.C.B.	60°	115°		310°

Obtain the length and bearing of CD.

Answer: 88 m; 239° 44'.

2. Using the data of a closed traverse given below, calculate the lengths of the lines BC and CD.

Line	Length	Whole circle beginning from N	Reduced Bearing	Latitude	Departure
AB	104·85 m	14° 31'	N 14° 31' E	+ 101·50	+ 26·29
BC		319° 42'	N 40° 18' W		
CD		347° 15'	N 12° 45' W		
DE	91·44	5° 16'	N 5° 16' E	+ 91·06	+ 8·39
EA	596·80	168° 12'	S 11° 48' E	− 584·21	+ 122·05

(*I.C.E.*, Oct., 1955)

Answer: $DC = 289\cdot13$ m; $BC = 143\cdot77$ m.

3. Readings of lengths and whole circle bearings from a traverse carried out by chain and theodolite reading to 1 minute of arc were as follows, after adjusting the angles—

Line	AB	BC	CD	DE	EF	FA
W.C.B.	0°00'	35° 40'	46° 15½'	156° 13'	180° 00'	270° 00'
Length (m)	146·16	161·58	134·67	88·62	289·17	226·86
Line	AG	GH	HD			
W.C.B.	64° 58'	346° 25'	37° 40'			
Length (m)	137·79	177·21	176·70			

Taking the direction AB as north, calculate the latitude and departure of each line. If A is taken as origin and the mean co-ordinates of D as obtained by the three routes are taken as correct, find the co-ordinates of the other points by correcting along each line in proportion to chainage. (*L.U.*, *B.Sc.*)

Answer: Mean co-ordinates of D are 370·56 N; 191·46 E.

4. Explain the meaning of the following terms: W.C.B.; reduced bearing; latitude; departure. In a closed traverse *ABCDA* made with chain and theodolite, the following readings were taken—

Angles	Lengths
$A\widehat{B}C$ 101° 32′ 30″	*AB* 73·33 m
$B\widehat{C}D$ 66° 54′ 30″	*BC* 108·54 m
$C\widehat{D}A$ 84° 58′ 30″	*CD* 122·32 m
$D\widehat{A}B$ 106° 40′ 30″	*DA* 85·44 m

Bearing of *B* from *A*, N 52° 23′ E, and the stations are lettered in an anticlockwise manner from *A*.

Prepare these results in a form suitable for plotting, showing how you would adjust any errors.

5. The following lengths, latitudes and departures were obtained for a closed traverse *ABCDEFA*—

	Length	Latitude	Departure
AB	183·79	0	+ 183·79
BC	160·02	+ 128·72	+ 98·05
CD	226·77	+ 177·76	− 140·85
DE	172·52	− 76·66	− 154·44
EF	177·09	− 177·09	0
FA	53·95	− 52·43	+ 13·08

Adjust the traverse by the Bowditch method. (*L.U., B.Sc.*)

6. The internal angles of a closed traverse *ABCDEF*, and the approximate lengths of the sides, are as follows—$A\widehat{B}C = 87°\ 49′$; $B\widehat{C}D = 144°\ 56′$; $C\widehat{D}E = 69°\ 33′$; $D\widehat{E}F = 120°\ 44′$; $E\widehat{F}A = 226°\ 51′$; $F\widehat{A}B = 70°\ 10′$. $AB = 323·0$ m; $BC = 366·3$ m; $CD = 338·3$ m; $DE = 259·0$ m; $EF = 239·3$ m; $FA = 215·0$ m. The magnetic bearing of the line *AB* is 365° 10′, and the magnetic variation is 8° 40′ W.

Calculate the reduced bearings (relative to the true meridian) and the true bearings of all the lines, and tabulate the results. *ABCDEF* lie in a clockwise direction round the traverse.

CHAPTER 5

AREAS AND VOLUMES

THE CALCULATIONS connected with the measurement of areas of land, etc., and of volumes and other quantities connected with engineering and building works, are dealt with in this chapter. Areas are considered first of all, since the computation of areas is involved in the calculation of volumes, and are dealt with under the following headings—

(1) Mechanical integration—the planimeter. (2) Areas enclosed by straight lines. (3) Irregular figures.

The Planimeter

Mechanical integration by the planimeter can be applied to figures of all shapes, and although the construction and application of the

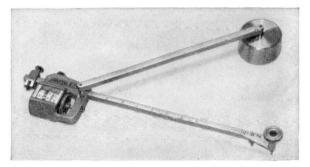

FIG. 5.1. AMSLER PLANIMETER TYPE 612a
(Courtesy of Hulden & Co., Ltd.)

instrument are simple, the accuracy achieved is of the highest degree, particularly when measuring irregular figures.

Fig. 5.1 shows a current sliding bar planimeter made by Alfred J. Amsler & Co., and Fig. 5.2 a fixed bar instrument from the Allbrit Co. The planimeter consists, essentially, of—

(a) The pole block, which is fixed in position on the paper by a fine retaining needle.

(b) The pole arm, which is pivoted about the pole block at one end and the integrating unit at the other.

(c) The tracing arm (which, as shown, may be either fixed or variable in length) attached at one end to the integrating unit and carrying at the other end the tracing point.

(d) The measuring unit, consisting of a hardened steel integrating

141

disc carried on cone pivots; directly connected to the disc spindle is a primary drum divided into 100 parts, a vernier opposite this drum enabling a reading of 1/1,000th of a revolution of the integrating disc to be made. Another indicator gives the number of complete revolutions of the disc.

Principles and Operation. It can readily be shown that if the pole is suitably placed relative to the figure to be measured and the tracing point moved round the outline of the figure, then the integrating disc will register an amount proportional to the area of the figure. (A rigorous

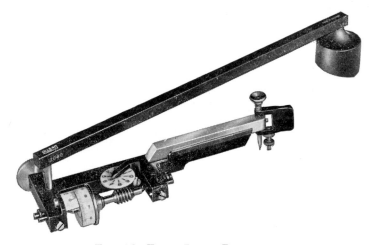

FIG. 5.2. FIXED INDEX PLANIMETER
(Courtesy of Stanley & Co., Ltd.)

proof of the principle may be found in Lamb's *Infinitesimal Calculus*.)*
In the fixed arm instrument, the drum is so graduated that the areas are given directly, and correction factors must be applied when working on plans.

EXAMPLE. What is the area of a piece of land which has a plan area of 1,613 mm² as measured by a fixed arm planimeter if the scale of the plan is 1/2,500.

On this scale 1 mm² represents 2,500² mm²

∴ 1,613 mm² ≡ 1,613 × 2,500 × 2,500 mm² or 1·008 hectare.

Where the length of the tracing arm is variable, the bar is graduated so that on being set at a particular graduation, direct readings of areas may be made on a plan of a particular scale.

A planimeter can be used in two ways—

    (*a*) with the pole *outside* the figure to be measured, or
    (*b*) with the pole *inside* the figure to be measured.

* Cambridge University Press.

The former method is the more convenient and should be used whenever possible.

(*a*) With the pole *outside* the figure the procedure for measuring any area, the plan being on a flat horizontal surface, is—

(i) Place the pole outside the area in such a position that the tracing point can reach any part of the outline.

(ii) With the tracing point placed on a known point on the outline, read the vernier.

(iii) Move the tracing point *clockwise* around the outline, back to the known point, and read the vernier again.

(iv) The difference between the two readings, multiplied by the scale factor, gives the area.

(v) Repeat until three consistent values are obtained, and the mean of these is taken.

(*b*) With the pole *inside* the figure—

The procedure is as above, but a constant, engraved on the tracing arm, must be added to the difference in readings in each case. This constant represents the area of the *zero circle* of the planimeter, i.e. of that circle which will be swept out when the plane of the integrating disc lies exactly through the pole and the integrating disc does not revolve at all.

Areas Enclosed by Straight Lines

Such areas include those enclosed inside the chain lines of a chain survey or theodolite traverse, and fields enclosed by straight-line boundaries.

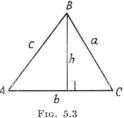

Fig. 5.3

Simple Triangles. Where the area is triangular in shape (Fig. 5.3), or is made up of a series of triangles, as for example in a chain survey, the following formulae are used—

$$\text{Area} = \tfrac{1}{2}(\text{base} \times \text{perpendicular height})$$
$$= \tfrac{1}{2}bh \quad . \qquad . \qquad . \qquad . \qquad . \qquad (5.1)$$

but $\qquad\qquad h = a \sin C$

Therefore $\qquad \text{area} = \tfrac{1}{2}ba \sin C \quad . \qquad . \qquad . \qquad . \qquad (5.2)$

Also $\qquad\qquad \text{area} = \sqrt{S(S-a)(S-b)(S-c)} \quad . \qquad . \qquad (5.3)$

where $\qquad\qquad S = \dfrac{a+b+c}{2}$

Irregular Straight-sided Figure. All such figures may be divided into triangles as shown in Fig. 5.4, so that the total area is derived by summing the parts. The calculation is somewhat simplified if pairs of triangles having common bases, such as X and Y, are considered. In this case, area $= \tfrac{1}{2}(h_1 + h_2)X + \tfrac{1}{2}(h_3 + h_4)Y$.

Area Enclosed by Survey Lines. In the particular case of traverse surveys which are plotted from co-ordinates, it is more convenient to calculate the area from the co-ordinates themselves, and this involves the use of the longitudes of the lines.

Longitudes. The longitude of a line is the perpendicular distance from a convenient meridian to the centre point of the line. The most

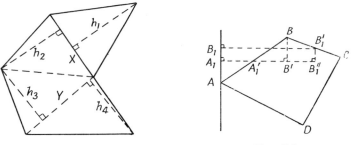

FIG. 5.4   FIG. 5.5

convenient meridian to take for any particular closed traverse is the North–South line through the most westerly point in the traverse, e.g. point A in Fig. 5.5.

Then the longitude of $AB = A_1A_1'(= \frac{1}{2} \text{ departure } AB)$

$$\text{the longitude of } BC = B_1B_1'$$

$$= A_1A_1' + A_1'B' + B'B''_1$$

$$= \text{longitude } AB + \tfrac{1}{2} \text{ departure } AB$$

$$+ \tfrac{1}{2} \text{ departure } BC$$

and in the general case, longitude of line n

$$= \text{longitude of line } (n-1)$$

$$+ \tfrac{1}{2} \text{ departure of line } (n-1)$$

$$+ \tfrac{1}{2} \text{ departure of line } n$$

or, the longitude of any line is equal to the longitude of the previous line plus half the departure of the previous line plus half its own departure, having due regard to sign, i.e. negative departures are summed algebraically. It will be seen now that by taking the meridian through the most westerly station no longitude is ever negative.

Area by Latitudes and Longitudes. Area equals algebraic sum of the products of the longitude of each line and the latitude of that line.

This is easily proved by considering some typical figure such as $ABCDEA$ in Fig. 5.6.

Area enclosed by $ABCDEA$ = trapezoid $bBCc$ + trapezoid $cCDd$

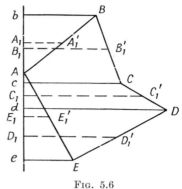

+ trapezoid $dDEe - \Delta bBA$
- ΔAEe
= $bc \times B_1B_1' + cd \times C_1C_1' + de$
$\times D_1D_1' - Ab \times A_1A_1' - eA$
$\times E_1E_1'$
= latitude BC × longitude BC
+ latitude CD × longitude CD
+ latitude DE × longitude DE
- latitude AB × longitude AB
- latitude EA × longitude EA.

FIG. 5.6

Thus the algebriac sum of the products of latitude and longitude gives the area enclosed by the lines.

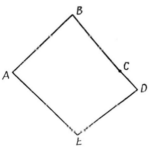

FIG. 5.7

It is often convenient to work in double longitudes and then to divide the answer by two.

Double longitude of a line = double longitude of previous line

　　　　+ departure of previous line
　　　　+ departure of line under con-
　　　　sideration.

EXAMPLE. Determine the area (in hectares) enclosed by the line of a closed traverse survey $ABCDEA$ (Fig. 5.7) from the following data—

Line	AB	BC	CD	DE	EA
Latitude (m)	+ 85·65	− 103·63	− 33·22	− 63·10	+ 114·30
Departure (m)	+ 106·98	+ 61·57	+ 24·38	− 101·19	− 91·74

If the chain, nominally 20 m long, used on the survey was later found to be 0·1 m too long, what would be the corrected value for the area?

Longitude of $AB = + \dfrac{106 \cdot 98}{2}$ m

Longitude of $BC = \dfrac{106 \cdot 98}{2} + \dfrac{106 \cdot 98}{2} + \dfrac{61 \cdot 57}{2}$

$\qquad = \dfrac{275 \cdot 53}{2}$ m

Longitude of $CD = \dfrac{275 \cdot 53}{2} + \dfrac{61 \cdot 57}{2} + \dfrac{24 \cdot 38}{2}$

$\qquad = \dfrac{361 \cdot 48}{2}$

Longitude of $DE = \dfrac{361 \cdot 48}{2} + \dfrac{24 \cdot 38}{2} - \dfrac{101 \cdot 19}{2}$

$\qquad = \dfrac{284 \cdot 67}{2}$

Longitude of $EA = \dfrac{284 \cdot 67}{2} - \dfrac{101 \cdot 19}{2} - \dfrac{91 \cdot 74}{2}$

$\qquad = \dfrac{91 \cdot 74}{2} = \tfrac{1}{2}$ departure EA (check)

Line	Latitude	Departure	Double Longitude	Double Longitude × Latitude	
				+	−
AB	+ 85·65	+ 106·98	+ 106·98	9,162·84	—
BC	− 103·63	+ 61·57	+ 275·53	—	28,553·17
CD	− 33·22	+ 24·38	+ 361·48	—	12,008·34
DE	− 63·10	− 101·19	+ 284·67	—	17,962·68
EA	+ 114·30	− 91·74	+ 91·74	10,485·88	—

$$19{,}648 \cdot 72 \qquad 58{,}524 \cdot 19$$
$$19{,}648 \cdot 72$$
$$2 \; | \; \overline{38{,}875 \cdot 47}$$

Area $= 19{,}437 \cdot 73$ m^2

Therefore uncorrected area $= 1 \cdot 9438$ hectares

$\qquad$ True area $=$ measured area $\times \dfrac{\text{(length of chain used)}^2}{\text{(nominal chain length)}}$

$\qquad\qquad = 1 \cdot 9438 \times \dfrac{20 \cdot 1^2}{20 \cdot 0^2}$

$\qquad\qquad = 1 \cdot 9438 \times (1 + 0 \cdot 005)^2$

$\qquad\qquad = 1 \cdot 9438 \times (1 + 0 \cdot 01)$, say,

$\qquad\qquad = 1 \cdot 9632$ hectares.

Area by Total Latitudes and Departures. Area equals half the sum of the products of the total latitude of each point and the sum of the departures meeting at that point, having regard to sign.

This may be proved by considering a closed traverse $ABCDEA$ (Fig. 5.8) with the points as co-ordinates x_A, y_A; x_B, y_B; etc., measured relative to two axes, one of which is the meridian and the other at right angles to it.

The ordinates y_A, y_B, y_C, etc., are called the total latitudes of these points, and are obtained by summing the latitudes algebraically, proceeding round the traverse in a constant direction. By making the axis

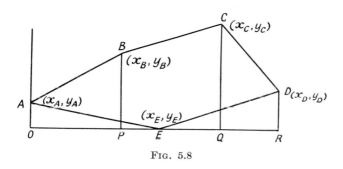

FIG. 5.8

at right angles to the meridian pass through the most southerly point (through station E in Fig. 5.8), the total latitude of this point thus becomes zero.

Area $ABCDEA$

$$= ABPO + BCQP + CDRQ - OAE - EDR$$

$$= \frac{y_A + y_B}{2} \times \text{dep. } AB + \frac{y_B + y_C}{2} \times \text{dep. } BC$$

$$+ \frac{y_C + y_D}{2} \times \text{dep. } CD - \frac{y_D + y_E}{2} \times \text{dep. } DE$$

$$- \frac{y_E + y_A}{2} \times \text{dep. } EA$$

$$= \tfrac{1}{2}\{y_A \cdot (\text{dep. } AB + \text{dep. } EA) + y_B \cdot (\text{dep. } AB + \text{dep. } BC)$$

$$+ y_C \cdot (\text{dep. } BC + \text{dep. } CD) + y_D \cdot (\text{dep. } CD + \text{dep. } DE)$$

$$+ y_E \cdot (\text{dep. } DE + \text{dep. } EA)\}$$

taking into account the signs of the departures.

EXAMPLE. Calculate the area of the traverse given in the last example by the method of total latitudes and departures.

Line	Latitude	Departure	Station	Total Lats.	Sum of Departures	Product +	Product −
AB	+ 85·65	+ 106·98					
			B	+ 199·95	+ 168·55	33,701·57	
BC	− 103·63	+ 61·57					
			C	+ 96·32	+ 85·95	8,278·70	
CD	− 33·22	+ 24·38					
			D	+ 63·10	− 76·81		4,846·71
DE	− 63·10	− 101·19					
			E	0	− 192·93	—	—
EA	+ 114·30	− 91·74					
			A	+ 114·30	+ 15·24	1,741·93	

$$\begin{array}{r} 43,722\cdot20 \\ 4,846\cdot71 \\ \hline 2\;|\;38,875\cdot49 \\ \hline 19,437\cdot74 \text{ m}^2 \end{array}$$

Therefore   Area = 19,437·74 m²
                                 = 1·9438 hectares.

Division of an Area by a Line of Known Bearing. It is required to divide the area $ABCDEA$ shown in Fig. 5.9 into two parts by a line XY. The procedure is as follows—

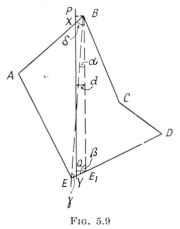

1. Calculate the total area enclosed by $ABCDEA$.

2. Draw a line at the known bearing through one of the stations which will divide the area approximately in the manner required, say BE_1.

3. Determine the bearings of lines EB, EE_1, and BA and so deduce the angles α, β, and γ.

4. Calculate the length of EB from the known co-ordinates, and thus find EE_1 and BE_1 using the sine rule relationship $\dfrac{EE_1}{\sin \alpha} = \dfrac{EB}{\sin \beta} = \dfrac{BE_1}{\sin \gamma}$.

5. Determine area ABE_1EA by the method previously described.

Fig. 5.9

6. The required area $AXYEA$ = area ABE_1EA − area XBE_1YX. Hence area XBE_1YX is found.

7. To locate line XY it will be necessary to calculate the distance separating the actual line XY and the trial line BE_1.

Area XBE_1YX = area PBE_1QP + area QE_1Y − area PBX
             = $d \cdot BE_1 + \tfrac{1}{2}d[d \cdot \tan (\beta - 90)] - \tfrac{1}{2}d[d \cdot \tan (90 - \delta)]$

All the terms in the above equation with the exception of d are known, so that the problem may be solved.

EXAMPLE. Locate a line running north-south which divides the area enclosed by the traverse given in the previous examples into two equal parts.

1. Calculate the total area as already shown. As this is 19,437·74 m², the required line will give areas of 9,718·87 m² on either side.
2. From the data (i.e. latitudes and departures) and using some

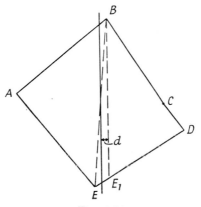

FIG. 5.10

convenient point—say A (Fig. 5.10)—as origin of co-ordinates, calculate the co-ordinates of the various points, which are

A	0	0
B	N 85·65	E 106·98
C	S 17·08	E 108·55
D	S 51·20	E 192·93
E	S 114·30	E 91·74

Take the trial line passing through B (it will be seen from the general shape that either B or E must be chosen to approximately halve the area).

3. Bearing of line $EB = \tan^{-1}\left(\dfrac{106·98 - 91·74}{85·65 + 114·30}\right)$

$= \tan^{-1}\dfrac{15·24}{199·95}$

$= $ N 4° 21′ E

Bearing of line $ED = \tan^{-1}\dfrac{101·19}{63·10}$

$= $ N 58° 03′ E

Bearing of line $AB = \tan^{-1}\dfrac{106·98}{85·65}$

$= $ N 51° 19′ E

4. BE_1 is parallel to the N–S meridian (Fig. 5.11

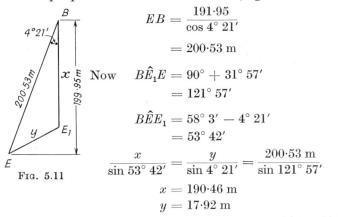

$$EB = \frac{191 \cdot 95}{\cos 4° 21'}$$

$$= 200 \cdot 53 \text{ m}$$

Now $\quad B\hat{E}_1E = 90° + 31° 57'$

$$= 121° 57'$$

$$B\hat{E}E_1 = 58° 3' - 4° 21'$$

$$= 53° 42'$$

$$\frac{x}{\sin 53° 42'} = \frac{y}{\sin 4° 21'} = \frac{200 \cdot 53 \text{ m}}{\sin 121° 57'}$$

$$x = 190 \cdot 46 \text{ m}$$

$$y = 17 \cdot 92 \text{ m}$$

FIG. 5.11

Therefore co-ordinates of E_1 are $190 \cdot 46 - 85 \cdot 65 = 104 \cdot 81$ S, and $106 \cdot 98$ E.

5. Calculate the area ABE_1EA.

Line	Latitude	Departure	Longitude	Double Longitude	Double Longitude × Latitude	
					+	−
AB	+ 85·65	+ 106·98	53·49	106·98	9,162·84	—
BE_1	− 190·46	0	106·98	213·96	—	40,750·82
E_1E	− 9·49	− 15·24	99·36	198·72	—	1,885·85
EA	+ 114·30	− 91·74	45·87	91·74	10,485·88	—

$$\begin{array}{rr} & 19{,}648{\cdot}72 & 42{,}636{\cdot}67 \\ & & 19{,}648{\cdot}72 \end{array}$$

$$\begin{aligned} \text{Double Area} &= 22{,}987{\cdot}95 \\ \text{Area} &= 11{,}493{\cdot}97 \\ \text{Required Area} &= 9{,}718{\cdot}87 \end{aligned}$$

Therefore $\qquad\qquad\qquad\qquad\qquad$ Residual Area $= 1{,}775{\cdot}10$

and $\quad 1{,}775 \cdot 10 = 190 \cdot 46d + \tfrac{1}{2}d \cdot d \tan 31° 57' - \tfrac{1}{2}d \cdot d \tan 31° 41'$

$$0 \cdot 0885d^2 - 190 \cdot 46d + 1{,}775 \cdot 10 = 0$$

Therefore $\qquad\qquad\qquad\qquad\qquad\qquad d = 9 \cdot 44 \text{ m}$

Thus the required dividing line would run N–S across the area at a distance of $9 \cdot 44$ m west of B, or $97 \cdot 54$ m from A.

Irregular Figures

The following methods may be used for determining the area of irregular curvilinear figures such as ponds, lakes, and the areas enclosed between chain lines and natural boundaries.

1. **"Give and Take" Lines.** The whole area is divided into triangles, or trapezoids, the irregular boundaries being replaced by straight lines so arranged that any small areas excluded from the survey by the straight line are balanced by other small areas outside the survey which are now included (*see* Fig. 5.12).

The positions of the lines are estimated by eye, using either a thin transparent straight-edge or a silk thread, and the lines are then drawn in faintly on the plan. The areas of the resulting triangles or trapezoids are calculated by the methods already described.

2. **Counting Squares.** An overlay of squared tracing paper or linen is laid on the drawing. The number of squares and parts of squares

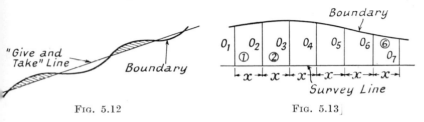

Fig. 5.12   Fig. 5.13

which are enclosed by the figure under consideration is now counted, and, knowing the scale of the drawing and the size of the squares on the overlay, the total area of the figure is computed.

3. **Trapezoidal Rule.** Fig. 5.13 shows an area bounded by a survey line and a boundary. The survey line is divided into a number of small equal intercepts of length x, and the offsets O_1, O_2, etc., measured, either directly on the ground or by scaling from the plan. If x is short enough for the length of boundary between the offsets to be assumed straight, then the area is divided into a series of trapezoids.

$$\text{Area of trapezoid (1)} = \frac{O_1 + O_2}{2} \, x.$$

$$\text{Area of trapezoid (2)} = \frac{O_2 + O_3}{2} \cdot x$$

$$\text{Area of trapezoid (6)} = \frac{O_6 + O_7}{2} \cdot x$$

Summing up, we get

$$\text{Area} = \frac{x}{2}\{O_1 + 2\,O_2 + 2\,O_3 + \ldots O_7\}$$

In the general case with n offsets, we get

$$\text{Area} = x\left\{\frac{O_1 + O_n}{2} + O_2 + O_3 + \ldots + O_{n-1}\right\}$$

6

If the area of a narrow strip of ground is required, this method may be used by running a straight line down the strip as shown in Fig. 5.14, and then measuring offsets at equal intercepts along this. By the same reasoning it will be seen that the area is given by

$$\text{Area} = x \left\{ \frac{O_1 + O_n}{2} + O_2 + O_3 + \ldots + O_{n-1} \right\}$$

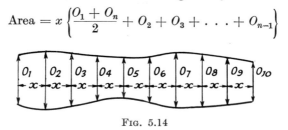

FIG. 5.14

EXAMPLE. Calculate the area of the plot shown in Fig. 5.14 if the offsets, scaled from the plan at intervals of 10 m, are—

Offset	O_1	O_2	O_3	O_4	O_5	O_6	O_7	O_8	O_9	O_{10}
Length (m)	16·76	19·81	20·42	18·59	16·76	17·68	17·68	17·37	16·76	17·68

$$\text{Area} = 10 \left\{ \frac{16 \cdot 76 + 17 \cdot 68}{2} + 19 \cdot 81 + 20 \cdot 42 + 18 \cdot 59 \right.$$

$$\left. + 16 \cdot 76 + 17 \cdot 68 + 17 \cdot 68 + 17 \cdot 37 + 16 \cdot 76 \right\}$$

$$= 1{,}622 \cdot 9 \text{ m}^2$$
$$= 0 \cdot 1623 \text{ hectare}$$

4. Simpson's Rule. This method, which gives greater accuracy than other methods, assumes that the irregular boundary is composed of a series of parabolic arcs. It is essential that the figure under consideration be divided into an *even* number of equal strips.

Referring to Fig. 5.13, consider the first three offsets, which are shown enlarged in Fig. 5.15.

The portion of the area contained between offsets O_1 and O_3

$$= ABGCDA$$

$$= \text{trapezoid } ABFCDA + \text{area } BGCFB$$

$$= \frac{O_1 + O_3}{2} \cdot 2x + \frac{2}{3} \text{ (area of circumscribing parallelogram)}$$

$$= \frac{O_1 + O_3}{2} \cdot 2x + \frac{2}{3} \cdot 2x \left(O_2 - \frac{O_1 + O_3}{2} \right)$$

$$= \frac{x}{3} \left\{ 3 O_1 + 3 O_3 + 4 O_2 - 2 O_1 - 2 O_3 \right\}$$

$$= \frac{x}{3} \left\{ O_1 + 4 O_2 + O_3 \right\}$$

For the next pair of intercepts, area contained between offsets O_3 and

$$O_5 = \frac{x}{3}\{O_3 + 4\,O_4 + O_5\}.$$

For the final pair of intercepts, area contained between offsets O_5 and

$$O_7 = \frac{x}{3}\{O_5 + 4\,O_6 + O_7\}.$$

Summing up, we get

$$\text{Area} = \frac{x}{3}\{(O_1 + O_7) + 2(O_3 + O_5) + 4(O_2 + O_4 + O_6)\}$$

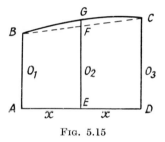

Fig. 5.15

In the general case,

$$\text{Area} = \frac{x}{3}\,(X + 2O + 4E)$$

where

X = sum of first and last offsets,

O = sum of the remaining odd offsets,

E = sum of the even offsets.

Simpson's Rule states, therefore, that the area enclosed by a curvilinear figure divided into an even number of strips of equal width is equal to one-third the width of a strip, multiplied by the sum of the two extreme offsets, twice the sum of the remaining odd offsets, and four times the sum of of the even offsets.

EXAMPLE. State Simpson's Rule for the determination of areas.

In a chain survey the following offsets were taken to a fence from a chain line—

Chainage (m)	0	20	40	60	80	100	120	140	160	180
Offset (m)	0	5·49	9·14	8·53	10·67	12·50	9·75	4·57	1·83	0

Find the area between the fence and the chain line.

It will be seen that there are ten offsets, and since Simpson's Rule can be applied to an *odd* number of offsets only, it will be used here to calculate the area contained between the first and ninth offsets. The residual triangular area between the ninth and tenth offsets is calculated separately. It is often convenient to tabulate the working.

Offset No.	Offset	Simpson Multiplier	Product
1	0	1	0
2	5·49	4	21·96
3	9·14	2	18·28
4	8·53	4	34·12
5	10·67	2	21·34
6	12·50	4	50·00
7	9·75	2	19·50
8	4·57	4	18·28
9	1·83	1	1·83

$$\Sigma = 185\cdot31$$

$$\text{Area } (O_1 - O_9) = \frac{20}{3} \cdot 185\cdot31 = 1{,}235\cdot40 \text{ m}^2$$

$$\text{Area } (O_9 - O_{10}) = \frac{20}{2} \cdot 1\cdot83 = \underline{\phantom{1{,}2}18\cdot30 \text{ m}^2}$$

$$1{,}253\cdot70 \text{ m}^2$$

$$= 0\cdot1254 \text{ hectare}$$

Volumes : Earthwork Calculations

The excavation, removal and dumping of earth is a frequent operation in building or civil engineering works. In the construction of a sewer, for example, a trench of sufficient width is excavated to given depths and gradients, the earth being stored in some convenient place (usually the side of the trench) and then returned to the trench after the laying of the pipe. Any material left over after reinstatement must be carted away and disposed of. In basement excavation probably all the material got out will require to be carted away, but for embankments the earth required will have to be brought from some other place.

In each case, however, payment will have to be made for labour, plant, etc., and this is done on the basis of the calculated volume of material handled. It is therefore essential that the engineer or surveyor should be able to make good estimates of volumes of earthwork.

There are three general methods for calculating earthworks: (I) by cross sections, (II) by contours, (III) by spot heights.

(I) Volumes from Cross Sections

In this method cross sections are taken at right angles to some convenient line which runs longitudinally through the earthworks, and although it is capable of general application it is probably most used on long narrow works such as roads, railways, canals, embankments, pipe excavations, etc. The volumes of earthwork between successive cross sections are calculated from a consideration of the cross-sectional areas, which in turn are measured or calculated by the general methods already given, i.e. by planimeter, division into triangles. counting squares, etc.

In long constructions which have constant formation width and side slopes, however, it is possible to simplify the computation of cross-sectional areas by the use of formulae. These are especially useful for

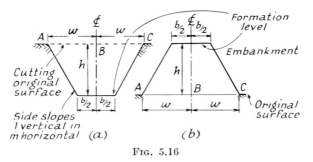

Fig. 5.16

railways, long embankments, etc., and formulae will be given for the following types of cross section—

(*a*) sections level across, (*b*) sections with a cross fall, (*c*) sections part in cut and part in fill, (*d*) sections of variable levels.

(*a*) **Sections Level Across** (Fig. 5.16). Depth at centre line (or height in case of embankment)

$$= h \text{ units}$$

$$\text{Formation width} = b \text{ units}$$

Then $$w = \text{side width}$$

$$= \frac{b}{2} + mh \qquad . \qquad . \qquad . \qquad (5.4)$$

Note: the sloping side has to fall (or rise a vertical height of h units from original level to final formation level. Since this side slopes in such a way that m units is the horizontal projection for every single unit vertical rise, then in h units the side gives a horizontal projection of mh units.

$$= 2 w$$

$$= b + 2 mh$$

$$\text{cross-sectional area} = h \cdot \left(\frac{b + b + 2 mh}{2} \right)$$

$$= h(b + mh) \qquad . \qquad . \qquad . \qquad (5.5)$$

EXAMPLE At a certain station an embankment formed on level ground has **a** height at its centre line of 3·10 metres. If the breadth of formation is 12·50 metres, find (*a*) the side widths, and (*b*) the area of cross section, given that the side slope is 1 vertical to $2\frac{1}{2}$ horizontal.

$$b = 12\cdot50 \text{ metres}, \quad m = 2\cdot5$$

$\therefore \qquad$ Area $= h(b + mh)$

$$= 3\cdot10(12\cdot50 + 2\cdot5 \times 3\cdot10)$$
$$= 62\cdot78 \text{ sq metres}$$

Sidewidths are both equal to $\dfrac{b}{2} + mh$

$$= 6\cdot25 + 2\cdot5 \times 3\cdot10$$
$$= 14 \text{ metres}$$

In setting out this section, for soil stripping and fixing the toe of each

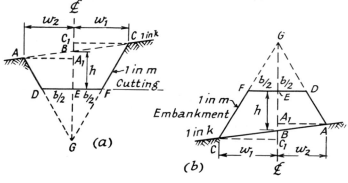

FIG. 5.17

side slope, pegs would be inserted at *A* and *C* at distances of 14 metres from centre peg *B* measured normal to the centre line.

(*b*) **Sections with Cross Fall** (Fig. 5.17). In this case the existing ground has a cross fall or transverse gradient relative to the centre line, and the sidewidths are not equal since the section is not symmetrical about the centre line.

Now $\qquad\qquad C_1B = \dfrac{w_1}{k}$

since this is the difference in level between *C* and *B* due to a gradient of 1 in *k* over a distance of w_1; similarly

$$A_1B = \frac{w_2}{k}$$

Also, if the side slopes intersect at *G*, then *GE* will be the vertical difference in level over a horizontal distance of $\dfrac{b}{2}$.

$\therefore \qquad\qquad GE = \dfrac{b}{2\,m}$

Since triangles C_1CG and EFG are similar,

$$\frac{CC_1}{EF} = \frac{GC_1}{GE}$$

$$\frac{w_1}{b/2} = \frac{\dfrac{b}{2\,m} + h + \dfrac{w_1}{k}}{b/2\,m}$$

$$w_1 = m\left[\frac{b}{2\,m} + h + \frac{w_1}{k}\right]$$

$$w_1\left(1 - \frac{m}{k}\right) = \frac{b}{2} + mh$$

$$\therefore \qquad w_1 = \left(\frac{b}{2} + mh\right)\left(\frac{k}{k - m}\right) \qquad . \qquad . \qquad . \quad (5.6)$$

Also
$$\frac{AA_1}{DE} = \frac{GA_1}{GE}$$

$$\frac{w_2}{b/2} = \frac{\dfrac{b}{2\,m} + h - \dfrac{w_2}{k}}{b/2\,m}$$

$$\therefore \qquad w_2 = \left(\frac{b}{2} + mh\right)\left(\frac{k}{k + m}\right) \qquad . \qquad . \qquad . \quad (5.7)$$

The area of the cutting or the embankment is the area $ACFDA$,

$$= \text{area } BCG + \text{area } ABG - \text{area } DFG$$

$$= \tfrac{1}{2}w_1\left(\frac{b}{2\,m} + h\right) + \tfrac{1}{2}w_2\left(\frac{b}{2\,m} + h\right) - \tfrac{1}{2}b\,\frac{b}{2\,m}$$

$$= \tfrac{1}{2}\left(\frac{b}{2\,m} + h\right)(w_1 + w_2) - \frac{b^2}{4\,m}$$

$$= \frac{1}{2\,m}\left\{\left(\frac{b}{2} + mh\right)(w_1 + w_2) - \frac{b^2}{2}\right\} \qquad . \qquad . \qquad . \quad (5.8)$$

Difference in level between C and F

$$= h + \frac{w_1}{k} \,. \qquad . \qquad . \qquad . \qquad . \quad (5.9)$$

Difference in level between A and D

$$= h - \frac{w_2}{k} \,. \qquad . \qquad . \qquad . \qquad . \quad (5.10)$$

This type of section is sometimes known as a "two-level section," since two levels are required to establish the cross fall of 1 in k.

In setting out pegs to locate A and C, the side widths may be scaled from the section, or, with greater accuracy, they may be located using level staff and tape (Fig. 5.18).

The readings on the staff at A and B are H_2 and H respectively. If h is the formation depth below B, it is evident that

$$h_2 + H_2 = H + h$$

or

$$h_2 = H - H_2 + h$$

but

$$h_2 = \frac{x}{m}$$

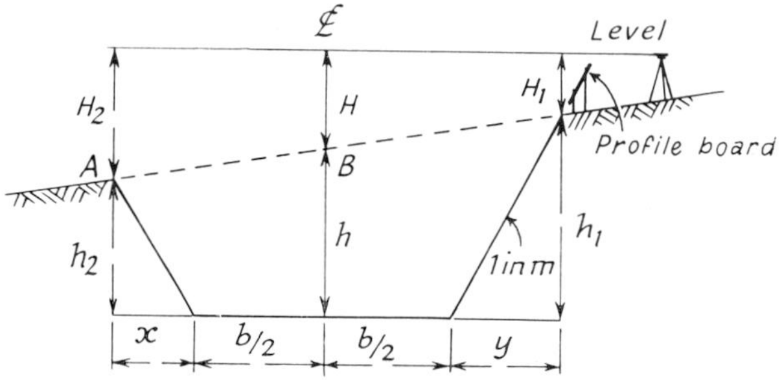

Fig. 5.18

Therefore

$$x = m(H - H_2 + h)$$

$$w_2 = \frac{b}{2} + x$$

$$= \frac{b}{2} + m(H - H_2 + h) \qquad . \qquad . \quad (5.11)$$

Similarly

$$w_1 = \frac{b}{2} + m(H - H_1 + h) \qquad . \qquad . \quad (5.12)$$

After taking a reading H on the staff at B, the staff man places the staff at some trial position which he estimates for A, and the tape reading from B to the staff is noted. The staff reading H_2 is now observed and the term $m(H - H_2 + h)$ computed. A value of w_2 is now deduced, as in equation (5.11), by adding $\frac{b}{2}$, and if this value agrees with the taped value, then the point A has been correctly located. If not, further trial positions for A must be taken until the deduced and taped values of w_2 agree.

This method will prove quite speedy after a short time spent in practice, and is, of course, more accurate than scaling. It may be somewhat simplified by utilizing the finished level of formation and the height of collimation of the instrument. The difference between these values is equal to $(H + h) = Y$ say. Then

$$w_2 = \frac{b}{2} + m(Y - H_2) \qquad . \qquad . \qquad (5.11a)$$

Profile boards are established to slopes of 1 in m so that the actual slopes can be realized during construction. For cuttings these would be placed outside the side width limits, but for embankments they could start just inside the toe, and could be set either to match the slope surfaces or to be in suitable positions at a known height above the finished surfaces. Formation levels can be controlled by sight rails as shown in Chapter 3 for sewers.

EXAMPLE. Calculate the side widths and cross-sectional area of an embankment to a road with formation width of 12·50 metres, and side slopes 1 vertical to 2 horizontal, when the centre height is 3·10 metres and the existing ground has a cross fall of 1 in 12 at right angles to the centre line of the embankment.

Referring to Fig. 5.17b,

$$w_1 = \left(\frac{b}{2} + mh\right)\left(\frac{k}{k - m}\right)$$

where $\qquad b = 12\cdot50,\ k = 12,\ m = 2,\ h = 3\cdot10\text{ m}$

$$= \left(\frac{12\cdot50}{2} + 2 \times 3\cdot10\right)\left(\frac{12}{10}\right)$$

$$= 14\cdot94\text{ metres}$$

$$w_2 = \left(\frac{b}{2} + mh\right)\left(\frac{k}{k + m}\right)$$

$$= \left(\frac{12\cdot50}{2} + 2 \times 3\cdot10\right)\left(\frac{12}{14}\right)$$

$$= 10\cdot67\text{ metres}$$

From equation (5.8),

$$\text{area} = \frac{1}{2\,m}\left\{\left(\frac{b}{2} + mh\right)\left(w_1 + w_2\right) - \frac{b^2}{2}\right\}$$

$$= \frac{1}{4}\left\{12\cdot45(14\cdot94 + 10\cdot67) - \frac{12\cdot50^2}{2}\right\}$$

$$= 60\cdot18\text{ sq metres}$$

EXAMPLE. In the previous example, when setting out peg A (see Fig. 5.17b) with a level, it was found that the staff reading at B was 3·636 m, and at the trial point the staff reading was 2·768 m with a tape reading of 10·67 m. Is this the correct position for A ?

$$\text{Staff reading at } B = H = 3\cdot636$$

Staff reading at trial position

$$= H_2 = 2\cdot768$$

Since we are now considering an embankment, equations (5.11) and (5.12) will require modification, and the student can easily prove that in this case

$$w_1 = \frac{b}{2} + m(H_1 + h - H) . \qquad . \qquad . \qquad . \quad (5.13)$$

$$w_2 = \frac{b}{2} + m(H_2 + h - H) . \qquad . \qquad . \qquad . \quad (5.14)$$

Substituting the trial values in equation (5.14),

$$w_2 = 6{\cdot}25 + 2(2{\cdot}746 + 3{\cdot}10 - 3{\cdot}636)$$
$$= 10{\cdot}67 \text{ metres}$$

which agrees with the taped values; thus the toe of the bank has been located correctly.

(c) **Sections Part in Cut and Part in Fill** (Fig. 5.19). $\triangle$'s GBA and EBJ are similar.

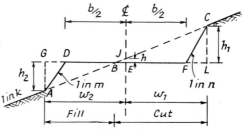

FIG. 5.19

Therefore
$$\frac{GB}{BE} = \frac{h_2}{h}$$

$$\frac{GB + BE}{BE} = \frac{h + h_2}{h}$$

but
$$GB + BE = w_2$$

and
$$BE = k \cdot h$$

Therefore
$$w_2 = k(h + h_2)$$

but
$$GD = mh_2$$
$$= GE - DE$$
$$= w_2 - \frac{b}{2}$$

Therefore
$$h_2 = \frac{w_2 - b/2}{m} = \frac{2w_2 - b}{2m}$$

and
$$w_2 = k\left(h + \frac{2\,w_2 - b}{2\,m}\right)$$

$$2\,mw_2 = 2\,mkh + 2\,w_2 k - bk$$

$$2\,w_2(k - m) = bk - 2\,mkh$$

i.e.
$$w_2 = \left(\frac{k}{k - m}\right)\left(\frac{b}{2} - mh\right) \qquad . \qquad . \ (5.15)$$

Similarly, the other side width may be deduced. From Fig. 5.19,

$$\frac{BL}{BE} = \frac{EL + BE}{BE} = \frac{h_1}{h}$$

Therefore
$$\frac{EL}{BE} = \frac{h_1 - h}{h}$$

$$BE = kh$$

$\therefore$
$$EL = w_1 = k(h_1 - h)$$

but
$$FL = nh_1 = w_1 - \frac{b}{2}$$

Therefore
$$h_1 = \frac{2\,w_1 - b}{2\,n}$$

$$w_1 = k\left(\frac{2\,w_1 - b}{2\,n} - h\right)$$

$\therefore$
$$w_1 = \left(\frac{k}{k - n}\right)\left(\frac{b}{2} + nh\right) \qquad . \qquad . \qquad . \ (5.16)$$

$$\text{Area of fill} = \tfrac{1}{2}\,h_2 \,.\, DB = \tfrac{1}{2}\,h_2\left(\frac{b}{2} - kh\right)$$

$$= \tfrac{1}{2}\left(\frac{2\,w_2 - b}{2\,m}\right)\left(\frac{b}{2} - kh\right)$$

$$= \tfrac{1}{2}\,\frac{(b/2 - kh)^2}{(k - m)} \qquad . \qquad . \qquad . \ (5.17)$$

$$\text{Area of cut} = \tfrac{1}{2}\,h_1 \,.\, BF = \tfrac{1}{2}\,h_1\left(\frac{b}{2} + kh\right)$$

$$= \tfrac{1}{2}\left(\frac{2\,w_1 - b}{2\,n}\right)\left(\frac{b}{2} + kh\right)$$

$$= \tfrac{1}{2}\,\frac{(b/2 + kh)^2}{(k - n)} \qquad . \qquad . \qquad . \ (5.18)$$

When the cross section is in fill at the centre line, instead of being in cut as in Fig. 5.19, then the following modified formulae obtain—

$$\text{Area of fill} = \tfrac{1}{2}\frac{(b/2 + kh)^2}{(k - m)} \qquad . \qquad . \qquad (5.17a)$$

$$\text{Area of cut} = \tfrac{1}{2}\frac{(b/2 - kh)^2}{(k - n)} \qquad . \qquad . \qquad (5.18a)$$

The student should confirm this as an exercise in the derivation of these formulae.

In setting out the limits of such earthwork part in cut, part in fill by means of a level set up above C (Fig. 5.19), a peg must be positioned at, say A, such that w_2 as measured by tape equals the value given by $\left(\dfrac{b}{2} + mh_2\right)$. If Y is the difference in level between the line of collimation of the instrument and formation level, then h_2 must equal the reading on a staff held at A, say H_2, minus Y,

i.e. $$h_2 = H_2 - Y$$

The value of w_2 given by the value of H_2 obtained with the staff in the trial position is computed from the expression

$$w_2 = \frac{b}{2} + mh_2 = \frac{b}{2} + m(H_2 - Y)$$

and compared with the taped value. Different staff positions, up or down the slope on the line of the cross section, are tried until agreement between the two values is achieved.

If the level is set up with the line of collimation below formation level, the value of Y is again, of course, given by the difference between the two, but Y must now be added to the staff readings at the trial positions.

EXAMPLE. A road has a formation width of 9·50 metres and side slopes of 1 vertical to 1 horizontal in cut, and 1 vertical to 3 horizontal in fill. The original ground had a cross fall of 1 vertical to 5 horizontal. If the depth of excavation at the centre line is 0·5 metre, calculate the side widths and the areas of cut and fill.

From formulae (5.15), (5.16), (5.17), and (5.18), we get—

$$w_2 = \left(\frac{k}{k - m}\right)\left(\frac{b}{2} - mh\right) = \left(\frac{5}{5 - 3}\right)(4\cdot75 - 3 \times 0\cdot5)$$

$$= 7\cdot88 \text{ metres}$$

$$w_1 = \left(\frac{k}{k - n}\right)\left(\frac{b}{2} + nh\right) = \left(\frac{5}{5 - 1}\right)(4\cdot75 + 1 \times 0\cdot5)$$

$$= 6\cdot56 \text{ metres}$$

Area of fill $= \dfrac{1}{2}\left(\dfrac{(b/2 - kh)^2}{(k - m)}\right) = \dfrac{1}{2}\left(\dfrac{(4\cdot75 - 5 \times 0\cdot5)^2}{(5 - 3)}\right)$

$\qquad\qquad = 1\cdot27$ sq metres

Area of cut $= \dfrac{1}{2}\left(\dfrac{(b/2 + kh)^2}{(k - n)}\right) = \dfrac{1}{2}\left(\dfrac{(4\cdot75 + 5 \times 0\cdot5)^2}{(5 - 1)}\right)$

$\qquad\qquad = 6\cdot57$ sq metres

(d) **Sections of Variable Level.** The type of section shown in Fig. 5.20 is sometimes referred to as a "three level section" since at least three levels are required on each cross section to enable the ground slopes to be calculated. The side width formulae are obtained as for two level sections (b), and are

$$w_1 = \left(\frac{b}{2} + mh\right)\left(\frac{k}{k - m}\right) \qquad . \qquad . \qquad . \qquad . \quad (5.19)$$

$$w_2 = \left(\frac{b}{2} + mh\right)\left(\frac{l}{l + m}\right) \qquad . \qquad . \qquad . \qquad . \quad (5.20)$$

If BA were falling away from the centre line,

$$w_2 = \left(\frac{b}{2} + mh\right)\left(\frac{l}{l - m}\right) \qquad . \qquad . \qquad . \qquad . \quad (5.20a)$$

Area of cross section

$$-\frac{1}{2}w_1\left(h + \frac{b}{2\,m}\right) + \frac{1}{2}w_2\left(h + \frac{b}{2\,m}\right) - \frac{1}{2}b\cdot\frac{b}{2\,m}$$

$$= \frac{1}{2}\left\{(w_1 + w_2)\left(h + \frac{b}{2\,m}\right) - \frac{b^2}{2\,m}\right\}$$

$$= \frac{1}{2\,m}\left\{(w_1 + w_2)\left(mh + \frac{b}{2}\right) - \frac{b^2}{2}\right\} \qquad . \qquad . \quad (5.21)$$

Note: The section shown in Fig. 5.20 is probably the most complex one to be worthy of analysis in the manner shown; it is worth using on long constructions, where many cross sections are involved, because the use of a formula means (a) that levels are taken at predetermined points on each section, and (b) that computation of areas and volumes can be tabulated and systematized.

In road construction, where it is quite possible that the formation will be cambered, it is generally preferable to take sufficient levels to give a true

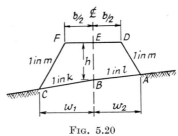

FIG. 5.20

representation of the existing ground profile and, by plotting to a suitable scale, obtain more accurate sectional areas by planimeter, counting squares, etc.

Computation of Volumes. Having determined the various areas of cross section, the volumes of earth involved in the construction are computed by one of the following methods: (1) mean areas, (2) end areas, (3) prismoidal formula.

1. VOLUMES BY MEAN AREAS. In this method the volume is determined by multiplying the mean of the cross-sectional areas by the distance between the end sections. If the areas are $A_1, A_2, A_3 \ldots A_{n-1}$, A_n and the distance between the two extreme sections A_1 and A_n is L, then

$$\text{Volume} = V = \frac{A_1 + A_2 + A_3 \ldots A_{n-1} + A_n}{n} \cdot L$$

The method is not a very accurate one.

2. VOLUMES BY END AREAS. If A_1 and A_2 are the areas of two cross sections distance D apart, then the volume V between the two is given by

$$V = D \cdot \frac{A_1 + A_2}{2} \qquad . \qquad . \qquad . \qquad . \quad (5.22)$$

This expression is correct so long as the area of the section mid-way between A_1 and A_2 is the mean of the two, and such can be assumed to be the case so long as there is no wide variation between successive sections. If there is such a variation, then a correction must be applied, and this refinement is dealt with in the next method. In general, however, in view of the usual irregularities in ground surface which exist between successive cross sections, and the problems of bulking and settlement normally associated with earthworks, it is reasonable to use the end areas formula for normal estimating.

For a series of consecutive cross sections, the total volume will be—

$$\text{Volume} = \Sigma V = \frac{D_1(A_1 + A_2)}{2} + \frac{D_2(A_2 + A_3)}{2}$$

$$+ \frac{D_3(A_3 + A_4)}{2} + \ldots$$

if $D_1 = D_2 = D_3$. etc. $= D$

$$\Sigma V = D \left\{ \frac{A_1 + A_n}{2} + A_2 + A_3 + \ldots + A_{n-1} \right\} \quad . \quad (5.23)$$

which is sometimes referred to as the trapezoidal rule for volumes.

EXAMPLE. An embankment is formed on ground which is level transverse to the embankment but falling at 1 in 20 longitudinally so that three sections 20 metres apart have centre line heights of 6·00, 7·60, and 9·20 metres respectively above original ground level. If side slopes of 1 in 1 are used, determine the volume of fill between the outer sections when the formation width is 6·00 m using the trapezoidal rule.

Using equation (5.5),

$$A = h(b + mh)$$
$$A_1 = 6·00 \ (6·00 + 6·00) = 72·00 \text{ sq metres}$$

$$A_2 = 7{\cdot}60(6{\cdot}00 + 7{\cdot}60) = 103{\cdot}36 \text{ sq metres}$$
$$A_3 = 9{\cdot}20(6{\cdot}00 + 9{\cdot}20) = 139{\cdot}84 \text{ sq metres}$$

Note that the mid-area A_2 is not the mean of $A_1 + A_3$.

$$V = \frac{20{\cdot}00}{2}\,[72{\cdot}00 + 2 \times 103{\cdot}36 + 139{\cdot}84]$$

$$= 4{,}185{\cdot}6 \text{ cu metres}$$

If the method of end areas is applied between the two extreme sections, we get

$$V = \frac{40{\cdot}00}{2}\,(72{\cdot}00 + 139{\cdot}84)$$

$$= 4{,}236{\cdot}8 \text{ cu metres}$$

an over-estimation of over 1 per cent compared with the first method.

As an extreme case of such discrepancy, consider the volume of a pyramid (Fig. 5.21) of base area A and perpendicular height D.

$$\text{True volume} - \tfrac{1}{3} \text{ base area} \times \text{height}$$
$$= \tfrac{1}{3}\,.\,D\,.\,A$$

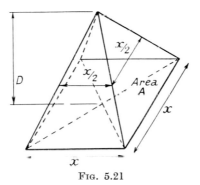

Fɪɢ. 5.21

But by end areas,
$$V - D\,.\,\frac{A \mid 0}{2}$$
$$= \tfrac{1}{2}\,D\,.\,A$$

3. Volumes by Prismoidal Formula. If the volume of earth between two successive cross sections be considered a *prismoid* then a more precise formula—the prismoidal formula—may be used. It is generally considered that, all things being equal, use of this formula gives the most accurate estimate of volume.

A *prismoid* is a solid made up of two end faces which must be parallel plane figures not necessarily of the same shape; the faces between them, i.e. sides, top and bottom, must be formed by straight continuous lines running from one end face to the other.

The volume of a prismoid is given by

$$V = \frac{D}{6}(A_1 + 4M + A_2) \qquad . \qquad . \qquad . \qquad (5.24)$$

where A_1 and A_2 are the areas of the two end faces distances D apart, M is the area of the section mid-way between.

The theory assumes that in calculating M each linear dimension is the average of the corresponding dimensions of the two end areas. (It is no use taking the area M as the mean of A_1 and A_2, for we should only get

$$V = \frac{D}{6}\left(A_1 + 4\frac{(A_1 + A_2)}{2} + A_2\right)$$

$$= D \cdot \frac{A_1 + A_2}{2}$$

i.e. the "end areas" formula.)

As an example of the application of the prismoidal formula, we may

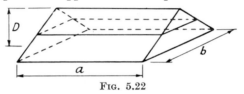

FIG. 5.22

derive the formula for the volume of a pyramid. Referring to Fig. 5.21,

$$\text{Base area} = A = x^2$$

$$\text{Mid-area} = M = \frac{x^2}{4} = \frac{A}{4}$$

$$\text{Top area} = A_2 = 0$$

$$\therefore \qquad V = \frac{D}{6}\left(A + 4 \cdot \frac{A}{4} + 0\right)$$

$$= \frac{D \cdot A}{3}$$

Similarly, for a wedge as in Fig. 5.22,

$$\text{Base area} = A_1 = a \times b$$

$$\text{Mid-area} = M = a \times \frac{b}{2} = \frac{A_1}{2}$$

$$\text{Top area} = A_2 = 0$$

$$\therefore \qquad V = \frac{D}{6}\left(A_1 + 4 \cdot \frac{A_1}{2} + 0\right)$$

$$= \frac{D \cdot A_1}{2}$$

The proof of the prismoidal formula will be found in any standard text-book on solid geometry.

There is a number of alternative ways in which the prismoidal formula may be used, and some of these are given below, assuming in each case that the basic longitudinal spacing of the cross sections is one chain of 100 links, i.e. D equals 100 links.

(*a*) Treat each cross section as the end area of a prismoid one chain long, and estimate the dimensions of the mid-areas at the 50-link points as the mean of the two corresponding dimensions in the end areas. This is difficult when the sections are irregular.

(*b*) Where estimation of the mid-area is difficult, arrange for extra sections to be levelled at the mid-area positions as required. This means, of course, a large increase in the amount of field work.

(*c*) Treat alternate sections as end areas, i.e. the length of the prismoid is 2 D. Unless the ground profile is regular both transversely and longitudinally, however, it is likely that errors will be introduced in assuming that a volume of earth is in fact prismoidal over such a length. Where this method is used, by taking A_1 and A_3, A_3 and A_5, A_5 and A_7, etc., as end areas of successive prismoids, we have

$$V_1 = \frac{2\,D}{6} \cdot (A_1 + 4\,A_2 + A_3)$$

$$V_2 = \frac{2\,D}{6} \cdot (A_3 + 4\,A_4 + A_5), \text{ etc.}$$

$$\therefore \quad V = \frac{D}{3}\,(A_1 + 4\,A_2 + 2\,A_3 + 4\,A_4 + \ldots$$

$$\vert\; 2\,A_{n-2} + 4\,A_{n-1} + A_n) \qquad . \qquad . \qquad . \qquad . \quad (5.25)$$

where n is an odd number and D is one chain.

This is Simpson's Rule for volumes. It can be applied with greater accuracy with method (*b*) in which mid-sections are levelled at the 50-link points, but note that D in equation (5.25) will then be 50 links instead of one chain.

(*d*) Calculate the volumes between successive cross sections by the method of end areas, and apply corrections to these volumes known as *prismoidal corrections*. Such corrections can be derived for regular sections only, e.g. consider sections level across—

If D = the spacing of cross sections, A_1 and A_2 = the two end areas, M = the mid-area, h_1 and h_2 = the difference in level between ground level and formation at A_1 and A_2, respectively, and b = formation width, then from equation (5.5),

$$A_1 = h_1(b + mh_1)$$
$$A_2 = h_2(b + mh_2)$$

$$\therefore \quad V_{end\ areas} = \frac{D}{2}\,(bh_1 + mh_1{}^2 + bh_2 + mh_2{}^2)$$

assuming $h_m = \dfrac{h_1 + h_2}{2}$ (difference in level between ground level and formation at the mid-section)

$$M = \left(\frac{h_1 + h_2}{2}\right)\left\{b + m\left(\frac{h_1 + h_2}{2}\right)\right\}$$

$$= \frac{bh_1}{2} + \frac{bh_2}{2} + \frac{mh_1{}^2}{4} + \frac{mh_2{}^2}{4} + \frac{mh_1h_2}{2}$$

$$\therefore \quad V_{prismoid} = \frac{D}{6}\left\{bh_1 + mh_1{}^2 + 4\left(\frac{bh_1}{2} + \frac{bh_2}{2} + \frac{mh_1{}^2}{4} + \frac{mh_2{}^2}{4}\right.\right.$$

$$\left.\left. + \frac{mh_1h_2}{2}\right) + bh_2 + mh_2{}^2\right\}$$

$$= \frac{D}{6}\{3\,bh_1 + 2\,mh_1{}^2 + 3\,bh_2 + 2\,mh_2{}^2 + 2\,mh_1h_2\}$$

$$= \frac{D}{2}\left\{bh_1 + \frac{2}{3}mh_1{}^2 + bh_2 + \frac{2}{3}mh_2{}^2 + \frac{2\,mh_1h_2}{3}\right\}$$

$$\therefore \quad V_{EA} - V_P = \frac{D}{2}\left\{\frac{mh_1{}^2}{3} - \frac{2\,mh\cdot h_2}{3} + \frac{mh_2{}^2}{3}\right\}$$

$$= \frac{D}{6}\cdot m(h_1 - h_2)^2 = \textit{prismoidal correction} \text{ for a leve}$$

section . . . (5.26

Since the term $(h_1 - h_2)^2$ must always be positive, the correctio must be *deducted* from the volume as calculated by the end area formula. The correction is simple to apply since it requires no additiona information to that which is already required for the end areas calcula tions.

Prismoidal corrections can be similarly derived for other types o section, and as a further example, the P.C. for the section with cros fall (the two-level section) will be evaluated.

The general expression for a prismoidal correction is given by—

$$V_{EA} - V_P = \text{prismoidal correction}$$

$$= \frac{D}{2}(A_1 + A_2) - \frac{D}{6}(A_1 + 4\,M + A_2)$$

$$= \frac{D}{3}(A_1 - 2\,M + A_2) \quad . \quad . \quad . \quad (5.2$$

For the two-level section, it has been proved that

$$A = \frac{1}{2\,m}\left\{\left(\frac{b}{2} + mh\right)(w_1 + w_2) - \frac{b^2}{2}\right\} \quad . \quad . \quad (5.$$

Substituting from equations (5.6) and (5.7) for w_1 and w_2,

$$A = \frac{1}{2\,m} \left(\frac{b}{2} + mh \right) \left(\frac{b}{2} + mh \right) \left(\frac{k}{k+m} + \frac{k}{k-m} \right) - \frac{b^2}{4\,m}$$

$$\therefore \quad A_1 = \frac{1}{2\,m} \left(\frac{b}{2} + mh_1 \right)^2 \left(\frac{k}{k+m} + \frac{k}{k-m} \right) - \frac{b^2}{4\,m}$$

$$A_2 = \frac{1}{2\,m} \left(\frac{b}{2} + mh_2 \right)^2 \left(\frac{k}{k+m} + \frac{k}{k-m} \right) - \frac{b^2}{4\,m}$$

$$A_m = \frac{1}{2\,m} \left\{ \frac{b}{2} + m \left(\frac{h_1 + h_2}{2} \right) \right\}^2 \left(\frac{k}{k+m} + \frac{k}{k-m} \right) - \frac{b^2}{4\,m}$$

$$\text{P.C.} = \frac{D}{3} \left(A_1 - 2\,A_m + A_2 \right)$$

$$= \frac{D}{6\,m} \left(\frac{k}{k+m} + \frac{k}{k-m} \right) \left\{ \left(\frac{b}{2} + mh_1 \right)^2 + \left(\frac{b}{2} + mh_2 \right)^2 \right.$$

$$\left. - 2 \left(\frac{b}{2} + m \left(\frac{h_1 + h_2}{2} \right) \right)^2 \right\}$$

$$= \frac{Dk}{6\,m} \left(\frac{1}{k+m} + \frac{1}{k-m} \right) \left(\frac{b^2}{4} + m^2 h_1^2 + mbh_1 + \frac{b^2}{4} \right.$$

$$+ m^2 h_2^2 + mbh_2 - \frac{2b^2}{4} - mbh_1 - mbh_2$$

$$\left. - \frac{mh_1^2}{2} - m^2 h_1 h_2 - \frac{m^2 h_2^2}{2} \right)$$

$$= \frac{Dk}{6\,m} \left(\frac{k-m+k+m}{k^2-m^2} \right) \left(\frac{m^2 h_1^2}{2} - m^2 h_1 h_2 + \frac{m^2 h_2^2}{2} \right)$$

$$= \frac{D}{6} \frac{k^2}{(k^2-m^2)} \cdot m(h_1 - h_2)^2 \qquad \qquad . \quad . \quad . \quad . \quad . \quad (5.28)$$

As with the P.C. for the one-level section, this expression will always be positive, since $k > m$.

As an exercise the student is advised to check that prismoidal corrections for two-level sections which are part in cut, part in fill, re—

Fill P.C. $= \dfrac{D}{12(k-m)} k^2 (h_1 - h_2)^2 \qquad . \quad . \quad , \quad . \quad (5.29)$

Cut. P.C. $= \dfrac{D}{12(k-n)} k^2 (h_1 - h_2)^2 \qquad . \quad . \quad . \quad (5.30)$

EXAMPLE. Using the data of the example previously solved by the end areas method, compute the volume by the prismoidal formula.

(*a*) Taking $\quad D = 200$ links $= 40$ metres

$$V = \frac{40}{6} (72 \cdot 00 + 4 + 103 \cdot 36 + 139 \cdot 84)$$

$$= 4{,}168 \cdot 5 \text{ cu metres}$$

(*b*) Taking $D = 20$ metres, and applying the prismoidal correction to the "end areas" volumes,

$$V_1 = \frac{20}{2} (72 \cdot 00 + 103 \cdot 36)$$

$$= 1{,}753 \cdot 6 \text{ cu metres}$$

$$\text{P.C.} = \frac{Dm}{6} (h_1 - h_2)^2$$

$$= \frac{20 \times 1}{6} (6 \cdot 00 - 7 \cdot 60)^2$$

$$= 8 \cdot 53 \text{ cu metres}$$

$$V_2 = \frac{20}{2} (103 \cdot 36 + 139 \cdot 84)$$

$$= 2{,}432 \cdot 0 \text{ cu metres}$$

$$\text{P.C.} = \frac{20}{6} (7 \cdot 60 - 9 \cdot 20)^2$$

$$= 8 \cdot 53 \text{ cu metres}$$

$$\text{Total volume} = 4{,}168 \cdot 5 \text{ cu metres}$$

The answers are the same because in this example the dimensions of the middle section are the exact mean of the two outer ones. In practice the second method is preferable, since it assumes prismoids of shorter length. A similar example, in which the volume contained between three cross sections is required, will now be worked; in this case however, the centre section dimensions are not the mean of the outer ones.

EXAMPLE. A cutting is to be made in ground which has a transverse slope of 1 in 5. The width of formation is 8·00 metres and the side slopes are 1 vertical to 2 horizontal. If the depths at the centre lines of three sections 20 metres apart are 2·50, 3·10, and 4·30 metres respectively, determine the volume of earth involved in this length of the cutting.

From equation (5.8),

$$A = \frac{1}{2\,m} \left\{ \left(\frac{b}{2} + mh \right) (w_1 + w_2) - \frac{b^2}{2} \right\}$$

From equations (5.6) and (5.7),

$$w_1 = \left(\frac{b}{2} + mh \right) \left(\frac{k}{k - m} \right)$$

$$w_2 = \left(\frac{b}{2} + mh \right) \left(\frac{k}{k + m} \right)$$

$$\therefore \qquad \frac{k}{k-m} = \frac{5}{3}, \frac{k}{k+m} = \frac{5}{7}, \text{ since } m = 2 \text{ and } k = 5$$

Tabulating,

Section	h	mh	$b/2 + mh$	w_1	w_2	$w_1 + w_2$	A (m²)
1	2·50	5·00	9·00	$9\!\cdot\!00 \times \frac{5}{3}$ $= 15\!\cdot\!00$	$9\!\cdot\!00 \times \frac{5}{7}$ $= \;\;6\!\cdot\!43$	21·43	40·24
2	3·10	6·20	10·20	$10\!\cdot\!20 \times \frac{5}{3}$ $= 17\!\cdot\!00$	$10\!\cdot\!20 \times \frac{5}{7}$ $= \;\;7\!\cdot\!29$	24·29	53·94
3	4·30	8·60	12·60	$12\!\cdot\!60 \times \frac{5}{3}$ $= 21\!\cdot\!00$	$12\!\cdot\!60 \times \frac{5}{7}$ $= \;\;9\!\cdot\!00$	30·00	86·50

(a) Treating the whole as one prismoid,

$$V = \frac{40}{6} (40\!\cdot\!24 + 4 \times 53\!\cdot\!94 + 86\!\cdot\!50)$$

$$= 2{,}283\!\cdot\!3 \text{ cu metres}$$

(b) Using end areas formula with prismoidal correction,

$$V_{EA} = \frac{20}{2} (40\!\cdot\!24 + 2 \times 53\!\cdot\!94 + 86\!\cdot\!50)$$

$$= 2{,}346\!\cdot\!2 \text{ cu metres}$$

From equation (5.28),

$$\text{P.C.} = \frac{D}{6} \frac{k^2}{k^2 - m^2} \cdot m(h_1 - h_2)^2$$

$$\therefore \qquad \text{Total P.C.} = \frac{20}{6} \cdot \frac{25}{21} \cdot 2[(2\!\cdot\!50 - 3\!\cdot\!10)^2 + (3\!\cdot\!10 - 4\!\cdot\!30)^2]$$

$$= 14\!\cdot\!3 \text{ cu metres}$$

Therefore
$$V_P = 2{,}346\!\cdot\!2 - 14\!\cdot\!3$$
$$= 2{,}331\!\cdot\!9 \text{ cu metres}$$

EXAMPLE. A road has a formation breadth of 9·00 m, and side slopes of 1 in 1, in cut, and 1 in 3 in fill. The original ground had a cross fall of 1 in 5. If the depth of excavation at the centre lines of two sections 20 metres apart are 0·40 and 0·60 metres respectively, find the volumes of cut and fill over this length.

From equations (5.17) and (5.18),

$$\text{Area of fill} = \tfrac{1}{2} \cdot \frac{(b/2 - kh)^2}{k - m}$$

$$\text{Area of cut} = \tfrac{1}{2} \cdot \frac{(b/2 + kh)^2}{k - n}$$

Section 1:

$$\text{Area of fill} = \tfrac{1}{2}\, \frac{(4\cdot50 - 5 \times 0\cdot40)^2}{(5 - 3)} = 1\cdot56$$

$$\text{Area of cut} = \tfrac{1}{2}\, \frac{(4\cdot50 + 5 \times 0\cdot40)^2}{(5 - 1)} = 5\cdot28$$

Section 2:

$$\text{Area of fill} = \tfrac{1}{2}\, \frac{(4\cdot50 - 5 \times 0\cdot60)^2}{(5 - 3)} = 0\cdot56$$

$$\text{Area of cut} = \tfrac{1}{2}\, \frac{(4\cdot50 + 5 \times 0\cdot60)^2}{(5 - 1)} = 7\cdot03$$

Fill:

$$V_{EA} = \frac{20}{2}\,(1\cdot56 + 0\cdot56)$$

$$= 21\cdot2 \text{ cu metres}$$

From equation (5.29),

$$\text{P.C.} = \frac{D}{12\,(k - m)}\, .\, k^2\,(h_1 - h_2)^2$$

$$= \frac{20}{12 \times 2} \times 25 \times 0\cdot2^2$$

$$= 0\cdot8 \text{ cu metres}$$

$$\therefore \qquad V_P = 20\cdot4 \text{ cu metres}$$

Cut:

$$V_{EA} = \frac{20}{2}\, .\, (5\cdot28 + 7\cdot03)$$

$$= 123\cdot1 \text{ cu metres}$$

From equation (5.30),

$$\text{P.C.} = \frac{D}{12\,(k - n)}\, .\, k^2\,(h_1 - h_2)^2$$

$$= \frac{20}{12 \times 4} \times 25 \times 0\cdot2^2$$

$$= 0\cdot4 \text{ cu metres}$$

Therefore $\qquad V_P = 122\cdot7 \text{ cu metres}$

i.e. there is an excess of 102·3 cu metres of cut over fill.

EXAMPLE. Obtain an expression for the total volume of a symmetrical embankment of length l with side slopes 1 in s horizontal and formation breadth b if the heigth varies uniformly from zero at $l = 0$ to $\dfrac{l}{n}$ at a distance l.

A new access road to an old quarry necessitates an incline crossing an existing vertical quarry face, 9 metres high, at right angles, the original ground being level above and below the face. A length of 120 metres available on the low side is to accommodate the embankment forming the lower portion of the incline, and it is desired to use spoil from cutting the higher ground to form this embankment in such a way that cut and fill balance. The cut has vertical sides and the material from each cubic metre fills 1·1 m³ of the bank which has side slopes of 1 in 3. The formation breadth is 6 metres. Calculate the gradient required and the length of cutting. (*B.Sc.* (*Ext.*) *Part II*, 1952.)

Referring to Fig. 5.23, volume of section of length δx

$$= \frac{1}{2} \frac{x}{n} \left\{ b + \left(b + \frac{2\,xs}{n} \right) \right\} \delta x$$

$$= \frac{x}{n} \left(b + \frac{xs}{n} \right) \delta x$$

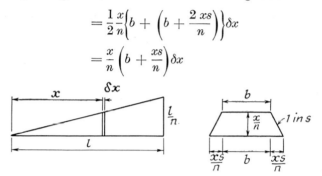

Fig. 5.23

Volume of embankment

$$= \int_0^l \frac{x}{n} \left(b + \frac{xs}{n} \right) dx$$

$$= \left[\frac{bx^2}{2\,n} + \frac{sx^3}{3\,n^2} \right]_0^l$$

$$= \frac{bl^2}{2\,n} + \frac{sl^3}{3\,n^2}$$

Or, using the prismoidal formula

$$A_1 = 0$$

$$A_2 = \frac{l}{n} \cdot \frac{\left(b + b + \dfrac{2\,ls}{n} \right)}{2} = \frac{l}{n} \left(b + \frac{ls}{n} \right)$$

$$M = \frac{l}{2\,n} \cdot \frac{\left(b + b + \dfrac{ls}{n} \right)}{2} = \frac{l}{4\,n} \left(2\,b + \frac{ls}{n} \right)$$

$$\therefore \qquad V = \frac{l}{6} \left\{ 0 + 4 \cdot \frac{l}{4\,n} \left(2\,b + \frac{ls}{n} \right) + \frac{l}{n} \left(b + \frac{ls}{n} \right) \right\}$$

$$= \frac{bl^2}{2\,n} + \frac{sl^3}{3\,n^2}$$

In the second part, if l_1 be the length of cutting, and the gradient required be 1 in n, then the volume of cut can be deduced from the above formula as

$$V_{cut} = \frac{bl_1^2}{2\,n}$$

(*Note :* the term $\dfrac{sl^3}{3\,n^2}$ disappears here because $s = 0$ for vertical sides.)

The embankment is to be 120 metres long, so that the height of embankment at the quarry face will be $\dfrac{120}{n}$ metres. Thus the depth of cut at the quarry face will be $\left(9 - \dfrac{120}{n}\right)$ metres.

Therefore
$$l_1 = n\left(9 - \frac{120}{n}\right)$$

$$= 9\,n - 120$$

Volume of material in embankment

$$= \frac{bl^2}{2\,n} + \frac{sl^3}{3\,n^2}$$

$$= \frac{6 \times 120^2}{2\,n} + \frac{3 \times 120^3}{3\,n^2}$$

Since cut and fill are to balance,

$$1\cdot1 \times \frac{6l_1^2}{2n} = \frac{6 \times 120^2}{2\,n} + \frac{3 \times 120^3}{3\,n^2}$$

$$\therefore \qquad 3\cdot3(9\,n - 120)^2 = 3 \times 120^2 + \frac{120^2}{n}$$

whence
$$n = 32\cdot8, \text{ say}$$

i.e.
$$\text{gradient} = 1 \text{ in } 32\cdot8$$

$$\text{Length of cut} = 9 \times 32\cdot8 - 120$$

$$= 175 \text{ metres, say.}$$

Effect of Curvature. In the previous calculations of volume by cross sections, it has been assumed that the sections are parallel to each other and normal to a straight centre line. When the centre line is curved, however, the sections will no longer be parallel to each other, since they will be set out radial to the curve; a correction for curvature is then necessary and to derive this use is made of Pappus's Theorem.

Pappus's Theorem states that the volume swept out by an area revolving about an axis is given by the product of that area and the ength of the path traced by the centroid of the area. The area is to be completely to one side of the axis and in the same plane.

Thus a sphere is swept out by a semicircle rotating about a diameter, and its volume is given by the area $(\frac{1}{2}\pi r^2)$ times the length of the path of the centroid in tracing one complete revolution. Since the centroid of a semicircle is $\frac{4}{3} \cdot \frac{r}{\pi}$ from the axis,

$$\text{Volume} = \frac{\pi r^2}{2} \times \left(2\pi \times \frac{4}{3}\frac{r}{\pi}\right)$$

$$= \frac{4}{3}\pi r^3$$

The volume of earth in cuttings and embankments which follow circular curves may therefore be determined by considering the solid

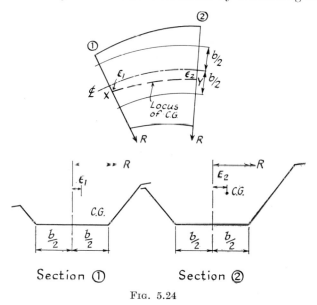

Section ①   Section ②

Fig. 5.24

s being formed by an area revolving about an axis at the centre of the curve. If the cross section is constant round the curve, then the volume will equal the product of that area and the distance traced out by its centre of gravity.

When the sections are not uniform, an approximate volume may be obtained by the method illustrated in Fig. 5.24.

If the mean distance of the centroids from the centre line is written s $\varepsilon = \frac{\varepsilon_1 + \varepsilon_2}{2}$, then the mean radius of the path of the centroid will be $R \pm \varepsilon$), the negative sign being adopted if the mean centroid lies on the

same side of the centre line as the centre of the curve, and positive if
on the other side.

If distance between sections is D, then the angle θ subtended at the
centre is equal to $\dfrac{D}{R}$.

$\therefore$ Length of path of centroid $= XY = \dfrac{D}{R} \cdot (R \pm \varepsilon)$

Then the volume is given approximately by

$$V = \tfrac{1}{2}(A_1 + A_2)D\left(1 \pm \dfrac{\varepsilon}{R}\right)$$

The position of the centroid may change from one side of the centre
line to the other if the transverse slope of the ground changes. I

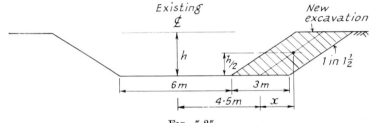

FIG. 5.25

regard is paid to sign, however, and ε_1 and ε_2 summed algebraically
the correct sign for ε will be obtained.

As an alternative, each cross-sectional area may be corrected for the
eccentricity of its centroid, so that instead of using A_1, an "equivalent
area" of $A_1\left(1 \pm \dfrac{\varepsilon_1}{R}\right)$ is used. Each area is thus corrected by

$\dfrac{A_1\varepsilon_1}{R}$, $\dfrac{A_2\varepsilon_2}{R}$, . . . etc., and these corrected values are then used in the
prismoidal formula adopted.

EXAMPLE. The centre line of a cutting is on a curve of 120 metres radius, the
original surface of the ground being approximately level. The cutting is to be
widened by increasing the formation width from 6 to 9 metres, the excavation to
be entirely on the inside of the curve and to retain the existing side slope of 1½
horizontal to 1 vertical. If the depth of formation increases uniformly from 2·4
metres to 5·10 metres over a length of 90 metres, calculate the volume of earthwork to be removed in this length. (*B.Sc.* (*Ext.*) London).

Referring to Fig. 5.25, area of new excavation $A = 3h$.
For any given depth at the centre line of h,

$$x = \frac{3}{2} \times \frac{h}{2} = \frac{3}{4}h \text{ metres}$$

Therefore eccentricity of centroid of excavation will then be $4·5 + \dfrac{3}{4}$
from the former centre line.

Distance (m)	h (m)	A (m²)	ε (m)	Mean distance of centroid from ℄
0	2·40	7·20	$4{\cdot}50 + 1{\cdot}80 = 6{\cdot}30$	6·64
30	3·30	9·90	$4{\cdot}50 + 2{\cdot}48 = 6{\cdot}98$	7·32
60	4·20	12·60	$4{\cdot}50 + 3{\cdot}15 = 7{\cdot}65$	7·98
90	5·10	15·30	$4{\cdot}50 + 3{\cdot}82 = 8{\cdot}32$	

Volume of excavation between 0 and 30 metres

$$= \left(\frac{7{\cdot}2 + 9{\cdot}9}{2}\right)\frac{30}{120}\Big(120 - 6{\cdot}64)\Big)$$

$$= 242{\cdot}31 \text{ cubic metres}$$

Volume of excavation between 30 and 60 metres

$$= \left(\frac{9{\cdot}9 + 12{\cdot}6}{2}\right)\frac{30}{120}\Big(120 - 7{\cdot}32\Big)$$

$$= 316{\cdot}91 \text{ cubic metres}$$

Volume of excavation between 60 and 90 metres

$$= \left(\frac{12{\cdot}6 + 15{\cdot}3}{2}\right)\frac{30}{120}\Big(120 - 7{\cdot}98\Big)$$

$$= 390{\cdot}67 \text{ cubic metres}$$

Total volume $= 949{\cdot}89$ cubic metres (950 cubic metres, say)

Alternatively, using equivalent areas and the prismoidal formula,

Distance	h (m)	A (m²)	ε (m)	$A\dfrac{\varepsilon}{R}$	$A\left(1 - \dfrac{\varepsilon}{R}\right)$ (m²)
0	2·40	7·20	6·30	0·38	6·82
15	2·85	8·55	6·64	0·47	8·08
30	3·30	9·90	6·98	0·57	9·33
45	3·75	11·25	7·31	0·68	10·57
60	4·20	12·60	7·65	0·80	11·80
75	4·65	13·95	7·99	0·93	13·02
90	5·10	15·30	8·32	1·06	14·24

$$\text{Volume} = \frac{30}{6}[7{\cdot}20 + 4 \times 8{\cdot}08 + 2 \times 9{\cdot}33$$

$$+ 4 \times 10{\cdot}57 + 2 \times 11{\cdot}80 + 4 \times 13{\cdot}02 + 14{\cdot}24]$$

$$= 950 \text{ cubic metres}$$

The correction for curvature is most important when the radius of

the curve is small compared to the width of the cross section. In th
above example, the uncorrected volume is

$$V = \frac{30}{6} [7\cdot20 + 4 \times 8\cdot55 + 2 \times 9\cdot90 + 4 \times 11\cdot25 + 2 \times 12\cdot60$$
$$+ 4 \times 13\cdot95 + 15\cdot30]$$

$$= 1{,}012\cdot5 \text{ cu metres}$$

Therefore the error caused by neglecting effect of curvature is abou
$6\frac{1}{2}$ per cent.

(II) Volumes from Contour Lines

As an alternative to the determination of volumes by means
vertical cross sections, it is possible to calculate volumes using th
horizontal areas contained by contour lines. Owing to the relative
high cost of accurately contouring large areas, the method is of limite
use, but where accurate contours are available, as, for instance,
reservoir sites, they may be conveniently used.

The contour interval will determine the distance D in the "end area
or prismoidal formula, and for accuracy this should be as small
possible, preferably 1 or 2 metres. The areas enclosed by individu
contour lines are best taken off the plan by means of a planimeter.
computing the volumes, the areas enclosed by two successive conto
lines are used in the "end areas" formula, whence

$$V = \text{volume of water, earth, tipped material, etc., between}$$
$$\text{contour lines } x \text{ and } y$$

$$= D \frac{A_x + A_y}{2}$$

where D = vertical interval.

If required, the prismoidal formula can be used, either by treati
alternate areas as mid-areas or by interpolating intermediate contou
between those established by direct levelling.

Fig. 5.26 shows by hatching the area enclosed by the 70 m conto
line. Note that the contour line is completed across the upstream fa
of the dam. The plan of the dam is determined as follows. Assume t
breadth at the top to be, say 8 m, and the height to be 73·0 m A.O.I
with side slopes 1 in 1 upstream and 1 in 2 downstream. Then at po
X the 72 m contour is cut by the toe of the bank, which has thus fall
1 m at 1 in 1, i.e. in 1 m horizontal distance from the top of the bar
Therefore XY, the distance from the centre of the dam, is $\frac{8}{2} + 1$

5 m. Similarly $XZ = \frac{8}{2} + 2 = 6$ m, since the slope here is 1 verti
to 2 horizontal. The points of intersection of the toes with other c
tours is found in like manner and the outline of the dam is thus obtain

EXAMPLE. The areas within the underwater contour lines of a lake are as follows—

Contour (m A.O.D.)	190	188	186	184	182
Area (m²)	3,150	2,460	1,630	840	210

Calculate the volume of water in the lake between the 182 and 190 m contours.

(a) By "end areas,"

$$V = \frac{2}{2}\{3,150 + 2(2,460 + 1,630 + 840) + 210\}$$

$$= 13,220 \text{ m}^3$$

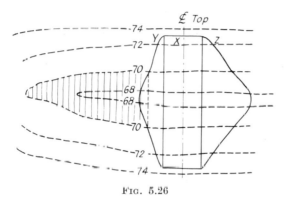

FIG. 5.26

(b) By prismoidal formula, using alternate sections as mid-areas,

$$V = \frac{1}{6}\{3,150 + 4(2,460 + 840) + 2 \times 1,630 + 210\}$$

$$= 13,213 \text{ m}^3$$

With reservoirs it is often required to know the volume of water contained, corresponding to a given height. This is done by calculating the total volumes contained below successive contours and then plotting volume against height to give a curve from which the volume at intermediate levels may be read. (It will be appreciated that a still water surface is a level surface, so that the water's edge is in fact a contour line.)

III) Volumes from Spot Levels

This is a method by means of which the earthworks involved in the construction of large tanks, basements, borrow pits, etc., and similar works with vertical sides may be calculated. The computation is simplified if the formation is to be to a fixed level or to fixed falls, but even basements with several levels present little difficulty.

Having located the outline of the structure on the ground, the engineer divides up the area into squares or rectangles, marking the corner points as described in the method of contouring by spot levels. Levels are taken at each of these corner points, and by subtracting from these the corresponding formation levels, a series of heights is obtained from which the mean heights of a series of vertical truncated prisms of earth can be found. (*Note:* The prisms are called truncated because, unless the ground and formation levels are parallel, the end planes are not parallel to each other.)

The volume of each prism is given by the plan area multiplied by the mean height of the prism. The prisms may of course be considered as either rectangles or triangles as is shown in the following example.

EXAMPLE. Fig. 5.27 shows a rectangular plot which is to be excavated to the given depths. Assuming the sides to be vertical, calculate the volume of earth to be excavated.

(*a*) Assume area is subdivided into four rectangles. Then, rather than evaluate each prism separately, since all the plan areas are the same, it is sufficient to sum the depths at the corners of all the rectangles and divide by four. Since some of the corners occur more than once we may tabulate thus—

Station	Depth of Exc. (h_n) (metres)	Number of rectangles in which it occurs (n)	Product $h_n \times n$
A	3·15	1	3·15
B	3·70	2	7·40
C	4·33	1	4·33
D	3·94	2	7·88
E	4·80	4	19·20
F	4·97	2	9·94
G	5·17	1	5·17
H	6·10	2	12·20
J	4·67	1	4·67

$$\Sigma h_n \times n = 73 \cdot 94$$

$$\therefore \quad \text{Volume} = 15 \cdot 0 \times 12 \cdot 5 \times \frac{73 \cdot 94}{4}$$

$$= 3,466 \text{ cubic metres}$$

(*b*) Assume the area is divided into triangles as shown by the dotted lines in Fig. 5.27.

The tabulation is as before, but the number of times each depth of excavation occurs is different, while to obtain the mean depth we now divide by three.

Station	Depth of Exc. (h_n)	Number of triangles in which it occurs (n)	$h_n \times n$
A	3·15	1	3·15
B	3·70	3	11·10
C	4·33	2	8·66
D	3·94	3	11·82
E	4·80	6	28·80
F	4·97	3	14·91
G	5·17	2	10·34
H	6·10	3	18·30
J	4·67	1	4·67

$$\Sigma h_n \times n = 111 \cdot 75$$

$$\text{Volume} = \frac{15 \cdot 0 \times 12 \cdot 5}{2} \times \frac{111 \cdot 75}{3}$$

$$= 3{,}492 \text{ m}^3$$

It often happens that the topsoil and vegetable matter is removed from the surface before work begins, such soil stripping normally being

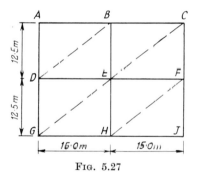

Fig. 5.27

paid for per square metre. If the levelling was done before the stripping and it is required to deduct this soil from the main excavation then the volume to deduct is given by the total area times the average depth of stripping.

Mass Haul Diagrams

In works where large volumes of earthwork have to be handled—more especially long works such as railways and arterial roads—a mass-haul diagram is of great value both in planning and construction. The diagram is plotted after the earthwork quantities have been computed, the ordinates showing aggregate volumes in cubic metres while the horizontal base line, plotted to the same scale as the profile, gives the points at which these volumes obtain. Cuttings are taken as positive and fills as negative when evaluating the aggregate volumes, and in plotting the mass-haul curve total positive volumes are plotted above the base line and total negatives below it.

Most materials are found to increase in volume after excavation ("bulking"), but after being re-compacted by roller or other means, soils in particular might be found to occupy less volume than originally, i.e. a "shrinkage" has taken place when compacted to the *in situ* volume.

If the shrinkage factor of such soils is known it may be used in the mass diagram to amend the volumes required for filling. For example if a certain material has undergone a volume shrinkage of 15 per cent on final consolidation compared with its pre-excavation volume, then 100 m³ of excavation produces 85 m³ of fill (or 118 m³ of excavation give 100 m³ of bank). Volumes of cut and fill may thus be related by multiplying the excavated volumes by 0·85 to give the equivalent volumes of fill. The following table gives typical swell or shrinkage factors for certain materials. It will be seen that rock is the only material which has a larger volume after consolidation in embankment than it had before excavation.

Volume before excavation = 1 m³

Material	Volume immediately after excavation	Volume after compaction
Rock (large pieces) .	1·5	1·4
Rock (small pieces) .	1·7	1·35
Chalk . . .	1·8	1·4
Clay . . .	1·2	0·90
Light sandy soil .	0·95	0·89
Gravel . . .	1·0	0·92

Haul. The cost of hauling excavated material will obviously depend to some extent on the distance it must be carried. In the Bill of Quantities for a job, the unit price of excavation will include for transporting the material a certain specified distance (say 0·5 km); this distance is known as the *"free haul."* If the material has to be moved a greater distance than the free haul, the extra distance is known as *"overhaul."* *Haul* is the total of the products of the separate volumes of cut and the distance they are transported to the embankment. This must equal the total volume of cut multiplied by the distance between the centroids of cutting and the embankment it forms.

Consider the mass haul diagram given in Fig. 5.28, which has been plotted from the detail contained in the quantities distribution table on p. 173. This curve is illustrative only, and it is assumed that earthworks prior to chainage 1,000 have balanced.

In plotting the curve the ordinate at *j* is plotted as the whole of the fill required between *A* and *J*, i.e. − 230 m³. The ordinate at chainage 1,100 is the aggregate volume up to this point, i.e. + 200 m³, so that the mass haul curve must cross the base line between chainage 1,040 and chainage 1,100. On plotting the remaining points it is seen that the

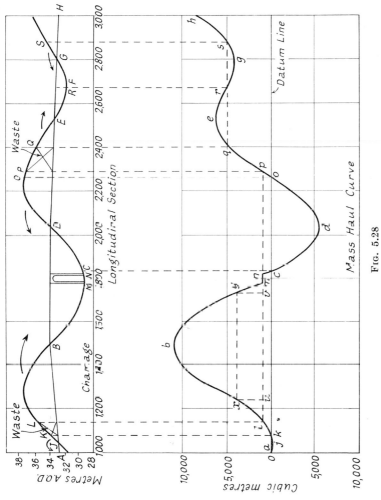

Fig. 5.28

curve crosses the datum line at four points, and the aggregate volume between pairs of these points is zero. The excavation and embankment between those points thus balance each other, i.e. between A and K, C and O.

The following points may also be noted in reading the mass haul curve—

(a) The mass haul curve rising indicates cut, since the aggregate volumes are increasing (i.e. J to B) and a maximum point occurs at the end of a cut (i.e. b, e).

(b) The mass haul curve falling indicates fill (i.e. B to M, N to D), and a minimum point occurs at the end of a fill (i.e. d).

(c) The difference between the ordinates at two points represents the volume of cut or fill between the two points so long as there is no maximum or minimum point between the two.

(d) If any horizontal line, e.g. lm, be drawn on the curve then the volumes of cut and fill balance because there is no difference in aggregate volume between l and m. Such a line is termed a balancing line and gives the distance of haul between the points. When the curve is above the balancing line, material must be moved to the right, i.e. LBM, and when it is below such a line the material must be moved to the left, i.e. RGS.

The base line on which the curve is drawn is a possible balancing line though not necessarily the most satisfactory one. In Fig. 5.28, using the base line as a balancing line, there is a surplus of 8,510 m³, though this may, of course, be used in the following section of the work. Any number of horizontal lines may be drawn on the curve to find balancing lines which enable the work to be done in the most economical manner, and they need not be continuous, e.g. lm and np are broken by the bridge, and np and qrs are not connected at all. When the balancing lines are not connected it means that the earthwork between those points on the profile not included by the lines will not be balanced, i.e. between K and L, P and Q, where the mass curve is rising. In these cases material will be carried to tip since it is surplus; if the mass curve were falling it would be necessary to borrow material, since the mass curve then indicates a fill.

In considering the ways in which balancing lines may be drawn on the mass haul curve it will be seen that to obtain the most economical scheme it is probable that the balancing lines will not be continuous. If the balancing lines were too long it would mean excessive and uneconomic haul distances. Thus a scheme involving balanced earthworks, over some lengths—having regard to the free haul—borrowing at some points and running to waste at others, is most likely to be used. It is better if material is hauled downhill, since this requires less power; thus in long hauls up steep gradients it may be worth while to waste material from the cutting and borrow for subsequent embankment. It is cheaper to borrow when it costs less to excavate in the cutting and run to waste, followed by excavation from a borrow pit to form the embankment, than it does to excavate in the cutting and transport this material to form the embankment.

As mentioned above, where the free haul is given, it may be plotted on the various parts of the mass haul curve, and the extra distance for overhaul may be estimated. If the free haul distance be 500 m, then between *KBM* this would be denoted by *xy* on the mass haul curve; elsewhere *np*, *qr*, and *rs* satisfy this requirement. The area of the portions of the curve cut off by the balancing lines, i.e. *lbm*, is the haul in that section, since haul is the product of volume and distance.

One square on the diagram represents $5,000 \times 200 = 10^6 \, \text{m}^4$. If the unit be taken as $1 \, \text{m}^3$ hauled 100 m, then 1 square represents 10,000 units. The area giving free haul is *uxbyvu*, and the overhaul is the area *lbmn* minus area *uxbyvu*, and is given by multiplying this net area by 10,000. *xu* is the volume on which overhaul is paid.

Chainage	Centre Height (m)	Volumes (m³)		Shrinkage Constant	Corrected Volume	Accumulated Volume
		Cut	Fill			
1,000	F 1·22					0
1,040	0		230			− 230
1,100	C 1·52	480		0·90	+ 430	+ 200
1,200	C 3·96	2,560		0·90	+ 2,300	+ 2,500
1,300	C 4·12	4,560		0·90	+ 4,100	+ 6,600
1,400	C 2·74	3,940		0·90	+ 3,550	+ 10,150
1,500	0	950		0·90	+ 850	+ 11,000
1,600	F 3·05		1,350		− 1,350	+ 9,650
1,700	F 4·27		4,010		− 4,010	+ 5,640
1,780	F 4·72		4,600		− 4,600	+ 1,040
1,820	F 4·72		BRIDGE			+ 1,040
1,900	F 3·51		4,130		− 4,130	− 3,090
2,000	F 1·22		2,370		− 2,370	− 5,460
2,035	0		60		− 60	− 5,520
2,100	C 1·98	510		0·90	+ 460	5,060
2,200	C 3·96	3,180		0·90	+ 2,860	− 2,200
2,300	C 3·66	4,055		0·90	+ 3,650	+ 1,450
2,400	C 2·44	3,860		0·90	+ 3,470	+ 4,920
2,500	C 0·61	1,320		0·90	+ 1,190	+ 6,110
2,530	0	100		0·90	+ 90	+ 6,200
2,600	F 1·07		− 350		− 350	+ 5,850
2,700	F 1·52		1,230		− 1,230	+ 4,620
2,800	0		− 420		− 420	+ 4,200
2,900	C 1·68	1,080		0·89	+ 960	5,160
3,000	C 3·66	3,730		0·89	+ 3,320	+ 8,480

It may be noted here that the prices in the Bill of Quantities are normally based on the unit volume of soil before excavation or the unit volume after compaction. The above analysis is mainly of use to the contractor when he is assessing plant requirements.

EXERCISES 5

1. State and prove Simpson's rule for areas.

Measurements made from a chain line to an irregular boundary were as follows—

Chainage (m)	0	10	20	30	40	50	60	70	80
Offset (m)	5·5	6·4	7·3	7·9	8·2	6·7	4·9	3·0	0

Calculate the area between the chain line and the boundary.

State a rule whereby the area enclosed by the chain lines in a chain and tape survey can be calculated from the measured lengths of the lines.

Answer: 4·84 ares.

2. Co-ordinates of corners of a polygonal area of ground are, taken in order, as follows, in metres—

A (0, 0); B (40, − 32); C (126, − 41); D (200, 14); E (144, 80); F (62, 108); G (− 19, 27), returning to A.

Calculate the area and the co-ordinates of the far end of a straight fence from A which just cuts the area in half. (*L.U., B.Sc.*)

Answer: 2·233 hectares; 160; 61.

3. The whole circle beatings of the straight sides of a plot of land $ABCD$ are—

AB 352° 26' BC 111° 04'
CD 195° 56' DA 242° 15'

The side CD is 175·82 m long and the area of the plot is 8·094 hectares. Calculate the length of side AB. (*L.U., B.Sc.*)

Answer: 412·4 m.

4. Derive an expression which will give the area of a piece of ground enclosed by straight lines joining a series of points having co-ordinates (N_0E_0); (N_1E_1); . . . (N_NE_N); (N_0E_0). Find the area enclosed by joining consecutively points A, B, C, D, E, F, A, whose co-ordinates are respectieely (0, 0); (600, 200); (900, 700); (850, 1,200); (400, 1,200); (− 200, 700); (0, 0); and find the position of a fence running due N/S which will divide it into two equal areas. (*L.U., B.Sc.*)

Answer: 99·162 hectares. Fence 645 m from A.

5. A closed traverse $ABCDA$ is run along the boundaries of a built-up area with the following results—

Side	W.C.B.	Length
AB	69° 55'	262 m
BC	166° 57'	155 m
CD	244° 20'	268 m
DA	347° 17'	181 m

Co-ordinate the stations B, C, and D on A as origin, and calculate the area of the traverse in hectares. (*Inst. Struct. E.*)

Answer: 4·38 hectares.

6. A straight embankment is made on ground having a transverse slope of 1 in 8. The formation width of the embankment is 10 m and the side slopes are 1 vertical to $1\frac{1}{2}$ horizontal. At three sections, 20 m apart, the heights of the bank, at the centre of the formation level are 3 m, 5 m and 6 m respectively.

Calculate the volume of the embankment and tabulate data required in the field for setting-out purposes. (*L.U., B.Sc.*)

Answer: 3,531 m³.

7. Describe three methods of carrying out the field-work for obtaining the volumes of earthworks.

Explain the conditions under which the "end area" and "prismoidal rule" methods of calculating volumes are accurate, and explain also the use of the "prismoidal correction."

The areas within the contour lines at the site of a reservoir are as follows—

Contour	Area in m²	Contour	Area in m²
158 m	476,000	148 m	164,000
156 m	431,000	146 m	84,000
154 m	377,000	144 m	10,000
152 m	296,000	142 m	1,000
150 m	219,000		

The level of the bottom of the reservoir is 142 m. Calculate (a) the volume of water in the reservoir when the water-level is 158 m, using the end-area method; (b) the volume of water in the reservoir when the water-level is 158 m, using the prismoidal formula (every second area may be taken as a mid-area); (c) the water-level when the reservoir contains 1,800,000 m³. (*L.U.*, *B.Sc.*)
Answer: (a) 3,639,000 m³; (b) 3,627 000 m³; (c) 153·6 m.

8. Cross sections at 20 metre intervals along the centre line of a proposed straight cutting are levelled at 10 metre intervals from − 30 m to + 30 m and the following information obtained—

Distances	− 30	− 20	− 10	0	+ 10	+ 20	+ 30
0	1·3	0·4	0	0	0	0·4	0·9
20	4·3	2·9	1·7	1·0	0·7	1·0	2·0
40	5·8	4·7	3·6	2·7	2·0	2·0	3·2
60	6·6	5·9	4·9	3·8	3·2	3·2	3·7
80	8·3	7·1	6·0	4·7	4·3	4·0	4·4

Tabulated figures are levels in metres relative to local datum.
The formation level is zero, its breadth 10 m and the side slopes 1 vertical in 2 horizontal. Find the volume of excavation over the section given.
Answer: 3,530 m³.

9. Fig. 5.29 shows the straight centre line *AB* for a road of formation width 6 metres, set out on a hillside which may be considered as a plane surface of uniform slope 1 in 5.

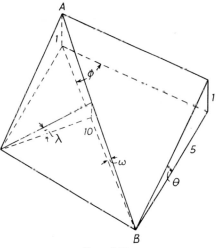

Fig. 5.29

Show that $$\tan \omega = \tan \theta \sin \phi$$
and $$\tan \lambda = \tan \theta \cos \phi$$

Hence determine the quantity of excavation obtained in a length of 100 m if side slopes of 1 vertical in 1½ horizontal are provided and there is neither cut nor fill along *AB*, which has a uniform gradient of 1 in 10. (*R.C.A.T.S.*)
Answer: 13 m³.

CHAPTER 6

OTHER SURVEYING INSTRUMENTS

I this chapter brief notes are given on those instruments which are not in the same general use as the basic equipment described in the first three chapters. The instruments to be dealt with are: (1) plane table, (2) Abney level, (3) sextant, (4) compass and dial, and (5) barometer.

1. Plane Table

The plane table differs from other surveying instruments in that the map or plan is prepared in the field without the direct measurement of

Fig. 6.1

any angles, and, except when contouring, without calculation. The equipment is simple, the most important piece being a drawing-board of convenient size mounted on a tripod, and it is due to this item that it owes its name. All the early English writers speak of it as a playne or plaine table, and in particular Aaron Rathborne* says of it: "This Instrument for the playnesse and perspicuitie thereof, and of his easie use in practice receiveth aptly the name and appelation of the Playne Table. A most excellent and absolute instrument for this our purpose in Survey." Fig. 6.1 shows Rathborne's plane table, and by comparing

* *The Surveyor*, by A. Rathborne (London, 1616).

188

it with modern plane table equipment it will be seen that little change has taken place.

Despite the enthusiastic praise of Rathborne and other sixteenth-century English surveyors, the plane table was not adopted for general use in England to anything like the same extent as on the Continent. Indeed, compared with the theodolite, which is essentially an English instrument from its first inception, the plane table had received little attention here until recent years, when its use in the Survey of India revealed its possibilities and led to the development of new techniques. In many parts of the world, e.g. the Continent, India, U.S.A., Canada, etc., the plane table is a frequently-used surveying instrument, and there is no doubt that, despite the English climate, greater use could be made of it here.

Figs. 6.2 to 6.5 show typical modern plane table equipment manufactured by Hilger & Watts, Ltd. The Plane Table consists of a drawing board mounted on a light, rigid tripod. Board sizes vary from, say, 450 mm × 450 mm to 750 mm × 600 mm, and soft woods are used where lightness is important. Tables made of teak are also available, but owing to their greater weight they require specially robust tripods. Also softwood strips are fitted to the under edges for drawing pins, though in this connexion it is always preferable to fasten the paper to the under surface of any board (by clips, tape or drawing pins) rather than damage the smooth upper surface.

FIG. 6.2. PLANE TABLE AND TRIPOD (SA12; SA2)
(Courtesy of Hilger & Watts, Ltd.)

Connexion of the board to the tripod in simple apparatus is by means of a single screw, levelling being carried out on the legs. Fig. 6.2 shows the Johnson quick-levelling head, which facilitates accurate levelling of the table, and this is fitted as a refinement. Alternatively a tribrach mounting may be used.

Accessories. These are shown in Figs. 6.3 and 6.4 and are as under.

(*a*) ALIDADE. This is a sight rule of either boxwood or metal and having folding sights which can be turned up at each end. One sight has a narrow vertical slit in it, while the other consists of a vertical wire stretched across an open frame.

A modification of this developed by the Survey of India is the *Indian clinometer*. On this the front vane is graduated in degrees and tangents and the rear vane has a single small aperture at the same height as the zero on the scale. The instrument is capable of being accurately levelled on the table so that when sighting a distant object through the small aperture, the angle of elevation or depression (or the tangent of same) can be read. The distance to the object is obtained from the map, and the difference in level can thus be calculated. This clinometer thus enables plane table surveys to include contouring, and has been widely used in this way.

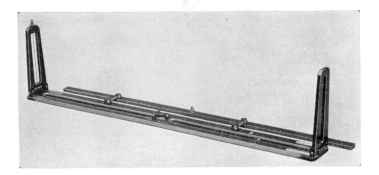

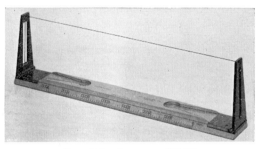

Fig. 6.3. Alidade (SA42)
(*Courtesy of Hilger & Watts Ltd.*)

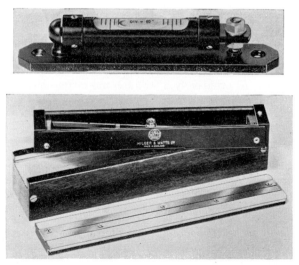

Fig. 6.4. Trough Compass and Adjustable Level
(*Courtesy of Hilger & Watts, Ltd.*)

The attachment of a telescope to the alidade will allow two-fold duty, and it has long been standard practice on the Continent, but the neat and well-proportioned Microptic alidade (Hilger & Watts, Ltd) shown in Fig. 6.5 marks a big advance in this field. The telescope is pillar-mounted on the base rule, and can be levelled independently of the table. Not only are clearer sights possible through the telescope, but distances and levels may be measured tacheometrically. Readings are taken through a special circle reading eyepiece which gives the angle of elevation or depression in degrees, and also the corresponding readings on vertical and horizontal Beaman scales. The use of Beaman

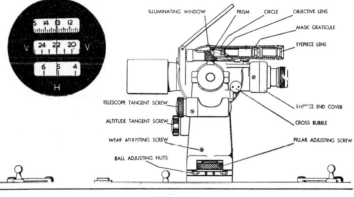

Fig. 6.5

stadia arcs, which is dealt with in Chapter 7, minimizes the work involved in reducing the telescope stadia readings.

(b) SPIRIT LEVEL, used for levelling the table.

(c) TROUGH COMPASS, used to orient the table in the magnetic meridian (Fig. 6.4).

(d) PLUMBING FORK. This consists of a U-shaped piece of metal with parallel arms of equal length, a plumb bob being attached to the free end of the lower arm. Thus, when the free end of the upper arm rests against a point on the drawing, the plumb bob must hang directly below the point; in this way the board can be accurately positioned over a station (Fig. 6.6).

(e) DRAWING MATERIALS. Finest quality paper should be used to withstand the rubbing of the alidade, and hard, well-pointed pencils of good quality.

Uses of Plane Table Surveying. These fall into two groups—(1) The production of complete maps with the plane table and its accessories alone; complete networks of control points may be arranged with it and the whole of the detail filled in. It is put to this use in the preparation of maps for exploratory surveys, and where speed is required.

(2) The filling in of detail after the control points have been fixed

by a theodolite. In this field it has much to commend it when compared with chain surveying. Under this heading comes the use of the plane table in locating contour lines and in revising existing plans.

The merits of plane tabling are—

(*a*) The plan is produced directly so that no measurements and bookings are required, rapid sketching in of detail is possible, and there is less chance of important features being omitted.

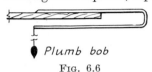

Plumb bob

Fig. 6.6

(*b*) The method is rapid; a good field technique is fairly easily acquired, and office work is cut to a minimum.

Disadvantages are—

(*a*) Work is impossible in a persistently windy or damp climate.

(*b*) The scale of the map or plan *must* be known before work begins, since the work is plotted in the field.

Plane Table Surveying Methods. The basic procedure in plane tabling consists in taking sights on to an object with the table, correctly oriented, at two separate stations. Rays drawn on the board along these sight lines must intersect at the plotted position of the object. Starting with two stations a known distance apart (or with a measured

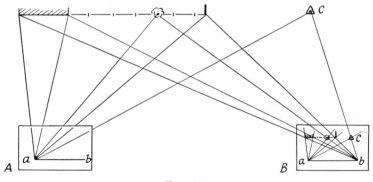

Fig. 6.7

base), it is possible to fix other stations and so build up a network of figures. Also, while the instrument is set up at each station, rays are drawn towards points of detail which require to be plotted (*see* Fig. 6.7). This is one of the basic (and earliest) methods of plane tabling and is known as either *triangulation* or *intersection*. It will be seen that it can be used for carrying out complete surveys, or for locating detail where certain fixed points are known.

In the diagrams, capital letters represent stations and the corresponding small letters represent their positions on the plan. In the simple example shown, *AB* is a base line measured by say chaining. Line *ab* is plotted to scale in a convenient position, and the board is then set up at *A* (with *a* as nearly over *A* as possible). With the alidade laid along *ab*, the board is turned until a sight can be obtained on station B, and then clamped: by means of the trough compass the direction of magnetic north is found and marked on the drawing. With the board thus oriented, rays are drawn as shown, after which the board is set up at B and reoriented by sighting back on station A with the alidade held along ba and a check is obtained by means of the trough compass. Another set of rays is drawn as required, and a new

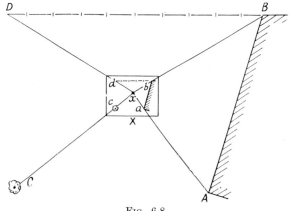

Fig. 6.8

station C could be located (by intersection of rays) over which the board could be set and oriented in the same way.

The choice of stations will be governed by the nature of the country, but in general,

(1) they must be in conspicuous positions,
(2) they must be clearly defined and capable of being occupied in their turn,
(3) they must give well-conditioned triangles.

The need for centring the board precisely so that each plotted point is over its station depends largely on the scale of the survey; e.g. if a single isolated tree is used as a station, and the instrument set up 10 m from the tree, the error in plotting at a scale of 1 in 50,000 is $\frac{1}{5}$ mm, which is negligible. At a scale of 1/500, the plotting error would be 20 mm; at this scale, therefore, centring must be accurate to within 0·25 m in order to give a plotting error of less than 0·5 mm.

RADIATION. Another method whereby complete surveys may be carried out is radiation.

Refer to Fig. 6.8, in which the board is set up at a convenient station X and a series of rays drawn through x towards the points to be surveyed. The lengths XA, XB, etc., are now measured, preferably using a telescopic alidade equipped with stadia hairs, and the distances scaled off along the rays xa, xb, etc. The method is mainly suitable for surveys of limited extent, so that the board must be centred with a plumbing fork to avoid excessive centring errors. Detail in close proximity to known points established by other methods may also be picked up by radiation.

TRAVERSING. This method is analogous to theodolite traversing, and is used only for fixing survey lines and stations on the plan. Detail is fixed by *intersection* or *radiation*. Fig. 6.9 illustrates the procedure. The table is set up at *B* and rays drawn towards adjacent stations *A* and *C*. *BA* and *BC* are measured and scaled off to give *ba* and *bc*, after which the instrument is moved to *C* and oriented by sighting back on *B* with the alidade held along *cb*. A ray is now drawn towards

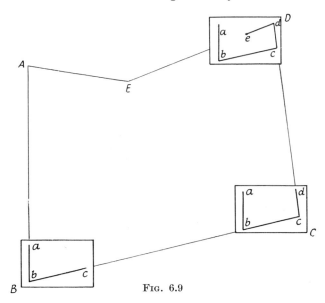

FIG. 6.9

D, *CD* measured, *cd* scaled off, and so on. Note that if intersection is used to fix the detail then rays will be drawn to the detail while the table is at, say *B*, and intersections obtained by rays from *C* or other stations.

Resection. The Three-point Problem. The above methods all involve setting the instrument up at well-marked points which have been previously located, but when we have a number of such points already fixed—either by plane table or otherwise, or because we intend to do revision, etc., on an existing survey—it is often more convenient to set up at some arbitrary point and to pick up the detail from there. Reflection will show that this adds enormously to the flexibility of plane-table surveying. Conspicuous points which have been located by triangulation etc., are usually church spires, mountain tops, etc., and as such are not suitable points from which to view low-lying detail such as rivers, bridges, roads, etc. Similarly, if we are revising an existing survey of low-lying land on which distant development (e.g. churches) can be identified, it is useless to set up the table near the known features and then try to draw rays towards distant hedges, gates, etc. These conspicuous known features will generally be visible from the area in which the detail lies, however, and if we set up in a position convenient to the detail, we can proceed to survey it so long as we can identify, by means of sights on the known points, the position on the plan corresponding to the instrument's position on the ground. This location and orientation is possible (with certain limitations) provided that three well-defined points are visible from the plane table, the process being known as *resection*. A point determined by resection is known as a *plane table fix*.

FIXING BY RESECTION. The three points used to give the fix must not (*a*) lie on, or nearly on, a circle passing through the point to be fixed, (*b*) be too close together, or diametrically opposed.

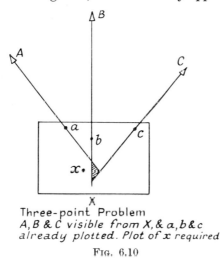

Three-point Problem
A, B & C visible from X, & a, b & c
already plotted. Plot of x required

FIG. 6.10

A rough orientation of the board is made either by guesswork or, more precisely, by means of the trough compass. With the alidade held so that its edge passes through point *a* (say) on the plan (Fig. 6.10), sight on the actual point *A* and draw a ray backwards. Repeat the process for the other points. If the board is correctly oriented, the three rays will meet in a point. Usually the preliminary orientation is faulty (and it must be borne in mind that even assuming complete absence of local attraction the trough compass will not give orientation to better than 20′) so that a small triangle of error results.

SOLUTION OF THE TRIANGLE OF ERROR. A number of methods are available, of which the one most used—*solution by trial and error*—is given here. (Other solutions to the three-point problem are given in Chapter 9.) The following rules are used to assist in determining the point *x* relative to the triangle of error—

Rule 1. If the plane table is set up in the triangle formed by the three points (i.e. *X* lies within the triangle *ABC*), then the position of the instrument on the plan will be inside the triangle of error; if not, it will be outside.

Rule 2.   Looking along the rays from the board towards the points to which they refer, the point x will lie either to the right of all three rays or to the left of all three rays, since the table is rotated in one direction to locate x.

Rule 3.   The perpendicular distances of x from the three rays are proportional to the distances from the instrument of the corresponding points, since the rotation of the table must have the same effect on each ray (displacement $= R\theta$ where θ is the rotation of the board to give correct orientation, and R is the distance from x to the station).

Referring to Fig. 6.10, therefore, by Rule 1, x is outside the triangle, by Rule 2 it cannot be in either of the sectors contained by the rays xA, xB, and xC, and by Rule 3, using the proportions for the perpendiculars given by scaling the distances XA, XB, and XC from the plan, it must be in the left-hand sector at the point shown. With the ruling edge of the alidade laid along, say xa, sight on A by turning the board, then reclamp. On repeating the resection for the other two points, no triangle of error (or at worst a very small one) should now be obtained.

FAILURE OF THREE-POINT FIX.   When the three points A, B and C and the instrument position X are so chosen that a circle can be drawn through the four points, the solution is indeterminate, because no matter how the board is oriented, the rays will always meet in a point, though not the same point. (This is because the two angles subtended by the three points at the instrument will always have the same values, no matter how the instrument is set up, since they are angles subtended at the circumference.) Hence the rays will always meet in a point. To avoid this possibility, and to increase what is called the *strength of fix*, the positions of points A, B and C are chosen according to the principles given in the last section but one.

2. Abney Level

This versatile little instrument can be used either as a hand level or for the measurement of vertical angles.

As shown in Fig. 6.11, which illustrates the instrument supplied by Hilger & Watts, Ltd., the Abney Level consists essentially of a sighting tube, with a draw tube extension to which is attached a graduated arc. An index arm, pivoted at the centre of the arc, carries a small bubble whose axis is normal to the axis of the arm, so that as the bubble is tilted the index moves over the graduated arc. By means of an inclined mirror mounted in one-half of the sighting tube, the bubble is observed on the right-hand side of the field of view when looking through the tube. The milled head in the centre of the arc is used to manipulate the bubble and a worm-screw provides fine adjustment.

The arc itself is graduated in degrees from 0 to $+ 90°$ and $- 90°$, and the scale is read by the vernier on the index arm to 10 min. Also engraved on the arc is a gradient scale giving gradients from 1:10 to 1:1. The hinged magnifier attached to the milled head assists the reading of the vernier.

Method of Use. (a) As a hand level. The index on the bubble arm is set at zero, and the instrument, held in the right hand, is sighted on the levelling staff. A horizontal sighting wire provides the axis of collimation, and the reading is taken when the sighting wire bisects the image of the bubble seen in the mirror. The height of the observer's eye is thus established for a given position of the observer by taking the reading with the staff on a point of known level. Readings are booked and reduced as for ordinary levelling. The degree of accuracy achieved with the hand level is not of a very high order, because (i) the bubble used is insensitive, to allow the instrument to be held in the

FIG. 6.11. ABNEY LEVEL

(Courtesy of Hilger & Watts, Ltd.)

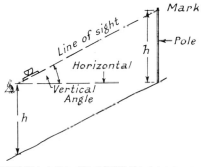

FIG. 6.12

hand, and (ii) changes occur in the height of collimation dependent on the stance of the observer. It is a compact, portable instrument, and is extremely useful for reconnaissance work.

As with any other type of level, the axis of the bubble tube must be parallel to the line of sight, and if the bubble tube is capable of being adjusted, then the two-peg test is used to check the adjustment.

(b) As a clinometer. In using the instrument for the measurement of vertical angles (as for instance when measuring mean ground slope) it will be seen from Fig. 6.12 that if the sight is taken on to a point whose height above the ground is the same as the observer's eye, then the line of sight will be parallel to the mean ground surface. A ranging pole with a mark on it at the required height makes a suitable target. To measure the angle, a sight is taken on to the mark, and the bubble brought into the field of view by means of the milled head. The fine adjustment can now be used to set the bubble so that it is bisected by the sighting wire at the same time as the wire is on the target. The angle is now read on the vernier with the aid of the magnifier.

The base of the sighting tube is machined flat so that the instrument can be set down and used to measure, for example, the dip of rock outcrops in geological surveying.

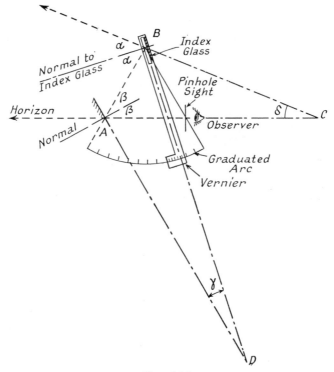

Fig. 6.13

3. The Sextant

There are two versions of this instrument (i) the nautical sextant, and (ii) the box sextant. The former instrument consists essentially of a 60° arc (i.e. $\frac{1}{6}$th of a circle), the periphery of which is graduated in degrees. An index arm, pivoted at the centre of the arc, carries a wholly-silvered index glass at the pivot end and at the other a vernier which moves along the graduated arc. A half-silvered glass A, known as the horizon glass, is attached to the sector as shown in Fig. 6.13, i.e. the plane of this glass is parallel to the radius through the zero of the graduated scale. When the index arm is set at zero the index glass should be parallel to the horizon glass, and adjusting screws are provided on the index glass to enable this condition to be complied with. A pinhole sight or a telescope is fitted opposite to the horizon glass, and to measure vertical angles the instrument is held upright and a sight taken on to the horizon. The index arm is now swung so that the image of the sun or star, seen in the mirror half of the horizon glass, cuts the horizon. The angle between the two glasses is half the angle of elevation of the object sighted, and to give direct reading the scale is graduated so that $\frac{1}{2}°$ reads as $1°$. Thus the scale reads to a total angle of $120°$, with perhaps an extra degree or two at each or one end.

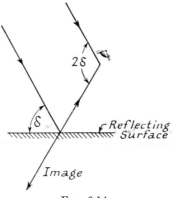

FIG. 6.14

Proof of the optical principle (which applies also to the box sextant) is as follows—

Referring to Fig. 6.13,

from $\triangle ABC$,
$$2\alpha = 2\beta + \delta$$
$\therefore$
$$\delta = 2(\alpha - \beta)$$
from $\triangle ABD$,
$$90 + \alpha = 90 + \beta + \gamma$$
$\therefore$
$$\gamma = \alpha - \beta$$
$\therefore$
$$\delta = 2\gamma$$

i.e. angle of elevation of the object δ is twice the angle swung out by the index arm as it travels round from its zero position when it is parallel to the horizon glass.

When the horizon is not visible, a so-called "artificial horizon" is used in which the observer views the image of the sun, etc. in a horizontal mirror—usually mercury. The principle is illustrated in Fig. 6.14, from which it will be seen that the angle measured is twice the required angle, assuming that the object being viewed may be considered to be at infinity. The mercury is contained in a rectangular flat tray on the ground in front of the observer, and is shielded from the wind by two glass plates which incline at $45°$ from the long sides of the tray. The short sides of the tray are screened by triangular-shaped end-pieces normal to the tray. The glass plates must be of equal thickness and have truly parallel faces. The observer sights through A (Fig. 6.13) the reflection of the object (say the sun) given by the mercury, and adjusts the index glass B by means of the arm until the image seen from the double reflections of the index and horizon glasses is coincident with the reflection in the mercury. For sights from aircraft, bubbles have been employed as artificial horizons, and also gyroscopes.

The sextant may also be used to measure vertical angles and horizontal angles subtended by two points in exactly the same way as described except that instead of sighting the horizon one of the points

FIG. 6.15. SEXTANT
(*Courtesy of Kelvin Hughes, Ltd.*)

would be observed as a referring object. Figs. 6.15 and 6.16 show a typical sextant manufactured by Kelvin Hughes (Marine) Ltd. and a box sextant manufactured by Stanley & Co., Ltd.

The Box Sextant. This is a compact version of the nautical instrument and is intended for use in land surveying as a pocket instrument. As may be seen from Fig. 6.16, it is similar in size and shape to the optical square described in Chapter 2, from which it is also a development. Fig. 6.17 gives the optical principles, and since they are identical with those of the nautical sextant (*see* Fig. 6.13) the student can readily ascertain that the same analysis as that given above shows that the angle between the index and horizon glasses is half the angle subtended at the eye by the two stations being viewed.

The instrument is held in one hand and Station X is viewed directly through the unsilvered half of the horizon glass, either using the small detachable telescope or, if this is removed, a simple sight hole. With the other hand, the arm on top of the box (to which the index glass is attached) is swung by means of the small actuating screw until the image of Station *Y* is seen coincident with Station *X*. The scale is graduated so that $\frac{1}{2}°$ reads as 1°, thus giving direct reading of the required angle, and the vernier is read through a small viewing glass.

Permanent Adjustments of a Sextant. The main adjustments involve—

1. Setting the index glass perpendicular to the plane of the arc.
2. Setting the horizon glass perpendicular to the plane of the arc.

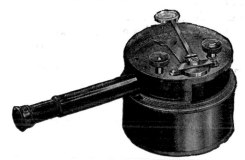

FIG. 6.16. BOX SEXTANT
(*Courtesy of W. F. Stanley & Co., Ltd.*)

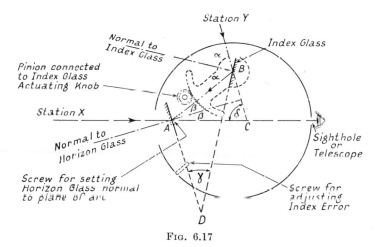

FIG. 6.17

3. Setting the index and horizon glasses parallel when the vernier reads zero (i.e. elimination of index error).

4. Setting the line of collimation parallel to the plane of the arc.

Of these, only Nos. 2 and 3 can be carried out on the box sextant and these are described below. In general, the adjustment of a nautical sextant, apart from ascertaining the index error, requires considerable skill and is best carried out by the makers.

SETTING THE HORIZON GLASS PERPENDICULAR TO THE PLANE OF THE ARC. With the vernier set at zero, a sight is taken on some clear distant object. If the reflected image overlaps the object viewed directly, or if there is a vertical gap, then the horizon glass is not truly normal to the plane of the arc. It is made normal by turning the small square-headed screw on top of the box (using the key provided) until the vertical gap or overlap is eliminated.

ELIMINATION OF INDEX ERRORS. To check the index error, the reflected image of some convenient and clearly defined point is brought into coincidence with the direct view of the object. The reading of the vernier, if not zero, gives the index error; if the reading is on the main scale the correction is negative, and if on the *arc of excess*, i.e. the negative graduations beyond the zero, the correction is positive. This correction is often made to subsequent angle readings, no adjustment being carried out to eliminate the error. The adjustment of the horizon glass is not then further disturbed.

Should it be required to remove the error, the vernier is set at zero and the object again sighted. The reflected image will now be laterally displaced relative to the object seen directly; by means of the key, the square-headed adjusting screw in the *side* of the box is turned until the horizon glass is brought truly parallel to the index glass, i.e. until the lateral displacement is eliminated.

Measurement of Horizontal Angles. Horizontal angles measured with a sextant are measured in the plane of the two objects sighted and the observer's eye, so that in land surveying, where angles in the horizontal plane are required, a solution of the spherical triangle has to be made to give the true answer. This adjustment becomes important when the two stations sighted differ considerably in level. Let θ be the angle as measured; the vertical angles α and β of the stations must also be measured. Then the corrected value of θ in Fig. 6.18, i.e. its value in the horizontal plane, is given by—

$$\cos \theta_h = \frac{\cos \theta - \cos \alpha \cos \beta}{\sin \alpha \sin \beta}$$

EXAMPLE. A sextant was used at station Q to measure the angle included between two other stations P and R. The horizontal distances PQ and QR were 373 m and 169 m respectively. The reduced levels of stations P, Q and R as determined by the levelling party were 147·90, 161·39 and 169·11 m respectively. The sextant observer sighted marks at stations P and R which were at the same height above the ground as that of the sextant at station Q, and recorded an included angle of 94° 20′. Deduce the horizontal angle PQR. (*L.U.*, *B.Sc.*, 1952.)

In the above equation (*see* Fig. 6.18)

$$\theta = 94° 20'$$

$$\alpha = 90 + \frac{\tan^{-1} 13·49}{373}$$

$$= 92° 04'$$

$$\beta = 90 - \frac{\tan^{-1} 7·72}{169}$$

$$= 87° 23'$$

$$\cos \theta_h = \frac{\cos \theta - \cos \alpha \cos \beta}{\sin \alpha \sin \beta}$$

$$= \frac{\cos 94° \ 20' - \cos 92° \ 4' \cos 87° \ 23'}{\sin 92° \ 4' \sin 87° \ 23'}$$

$$= \frac{- \ 0 \cdot 0756 + 0 \cdot 0016}{+ \ 0 \cdot 9984}$$

$$= - \ 0 \cdot 0739$$

Therefore $\theta_h = 94° \ 15'$

The Sextant as a Surveying Instrument. The sextant was developed by Sir Isaac Newton and John Hadley during the eighteenth century,

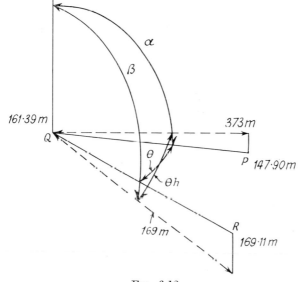

Fig. 6.18

for use by navigators, and it performs its function as an aid to navigators admirably. The vernier enables readings to be made to 10″, and although the accuracy is probably more like 30″, this is still good enough for locating position at sea. Also the readings are taken at the instantaneous coincidence of two signals, so that movement of the observer does not matter.

On land, however, although both versions of the instrument are portable, they must be reckoned as being much inferior to the theodolite for the determination of latitude and for the measurement of horizontal angles. The reasons for this are—

1. Because the reading is made at one pointing, the observer has to look directly at one station, so that in measuring a horizontal angle, for instance, the second station cannot be much further round than the

observer's shoulder, giving a maximum angle of, say 125°. Larger angles have thus to be measured in two or more parts. Angles of elevation of heavenly bodies are also affected by this limitation on land; the artificial horizon generally has to be used, and the limiting elevation is thus 60°–65°.

2. Angles are read on one side of the circle only (compare the theodolite in which two opposite points are read) so that a systematic error due to eccentricity of the graduated arc can occur. This error cannot be eliminated.

3. The sextant is cumbersome when used to measure azimuths, unless the stations concerned are absolutely on the same horizontal plane.

For these reasons, therefore, the lightweight theodolites now available are to be preferred for land surveying. In hydrographic surveying, however, the sextant is used to give three point fixes of boat positions during in-shore sounding operations*† (see Chapter 10), since it is admirably suited to this class of work.

4. The Compass

The fact that a magnetized needle, pivoted so that it can rotate freely in the horizontal plane, will come to rest in the magnetic meridian, has been used for centuries by navigators. Compasses are still used in surveying, the main instruments being (a) the prismatic compass, of which the best-known type is perhaps the military prismatic, widely used in rough and exploratory surveys, and (b) the compass theodolite, used also for such low-order surveys. Before dealing briefly with these instruments and their use, however, it is necessary to consider one or two aspects of terrestrial magnetism insofar as they affect the accuracy of compass work.

The *magnetic* meridian, which is the line taken by the compass needle outside the range of any uncompensated masses of magnetized material, lies approximately N–S, and the angle between it and the *true* meridian (i.e. the line passing through the compass and the *true* N and S poles) is known as the *declination*. This varies from one locality to another and in any given locality it also varies with time. Lines joining points of equal declination are known as *isogonals*, and in England they run across the country in a north-east, south-west direction, declination being greater in the north than in the south.

Variations in the strength of the magnetic field at any one place produce the following variations in the declination on a time basis—

(a) *Secular variation*, the full cycle of which takes several centuries. The annual amount of this change is 5′–9′ at Greenwich. In 1657, the declination at this point was nil (i.e. the magnetized needle pointed to true north); in 1819 it had reached its maximum westerly value and it is now moving eastwards again. Thus when magnetic bearings are

* "U.S. Coast and Geodetic Survey," Robert F. A. Stubbs. *Empire Survey Review*, July, 1951.
† *Hydrographic Surveying.* Lieut.-Cmdr. A. D. Margrett, R.D., R.N.R. F.R.G.S. (Kelvin Hughes, London.)

given, the date, the declination, and the annual rate of change for the locality should also be given.

(*b*) *Diurnal variations*, which are more or less regular changes in the needle about its mean position during the course of the day. The maximum change on the daily basis is about 10'.

(*c*) *Annual variations*, in which the period of variation is a year; these are so small that they may be ignored.

In addition, *magnetic storms* can cause sudden variations of as much as 1°. Of these various effects, the diurnal variation and magnetic

FIG. 6.19. SC II PRISMATIC COMPASS
(*Courtesy of Hilger & Watts, Ltd.*)

storms are the most serious in reducing the accuracy of compass bearings. Interference with the magnetic field caused by electric cables, small masses of iron, or iron ore is known as *local attraction*, and if it is suspected at any one station then back bearings are taken from adjoining stations free from attraction, and a correction applied. (Note here that this applies only where the effect is localized. In some parts of the globe, notably South Africa, the effect of magnetic rocks is so widespread as to make the compass useless.)

The Prismatic Compass. Fig. 6.19 shows a prismatic compass, supplied by Hilger & Watts, Ltd. It consists essentially of a glass-topped case of diameter 114 mm, within which an engine-divided aluminium ring (to which the needle is attached) rotates on a jewelled centre. The scale is divided round the ring from 0° to 360° in a clockwise manner, and a damping device, finger actuated, causes the needle to take up its position quickly. The object whose bearing is required is sighted by means of the fine slit at the top of the prism holder and the sighting wire in the hinged frame. At the same time the prism (which has a protective hinged cover) allows the eye to read the graduation on the circle. The sighting wire is seen superimposed on the compass dial, thus forming an index against which to read. The ring, which has a diameter of 102 mm, is graduated in degrees and half degrees, but a closer estimation is possible. Hinged shades are available for viewing bright objects. The instrument must be held in a horizontal position when observing and it may be tripod-mounted. When the observer looks through the prism with the needle in the meridian he is actually looking at the south pole of the compass; the scale is engraved so as to allow for this and a zero reading is given through the prism. Thus when a bearing is being taken a direct reading is obtained.

TRAVERSING WITH THE PRISMATIC COMPASS. As stated earlier, this instrument is normally used for rapid and exploratory surveys. It is

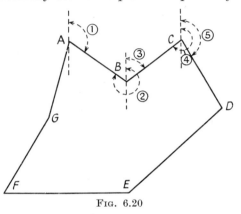

FIG. 6.20

generally held in the hand so that the length of the legs in the traverse must be sufficiently long to reduce the effects of centring. Under these conditions, the *free-needle method* of surveying is used in which the needle is floated before each reading so as to establish the magnetic meridian relative to which the actual reading is taken. Referring to the traverse *ABCDEFGA* in Fig. 6.20, and starting at *A*, the instrument is held horizontally and a sight taken on *B*; the bearing, which is the *forward bearing* of *AB*, is taken (1). At *B*, angle (2), which is the *back bearing* of *AB* (or the forward bearing *BA*) is read, and also angle (3), the forward bearing of *BC*. This procedure is repeated round the traverse.

Readings may be booked as *magnetic bearings*, i.e. in the form $Nx°E$, $Sy°E$, etc., the bearings then being directly applicable when calculating latitudes and departures, or as *whole circle bearings*; the latter booking being preferable.

The back bearing of each line should differ from the forward bearing by 180°, so that by comparison of the two, errors can be detected, including, in particular, the presence of local attraction. In rough work, forward bearings only are read, or even forward and back bearings at alternate stations, but this is not normally to be recommended since gross errors can escape undetected. If local attraction is found to be present at, say one station, it is corrected for; reduced bearings are then determined and latitudes and departures computed as with theodolite traverses. Errors are best adjusted using Bowditch's rule, i.e.

correction to latitude = total correction in latitude
  (or departure)   (or departure)

$$\times \frac{\text{length of corresponding side}}{\text{total length of traverse}}$$

Bowditch's method is suitable for use in correcting a compass traverse; it gives greater angular distortion of the framework than the transit rule, and this is required owing to the relatively large errors in angle measurement which occur in compass work. A further treatment of traverse correction is given in Chapter 4.

EXAMPLE. The following bookings were taken during a rough compass traverse round stations *ABCDEA*. Since magnetic interference was suspected, forward and back bearings were taken at each station. If the magnetic declination was 12°W, plot the traverse to a scale of 1 in 20,000, making any adjustments required and using co-ordinates relative to the true meridian.

Line	Length (m)	Forward Bearing	Back Bearing
A B	1,500	32°	212°
BC	540	77°	262°
CD	1,580	112°	287°
DE	1,060	122°	302°
EA	3,750	265°	85°

Considering each line in turn, the forward and back bearings should differ by 180°. This is true for lines *AB*, *DE*, and *EA*, and we can assume therefore that stations *A*, *B*, *D* and *E* are free from local attraction. The bearings for lines *BC* and *CD* do not agree, indicating interference at *C*, and the bearings for these lines are thus taken from the readings at *B* and *D*. The corrected forward bearings are now tabulated, and after correcting for declination the appropriate reduced bearing derived and the latitudes and departures calculated. (Note that where the forward and back bearings differ by only a few *minutes*, this probably only indicates diurnal variations and errors in reading; in such cases the mean of the two bearings is used.)

Line	Corrected Forward Bearing	True Bearing	Reduced Bearing	Length (metres)	Latitude (metres) +	Latitude (metres) −	Departure (metres) +	Departure (metres) −
AB	32°	20°	N 20° E	1,500	1,410	—	510	—
BC	77°	65°	N 65° E	540	230	—	490	—
CD	107°	95°	S 85° E	1,580	—	140	1,570	—
DE	122°	110°	S 70° E	1,060	—	360	1,000	—
EA	265°	253°	S 73° W	3,750	—	1,100	—	3,590
				8,430	1,640	1,600	3,570	3,590

Total error in latitude = + 40 m.

Total error in departure = − 20 m.

Apportioning out the error in latitude by Bowditch's rule (to slide rule accuracy) gives the following table of adjusted latitudes and departures which may be plotted.

Line	Adjusted Lat. (m) +	Adjusted Lat. (m) −	Adjusted Dep. (m) +	Adjusted Dep. (m) −	Point	Co-ordinates (m)	Co-ordinates (m)
A B	1,405	—	515	—	A	0 N	0 E
B C	230	—	490	—	B	1,405 N	515 E
C D	—	150	1,575	—	C	1,635 N	1,005 E
D E	—	365	1,000	—	D	1,485 N	2,580 E
E A	—	1,120	—	3,580	E	1,120 N	3,580 E
	1,635	1,635	3,580	3,580			

At the scale of 1:20,000 an error of 5 m is represented by 0·25 mm and, accordingly, the corrections have been broadly taken to 5 metres.

EXAMPLE. Enumerate the principal sources of error in compass traversing.

The following record was obtained in a compass traverse by the free-needle method. Apply any necessary corrections to the readings and tabulate the corrected bearings.

Line	Observed bearing
A E	123° 50′
A B	55° 30′
B A	236° 50′
B C	165° 20′
C B	344° 20′
C D	116° 30′
D C	294° 20′
D E	236° 30′
E D	58° 40′
E A	303° 50′

(*L.U.*, *B.Sc.*, 1950)

Line	Forward Bearing	Back Bearing	Difference	Error
A B	55° 30′	236° 50′	181° 20′	+ 1° 20′
B C	165° 20′	344° 00′	178° 40′	− 1° 20′
C D	116° 30′	294° 20′	177° 50′	− 2° 10′
D E	236° 30′	58° 40′	177° 50′	− 2° 10′
E A	303° 50′	123° 50′	180° 00′	0

Considering each line in turn, as in the previous example, it is seen that stations *E* and *A* are free from interference. Using the back bearing *DE* and forward bearing *AB* as being free from error, corrections can be applied to the other bearings taken from *D* and *E* which are affected by interference. At the same time a study of the amounts of error due to interference at *D* and *E* shows that station *C* is free from interference. (A quick sketch will show how the interference, by altering the declination, produces the error.) The corrected bearing can thus be tabulated:

Line	Corrected Forward Bearing	Corrected Back Bearing
AB	55° 30′	235° 30′
BC	164° 00′	344° 00′
CD	116° 30′	296° 30′
DE	238° 40′	58° 40′
EA	303° 50′	123° 50′

This adjustment, however, does not mean that the sum of the internal angles is necessarily equal to $(2n - 4)$ right angles.

Wild T0 Compass Theodolite. This instrument, shown in Fig. 6.21*a*, reads horizontal angles directly to 1′ and vertical angles to 1°. It is fitted with a swinging balanced graduated circle having a north-seeking needle beneath, and this orients the zero of the circle to magnetic north. Diametrically opposite graduations of the circle are seen in the field of view of the reading eyepieces on each side of the standard carrying the vertical circle, and they are brought into coincidence by turning a micrometer drum found on the other standard. A change-over knob is provided to obviate the need for moving round the instrument to read through the other eyepiece.

The circle is divided at intervals of 2° and numbered every 10°, and it is read in a somewhat similar manner to that of the T2 instrument. The first upright number of the lower graduations to the left of the field of view gives the tens of degrees, and one then counts the intervals to the inverted number of the upper graduation differing by 180°; these indicate the number of single degrees to be added on, and finally the micrometer drum reading is included. It will be noted in Fig. 6.21*b* that some of the lines are convergent but those in the middle of the field of view are the main ones used to obtain coincidence. The bearing of the compass circle rests on a sharp pivot, from which it can be lifted when being transported: in fact the instrument can be used as a normal theodolite when the circle does not rest on the pivot. The verticle circle is read through another eyepiece, but there is no provision for coincidence setting and estimation to 1′ has to be made.

One interesting feature of the telescope, which has a magnification of ×20, is that two sets of stadia lines are provided giving multiplying constants of 50 and 100 (*see* Chapter 7).

For compass traversing the leap-frog method allows rapid progress to be made since alternate stations only are occupied by the instrument. Say the position of *A* in Fig. 6.20 is known, then the instrument can be set up at *B* and the magnetic bearings of *BA* and *BC* can be established. If the instrument be now taken to *D* the bearings of *DC* and *DE* can be found. Any error which might be present in the bearing of *BC* is not therefore transferred to *CD*, and the whole traverse is not swung round but is displaced more or less parallel by that error.

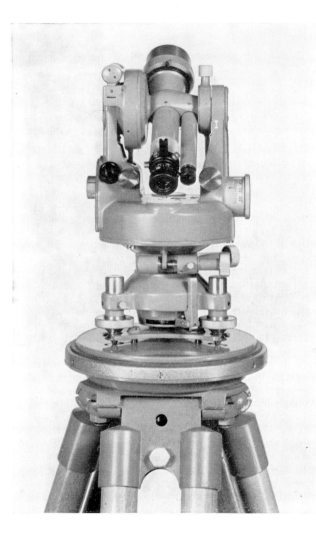

Fig. 6.21*a*. T0 Compass Theodolite
(Courtesy of Wild Heerbrugg, Switzerland)

Alternatively one could start at *A* and perhaps obtain the magnetic bearing of a point whose true bearing from *A* is known. Station *C* would then be occupied, and then *E*, but the whole traverse could later be oriented to the true meridian at A. The remarks on page 206 in respect of these methods must be remembered however.

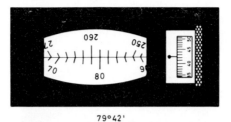

79°42'

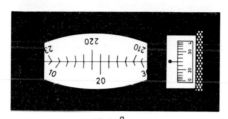

19. 24 9

FIG. 6.21*b*. COMPASS CIRCLE READING
(Couretsy of Wild Heerbrugg, Switzerland)

Barometric Levelling

Atmospheric pressure decreases with increase in altitude, and, since relationships between pressure and altitude can be derived, the accurate measurement of atmospheric pressures at a series of stations affords a means of calculating the relative altitudes of the stations. The pressures are expressed in terms of the height of a column of mercury, and are measured with a barometer which may be either (more accurately) of the mercury column type, or (less accurately but more conveniently) of the aneroid type. Analysis of the pressure/height relationship was originally carried out on the assumption that the atmosphere has a constant temperature between points of different height, i.e. that it is isothermal, and many conversion tables have been drawn up on this basis. It is a fact, however, that a temperature gradient exists between points of different height, and after the First World War new formulae were derived which take into account the fall in temperature with altitude. This new interest was due of course to the use of aneroids as aircraft altimeters, and attempts are being made to obtain international agreement on a "standard atmosphere" for use in drawing up altitude tables.* Both methods of analysis are given in the following pages.

* E.g. B.S. 2520: 1967. *Barometer Conventions and Tables* which gives the requirements of the International Standard Convention regarding (i) density of mercury, (ii) value of g, and (iii) standard calibrating temperature of 0°C.

The Isothermal Atmosphere. From the laws of Boyle and Charles, and treating dry air as a perfect gas, we have—

$$pv = r\theta$$

where p and v are the pressure and volume of a *unit* mass of air a absolute temperature θ.

Thus $\qquad\qquad\qquad \theta = t°C + 273°$

and also $\qquad\qquad\qquad r = $ gas constant.

If the density of air under these conditions of temperature and pressure is ρ, then since we have unit mass,

$$\frac{p}{\rho} = r\theta \qquad . \qquad . \qquad . \qquad . \qquad . \qquad (6.1)$$

Further, from a consideration of the equilibrium of a column of air under the action of gravity, we have from Fig. 6.22,

$$(p + \delta p) A + \rho g A \delta h = pA$$
$$\delta p . A = - \rho g A \delta h$$

$\therefore$ In the limit, $\qquad\qquad \dfrac{dp}{dh} = - \rho g \qquad . \qquad . \qquad . \qquad . \qquad (6.2)$

Now from (6.1) $\qquad\qquad \rho = \dfrac{p}{r\theta}$

$\therefore$ From (6.2)

$$\frac{dp}{p} . \frac{r\theta}{g} = - dh$$

$$\int_{h_2}^{h_1} dh = - \int_{p_2}^{p_1} \frac{r\theta}{gp} dp$$

$\therefore \qquad\qquad h_2 - h_1 = \dfrac{r\theta_0}{g} \log_e \left[\dfrac{p_1}{p_2}\right] \qquad . \qquad . \qquad . \qquad (6.3)$

assuming a constant absolute temperature of θ_0. The expression is usually given in the form

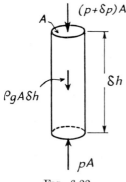

$$h_2 - h_1 = \bar{K} . \log_{10} \left[\frac{P_1}{P_2}\right] \qquad . \qquad (6.4)$$

where h_1, h_2 are the altitudes at lower and upper stations 1 and 2, and P_1 and P_2 are the corresponding pressures expressed in millimetres of mercury, $\bar{K}$ is a parameter dependent on the value of θ_0, and on the value of g. With θ_0 equal to the temperature of melting ice (273°K) equation (6.4) may be written as

$$h_2 - h_1 = 18,336 \cdot 5 \log_{10} \left[\frac{P_1}{P_2}\right] \text{ metres}$$

$(p+\delta p)A$

A

δh

$\rho g A \delta h$

pA

FIG. 6.22

CORRECTION FOR TEMPERATURE. This is only true for the particular value of θ_0 chosen, and correction is required if the absolute air temperature is different from this. Thus if the actual mean absolute air temperature between stations 1 and 2 is θ_m, equation (6.4), with temperature correction, becomes

$$h_2 - h_1 = \bar{K} \log_{10} \left[\frac{P_1}{P_2} \right] \cdot \frac{\theta_m}{\theta_0}$$

Now if the temperatures at stations 1 and 2 are $t_1 °C$ and $t_2 °C$, then)

$$\theta_m = \tfrac{1}{2}(t_1 + 273° + t_2 + 273°)$$
$$= \tfrac{1}{2}(t_1 + t_2 + 2\theta_0)$$

Therefore $\quad \dfrac{\theta_m}{\theta_0} = 1 + \dfrac{t_1 + t_2}{2\theta_0}$

Instead of writing $2\theta_0$ as 546°, however, 500° may be used to allow for humidity, the equations being based on dry air, which rarely obtains in practice. Thus, writing H as the difference in level in metres,

$$H = 18{,}336{\cdot}5 \log_{10} \left[\frac{P_1}{P_2} \right] \cdot \left(1 + \frac{t_1 + t_2}{500} \right) \quad . \qquad . \quad (6.5)$$

Although θ_0 is usually 273°, manufacturers of surveying aneroids (which have engraved altitude scales) often use different calibration temperatures, and this of course will give different values for the parameter $\bar{K}$.

The Troposphere and the Lapse-rate Formula. The first seven miles thickness of atmosphere is known as the troposphere (Greek: *tropos*, to turn) because it is continually in motion due to wind and storm. There exists a temperature gradient with respect to altitude, and under these conditions when a bulk of air is displaced vertically, its temperature, unlike its pressure, has insufficient time to adjust itself to the conditions obtaining at the new level before the air is moved again. The pressure and density at different levels thus obey the polytropic law—

$$\frac{p}{\rho^n} = k \quad . \qquad . \qquad . \qquad . \qquad . \quad (6.6)$$

Substituting for ρ from equation (6.2),

$$k^{\frac{1}{n}} \int \frac{dp}{p^{\frac{1}{n}}} = - g \int dh$$

Therefore $\quad \dfrac{n k^{\frac{1}{n}}}{n - 1} \cdot p^{\frac{n-1}{n}} = - gh + A$

Putting $\quad p = p_0$ when $h = 0$, $A = \dfrac{n k^{\frac{1}{n}}}{n - 1} \cdot p_0^{\frac{n-1}{n}}$

Therefore
$$p^{\frac{n-1}{n}} = p_0^{\frac{n-1}{n}} - \frac{n-1}{nk^{\frac{1}{n}}} \cdot gh \qquad . \qquad . \quad (6.7)$$

To find k, let ρ_0, θ_0 be the density and absolute temperature at $h = 0$.

From (6.6)
$$\frac{p}{\rho^n} = \frac{p_0}{\rho_0^n} = k$$

From equation (6.1),
$$\frac{p_0}{\rho_0} = r\theta_0$$

$\therefore$
$$\frac{1}{\rho_0^n} = \left(\frac{r\theta_0}{p_0}\right)^n$$

$\therefore$
$$k = (r\theta_0)^n p_0^{1-n}$$

and
$$\frac{1}{k^{\frac{1}{n}}} = \frac{p_0^{\frac{n-1}{n}}}{r\theta_0}$$

Substituting in (6.7),
$$\frac{p}{p_0} = \left(1 - \frac{n-1}{n}\frac{gh}{r\theta_0}\right)^{\frac{n}{n-1}} \qquad . \qquad . \qquad . \quad (6.8)$$

The *temperature gradient* is found as follows—

From equations (6.1) and (6.6),
$$\frac{\theta}{\theta_0} = \frac{p}{\rho} \cdot \frac{\rho_0}{p_0} = \frac{p}{p_0}\left(\frac{p_0}{p}\right)^{\frac{1}{n}}$$

$$= \left(\frac{p}{p_0}\right)^{\frac{n-1}{n}} \qquad . \qquad . \qquad . \qquad . \quad (6.9)$$

Substituting in equation (6.8),
$$\frac{\theta}{\theta_0} = 1 - \frac{n-1}{n} \cdot \frac{gh}{r\theta_0}$$

i.e.
$$\theta - \theta_0 = t - t_0 = -\frac{n-1}{n} \cdot \frac{gh}{r} \qquad . \qquad . \quad (6.9a)$$

where t and t_0 are the corresponding Celsius temperatures.
Since n and r are constant, the temperature gradient is linear.

$$\frac{d\theta}{dh} = \frac{dt}{dh} = -\frac{n-1}{nr} \cdot g \qquad . \qquad . \quad (6.10)$$

This gradient is known as the lapse-rate, since it gives the rate of lapse of temperature with altitude, and writing it as s we have

$$s = \frac{n-1}{n} \cdot \frac{g}{r}$$

$$\frac{n-1}{n} = \frac{sr}{g} \qquad . \qquad . \qquad . \qquad . \qquad . \qquad (6.11)$$

Substituting in (6.8), $\dfrac{p}{p_0} = \left(\dfrac{\theta_0 - sh}{\theta_0}\right)^{\frac{g}{sr}} \qquad . \qquad . \qquad . \qquad . \qquad (6.12)$

The gradient, or lapse-rate, has been obtained by international agreement on the basis of a theoretical standard atmosphere which represents average conditions in western Europe, and is given by the following formula—

$$t = 15 - 0 \cdot 0065\, h \qquad . \qquad . \qquad . \qquad . \qquad . \qquad . \qquad (6.13)$$

where t = temperature (°C) at height h above sea level.
The lapse-rate is thus virtually 6·5°C per 1,000 m. Assuming that n is 1·235 (the value used for the standard atmosphere , that g is 9·8066 m/sec², that at mean sea level the temperature is 15°C, that the barometric height reduced to 0°C is 760 mm of mercury, and using the value of r appropriate to the temperature scale used, we get,

$$\frac{p}{p_0} = (1 - 0 \cdot 00002256\, h)^{5 \cdot 256} \qquad . \qquad . \qquad (6.14)$$

The following table gives some numerical results obtained from applying this formula.

Height (m above S.L.)	$\dfrac{p}{p_0}$	Pressure (mm of mercury at 0°C)	Temp. (°C)
0	1·000	760·0	15·00
500	0·942	715·9	11·75
1,000	0·887	673·1	8·50
1,500	0·835	634·6	5·26

B.S. 2520:1967 lists the following pressure units (in order of preference)—

(a) the basic unit of pressure, N/m² or its decimal multiple,
(b) the millibar, mb,
(c) the millimetre of mercury, mm Hg, at 0°C and standard gravity 9·80665 m/s². This is included to cater for those barometers in current use which could become obsolete in the future.

Conversions are as follows—

mm Hg	kN/m²	mb
600	79·993	799·93
700	93·326	933·26
800	106·658	1,066·58
900	119.990	1,199·90

Thus 760 mm Hg is equivalent to 101·325 kN/m².

CORRECTION FOR TEMPERATURE ON STANDARD ATMOSPHERE SCALE. Readings require correction (as with the isothermal atmosphere) for casual variation of temperature, since it is unlikely that the ambient temperature at the time any one reading is made will equal that corresponding to the standard atmosphere for the particular altitude of the pressure measuring device.

Let H = actual height above sea level

θ_H = actual temperature in °K

t_{obs} = actual temperature in °C at the station from which θ_H is derived

p = pressure at height H

h = height above sea level on the standard atmosphere scale corresponding to p

θ = temperature in °K on the standard atmosphere scale corresponding to height h

t = temperature in °C at height h in the standard atmosphere from which θ is derived.

Then, using the suffix 0 for sea level,

$$\frac{\theta}{\theta_0} = \left(\frac{p}{p_0}\right)^{\frac{h-1}{h}} = \frac{\theta_H}{\theta_{H_0}}$$

$$\therefore \quad \frac{\theta - \theta_0}{\theta_0} = \frac{\theta_H - \theta_{H_0}}{\theta_{H_0}}$$

$$\therefore \quad \frac{sh}{\theta_0} = \frac{sH}{\theta_{H_0}}$$

$$\therefore \quad H = h \cdot \frac{\theta_{H_0}}{\theta_0}$$

$$\therefore \quad H - h = \frac{h(\theta_{H_0} - \theta_0)}{\theta_0} = \text{Correction for temperature}$$

But $\theta_{H_0} - \theta_0 = t_{obs} - t$

and if $\theta_0 = 288°\text{K}$

then $H = h(1 + 0{\cdot}003472\,(t_{obs} - t))$. . . . (6.15)

CORRECTION FOR GRAVITY. Barometer calculations are based on the following constants (B.S. 2520: 1967)—

$$\text{Density of mercury} = 13\cdot5951 \text{ kg/m}^3$$
$$\text{Temperature} = 0°\text{C}$$
$$g = 9\cdot80665 \text{ m/s}^2$$

The value of g is not constant but varies with latitude and altitude. The following expressions are quoted in B.S. 2520—

$$g_{\phi,0} = 9\cdot80616(1 - 0\cdot0026373 \cos 2\phi + 0\cdot0000059 \cos^2 2\phi) \text{ m/s}^2$$
$$g_{\phi,H} = g_{\phi,0} - 0\cdot0003086 H \text{ m/s}^2$$

where

ϕ = latitude and H = altitude above m.s.l.

The first formula, therefore, gives correction for latitude, and the second for elevation.

Measurement of Pressure. The most accurate method of measurement is undoubtedly a barometer of the Fortin type, in which the height of a vertical column of mercury is measured relative to the free surface of the mercury reservoir at the lower end. The particular feature of this instrument is the adjusting screw, operating on a leather bag forming part of the reservoir, whereby the free surface is brought to a constant level prior to each reading. The top of the column can thus be read against a fixed brass scale with a vernier. In accurate work, corrections are applied for thermal expansion of the mercury column and the brass scale when readings are taken at other than the calibration temperatures. For a fuller description, however, the student is referred to any physics text-book.

Despite its accuracy as a pressure measuring device, the mercury barometer is still much inferior to the surveying level in the measurement of altitude. As a consequence, the main use of barometric levelling is in exploratory surveys, where portability and compactness are important, and on these grounds the aneroid barometer is much more convenient, although less accurate.

The Aneroid. The main feature of the aneroid barometer is a shallow cylindrical box of about 75 mm diameter from which air has been evacuated, but which is prevented from collapsing by a sheet metal spring connected at one edge to the base via a bridge piece, and at the other to the centre of the box via a knife-edge (*see* Fig. 6.24). Any change in atmospheric pressure causes movement in the ends of the box, and this movement, magnified by a lever system, is transferred to a pointer. Fig. 6.23 shows a typical surveying aneroid instrument made by Short & Mason Ltd. of London.

The type of aneroid illustrated is designed for use in approximate and exploratory surveys and levelling in connexion with roads, railways, canals, water courses, and mines. The altitude scale, derived by a formula similar to equation (6.4), is evenly divided, and by means of a revolving vernier and a rotating reading lens, readings can be made to 1 m. It must be borne in mind, however, that the error of a single reading on an aneroid may be considerably greater than this.

To obtain the difference of level of any two stations, the aneroid is placed horizontally at one of the stations and the zero of the vernier is brought underneath the pointer by means of the adjusting nut on the top, and the reading taken. This operation is repeated at the next station, and the difference of level obtained by subtraction of the two readings. If the temperatures have been taken at the two stations, correction for temperature is applied as indicated earlier (equation (6.5) or equation (6.15)).

FIG. 6.23. ANEROID BAROMETER
(*Courtesy of Short & Mason, Ltd.*)

Paulin Aneroids. Aneroids of the type shown in Fig. 6.24 tend to stick due to friction in the lever system and at the pivots so that gentle tapping or vibration is necessary before taking a reading. Also, when the vacuum capsule is subjected to a pressure cycle, hysteresis often occurs and different readings are obtained for a given pressure according to whether the pressure is increasing or decreasing. To overcome these defects the American Paulin Co. manufacture an aneroid in which the pressure on the diaphragm of the vacuum box is in effect weighed by an accurate tension spring. There are very few moving parts in the Paulin-type instruments and the above defects are largely eliminated.

The principle is illustrated schematically in Fig. 6.25. The limit stops prevent excessive deflection of the vacuum box diaphragm, reducing elastic hysteresis, and all readings are made with the diaphragm in a constant position achieved by bringing the pointer to the null position. This is done by means of the spring actuating screw which is turned until the tension in the spring balances the atmospheric load on the

:apsule diaphragm. The reading on the height (or pressure) scale is now ι̇oted. If the atmospheric pressure changes due to the instrument ϸeing moved to a point of different altitude the arm will move out of ϸalance until it hits one of the stops.

The actuating screw is again turned and the spring tension adjusted ι̇ntil the null position is reached. The change in tension required to ᵗemove the out-of-balance caused by the change in atmospheric ϸressure is by Hooke's law proportional to the extension of the spring, ᵗnd hence to the movement of the gearing (or cam system) containing ᴛhe height scale. Thus by reading the height scale and subtracting the

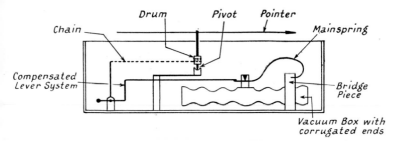

FIG. 6.24. SCHEMATIC SECTION THROUGH ANEROID BAROMETER

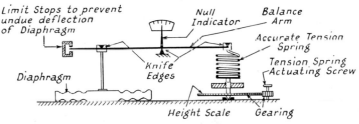

FIG. 6.25. SCHEMATIC SECTION THROUGH PAULIN-TYPE INSTRUMENT

ι̇nitial reading, the change in altitude is obtained; if the known base ᴋeight is not registered due to changing air pressure, a reset knob ᵃllows the correct reading to be set, putting the instrument into ᵃdjustment. The scale is calibrated for a temperature of +10°C.

CORRECTIONS FOR WEATHER. All altitude barometers are affected ϸy weather changes and some means of correcting for these must be ᴨade. With a single barometer used as described in the last section, t is clear that any change in atmospheric pressure due to weather and ᴨot to change in altitude, occurring during the journey from the ᵃrst station to the second would not be detected. Such a change, ᴋowever, could have a considerable effect on the estimated dif-ᵉrence in level. In certain parts of the world, notably the tropical ᵗegions such as Southern Nigeria, it is found that changes in atmospheric pressure throughout the day are so regular that they may be studied, and in a particular locality corrections may be derived for different times of the day. Even so, unchecked single-barometer readings are always of doubtful accuracy, especially when stations have considerable differences in elevation. Hinks* reports that heights of mountains in Africa determined by early single-barometer readings are often found to be excessively in error when compared with the results of more precise surveys.†

There are two methods of checking and correcting for weather: (i) by comparing the aneroid barometer with a known height, which will

FIG. 6.26. MICROBAROGRAPH
(*Courtesy of Short & Mason, Ltd.*)

give the difference in altitude reading between them. To be effective however, such checks must be made at frequent intervals, e.g. every hour, and this may not always be feasible.

(ii) By the use of a second barometer at the base. If an instrument such as the Short & Mason Microbarograph (Fig. 6.26) is used, a chart record of all weather changes is obtained, and, if the times of reading the other barometer are noted, corrections can be made for the effect of weather. This method is the more useful of the two, and should be used if possible. For full treatment of this subject see "The Use of Sensitive Pressure Devices in Surveying" by D. O'Connor B.E.(Syd)., L.S., F.R.G.S., *Empire Survey Review*, Vol. XIV, No. 108 *et seq.*, London, 1957.

EXERCISES 6

1. The angles of elevation of two points A and B were found to be 2° 11' and 27° 15' respectively. The angle subtended at the observer by A and B was measured by sextant in the inclined plane and found to be 41° 04'. What is the azimuthal angle subtended at the observer by A and B?

Describe the method of use of the artificial horizon with a sextant and prove the relation between the angle measured and the angle of elevation. (*L.U., B.Sc.*)

Answer: 33° 55'.

* *Maps and Survey*, A. R. Hinks (C.U.P.)

† *See also* "Height of Lake Tana, Ethiopia," by Col. D. R. Crone; *Empire Survey Review*, No. 99, Vol. XIII, London, 1956.

2. The (non-azimuth) angle between two signals A and B was measured by sextant from a point O, while the angular elevations of A and B above O were measured by clinometer, with the following results—

> Non-azimuth angle ACB $74° 25'$
> Angular elevation of A above O $15° 22'$
> Angular elevation of B above O $4° 06'$

Calculate the (azimuth) angle ACB. (*L.U., B.Sc.*)

Answer: $74° 57'$.

3. What is the lapse rate, used in connexion with barometrical levelling? Derive the simple lapse-rate formula

$$P = P_{\varepsilon}\left(\frac{T_{\varepsilon} - Kh}{T_{\varepsilon}}\right)^{\frac{g}{cK}}$$

in which P_{ε} is the atmospheric pressure at sea level, P the corresponding pressure at elevation h m above sea level, T_c is the absolute temperature at sea level, K is the lapse rate in deg. C per m rise, g is the acceleration of gravity, and c is the gas constant. (*L.U., B.Sc.*)

4. Describe briefly the principle of a surveyor's aneroid, commenting critically on its value as a surveying instrument. State, with reasons, whether in your opinion it is better to have the graduations of the instrument in units of pressure or height.

What particular precautions would you take in making aneroid surveys for the following purposes—

(i) To determine spot heights between a network of trigonometrical control points in order to provide height control for plotting 20 m contours from air photos.

(ii) To provide section levels in a reconnaissance survey for a proposed road through thick bush in a tropical country.

(iii) To provide cross sections in the preliminary reconnaissance of a large reservoir site in an area where the weather regime is most irregular and unstable. (*I.C.E.*)

5. The following data refer to a small reconnaissance compass traverse made through rough rocky country. Plot the traverse to a scale of 5 paces to the millimetre on squared paper, having first meaned the magnetic bearings and adjusted them, where necessary, for any discrepancy in bearing due to local attraction.

Adjust the traverse graphically and state by how much and in what direction the corrected point F differs from its plotted position.

Side	Distance	Magnetic Bearings	
AB	300 paces	AB 69°	BA 248°
BC	200 paces	BC 56°	CB 232°
CD	400 paces	CD 121°	DC 305°
DE	150 paces	DE 180°	ED 360°
EF	350 paces	EF 265°	FE 89°
FG	200 paces	FG 313°	GF 129°
GH	150 paces	GH 245°	HG 67°
HA	200 paces	HA 301°	AH 121°

(*I.C.E.*)

Answer: Local attraction at C and F, 35 paces at a bearing N 60° E.

6. It is proposed to survey a tract of land by plane table, triangulation points having previously been established on prominent points in the area. Describe in detail the equipment required and the field procedure to be adopted.

CHAPTER 7

TACHEOMETRY—THE OPTICAL MEASUREMENT OF DISTANCE

IN THIS BRANCH OF SURVEYING, heights and distances are determined from the instrumental readings alone, these usually being taken with a specially adapted theodolite known as a tacheometer. The chaining operation is eliminated, and tacheometry is therefore very useful in broken terrain, e.g. land cut by ravines, river valleys, over standing crops, etc., where direct linear measurement would be difficult and inaccurate. All that is necessary is that the assistant, who carries a staff on which the tacheometer is sighted, shall be able to reach the various points to be surveyed and levelled, and that a clear line of sight exists between the instrument and the staff. An additional limitation is imposed in some branches of tacheometry in that the distance between staff and instrument must not exceed a maximum, beyond which errors due to inaccurate reading become excessive.

The field work in tacheometry is rapid compared with direct levelling and measurement (the name derives from the Greek Ταχνσ—swift, and μετρον—a measure), and it is widely used therefore to give contoured plans of areas, especially for reservoir and hydro-electric projects, tipping sites, road and railway reconnaissance, housing sites, etc. With reasonable precautions the results obtained can be of the same order of accuracy as, or better than, those obtainable by direct measurement in some cases.

Systems of Tacheometry. Present-day methods of tacheometry can be classified in one of the following three groups—

1. The theodolite, with the measuring device inside it, is directed at a levelling staff which acts as target. This is usually known in England as the *stadia system*. One pointing of the instrument is required for each set of readings.

2. An accurate theodolite, reading to 1″ of arc, is directed at a staff, two pointings being made, and the small subtended angle measured. There are two variants, depending on the staff used,

    (a) an ordinary levelling staff, held vertically, is used—known as the *tangential system*, or

    (b) a bar of fixed length, usually held horizontally, is used—known as the *subtense system*.

3. A special theodolite with a measuring device in front of the telescope is directed at a special staff. One pointing of the instrument is required for each set of readings—the *optical wedge system*.

The Stadia System

There are two types of stadia instrument, (A) those in which the distance between the two hairs is fixed, and (B) those in which the distance is variable, being measured by means of a micrometer. These latter, which are sometimes described as *subtense* tacheometers, are not so common as the fixed-hair types, and will be dealt with only briefly. Fixed-hair tacheometry, or *stadia surveying* as it is often called, is dealt with at some length.

The invention of the stadia principle has been variously ascribed to the Englishman, William Green, who described in 1778 the method of stadia measurement using two fixed wires; to the Dane, Brander, who, between 1764–73 constructed the first glass diaphragms with fine lines cut on them, and applied them to the measurement of

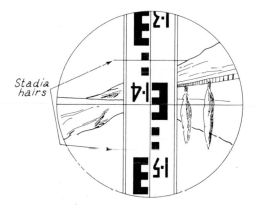

FIG. 7.1. VIEW OF STAFF THROUGH TACHEOMETER

distances; and to the Scot, James Watt, who, in 1771, used a tacheometer of his own construction in Scotland. There is reason to believe that they worked independently and evolved substantially similar methods of surveying, though it is interesting to note that working in the field of astronomy the great Dutch scientist, Huygens, had constructed a simple type of micrometer eyepiece as long ago as 1659.

(*A*) The Fixed-hair Tacheometer

The field of view through the telescope of a typical fixed-hair tacheometer is shown in Fig. 7.1. It will be appreciated by the student from the previous work that most modern theodolites and levels can also be used, in fact, as fixed-hair tacheometers. The two hairs (or lines) used to give an intercept on the staff are generally called *stadia* lines. The name was originally applied to the graduated staff or stadia rod (the stadia being a Greek measure of distance), but is now used exclusively when referring to the top and bottom hairs or lines in the diaphragm. In addition to the normal levelling staff, there are other types of stadia staves designed to give easier reading on long sights.

THEORY. The basic principles are shown in Fig. 7.2, and since a simple and exact theory is theoretically only possible for external focusing instruments, this case will be dealt with first. Taking the particular instance in which the telescope is level and the staff is vertical, then considering only the extreme rays which pass through the optical centre of the object glass,

triangles AOB, aOb are similar.

$$\therefore \qquad \frac{OX}{Ox} = \frac{u}{v} = \frac{AB}{ab} \qquad . \qquad . \qquad . \qquad . \qquad (7.1)$$

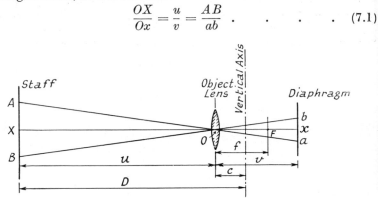

FIG. 7.2

Also, if $OF = f =$ focal length of object lens,

$$\frac{1}{v} + \frac{1}{u} = \frac{1}{f} \qquad . \qquad . \qquad . \qquad . \qquad (7.2)$$

Multiply both sides by uf,

$$u = \frac{u}{v} \cdot f + f$$

Substituting from equation (7.1),

$$u = \frac{AB}{ab} \cdot f + f$$

AB, which is obtained by subtracting the reading given on the staff by the lower stadia hair from that given by the top one, is usually denoted by s, and ab, the distance apart of the stadia lines, is denoted by i. This latter distance is, of course, constant in the fixed-hair tacheometer. Thus we get

$$u = \frac{f}{i} \cdot s + f$$

Thus the horizontal distance D from the vertical axis of the tacheometer to the staff is obtained by adding the small distance c between the object glass and the vertical axis.

$$D = \frac{f}{i} \cdot s + (f + c) \qquad . \qquad . \qquad . \qquad . \qquad (7.3)$$

In Fig. 7.1, $\qquad\qquad s = 1{\cdot}49 - 1{\cdot}37 = 0{\cdot}12$

The Anallatic Lens. The reduction of formula (7.3) would be simplified considerably if (a) the term $\frac{f}{i}$ is made some convenient figure, and (b) if the term $(f + c)$ can be made to vanish. Requirement (a) is solved by using values of f and i such that $\frac{f}{i}$ equals 100. The second requirement was solved by the Italian instrument-maker, Porro, in 1823 with a telescope incorporating an additional convex lens between the object lens and the diaphragm. (Note that this lens, which Porro called the anallatic* lens, has not the same function as the concave

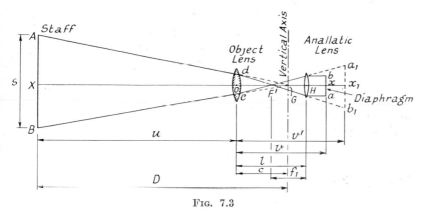

Fig. 7.3

lens used to focus the telescope in *internal*-focusing theodolites and levels.)

Referring to Fig. 7.3, consider those rays of light from points A and B on the staff which, after refraction through the objective and the anallatic lens, emerge parallel to the optical axis of the telescope to form an image abx, i.e. they pass through F^1, the focal point of the anallatic lens. Had the anallatic lens not been present, these rays would have formed an image at $a_1b_1x_1$.

The rays under consideration strike the objective at d and e, and if carried forward will, from symmetry, intersect at G on the axis. They will strike the objective at a fixed angle because they have to pass through the point G, so that G can be considered as a fixed point.

i.e. $\qquad\qquad XG = \tfrac{1}{2}\cot\theta \,.\, AB \quad$ where $\theta = A\hat{G}B$

Therefore $\qquad\qquad\qquad XG \propto AB$

Thus if G can be made to fall on the vertical axis of the instruments the term $(f + c)$ will have been eliminated from the tacheometric formula. Using the lens formulae and sign convention already given

---

* Referred to as an anallactic lens by some writers.

in Chapter 2, the necessary relationship between f, f_1, d and c can be deduced to give the condition that $XG(=D)$ is proportional to s.

Take first the object lens with the object AB and the image a_1b_1.

$$\frac{1}{v'} + \frac{1}{u} = \frac{1}{f} \qquad . \qquad . \qquad . \qquad . \quad (7.4)$$

$$\frac{u}{v'} = \frac{s}{a_1b_1} \qquad . \qquad . \qquad . \qquad . \quad (7.5)$$

Now applying the lens formulae to the anallatic lens, with virtual object a_1b_1 and image ab.

$$\frac{1}{(v-l)} - \frac{1}{(v'-l)} = \frac{1}{f_1} \qquad . \qquad . \qquad . \quad (7.6)$$

$$\frac{v-l}{v'-l} = \frac{ab}{a_1b_1} \qquad . \qquad . \qquad . \quad (7.7)$$

Note that from a consideration of the sign convention a negative sign is required in equation (7.6). From these four equations an expression for D can be obtained.

From equation (7.5),

$$v' = a_1b_1 \cdot \frac{u}{s}$$

from equation (7.6),

$$v - l = \frac{f_1(v'-l)}{f_1 + v' - l}$$

Therefore from equations (7.4) and (7.7),

$$a_1b_1 = \frac{ab\left(f_1 + \dfrac{fu}{u-f} - l\right)}{f_1}$$

Also from equation (7.5),

$$a_1b_1 = \frac{v's}{u}$$

Substitute in equation (7.4),

$$a_1b_1 = \frac{fs}{(u-f)}$$

Therefore
$$\frac{fs}{(u-f)} = \frac{ab\left(f_1 + \dfrac{fu}{u-f} - l\right)}{f_1}$$

Writing $\quad ab = i$, the stadia interval,

Then $\quad ff_1s = i[(u-f)f_1 + fu - l(u-f)]$

$$ui(f + f_1 - l) = ff_1s + iff_1 - lfi$$

$$\therefore \qquad u = \frac{ff_1s}{i(f+f_1-l)} - \frac{f(l-f_1)}{(f+f_1-l)}$$

but
$$D = u + c$$

$$= \frac{ff_1s}{i(f+f_1-l)} - \frac{f(l-f_1)}{(f+f_1-l)} + c \qquad . \qquad . \quad (7.8)$$

i.e.
$$D = Cs - \frac{f(l-f_1)}{(f+f_1-l)} + c$$

If
$$c = \frac{f(l-f_1)}{f+f_1-l}$$

then
$$D = Cs$$
$$= 100s \text{ (usually)}$$

and f_1 must be less than l since

$$l = f_1 + \frac{fc}{(f+c)} \qquad . \qquad . \qquad . \qquad . \qquad . \qquad . \quad (7.9)$$

The distance l between the lenses is made such that the constant C may be set, and kept, at 100,

i.e.
$$C = 100$$

$$= \frac{ff_1}{i(f+f_1-l)}$$

and
$$l - f + f_1 - \frac{ff_1}{100\,i} \qquad . \qquad . \qquad . \quad (7.10)$$

An inspection of equation (7.10) shows that with reasonable use adjustment should rarely be required, for f and f_1 are constant, and if the stadia lines are engraved on glass, then i is constant. Adjustment would then be required only when the diaphragm was replaced.

EXAMPLE. The focal lengths of object glass and anallatic lens are 127 mm and 114 mm respectively. The stadia interval was 2·5 mm. A field test with vertical staffing yielded the following—

Inst. Station	Staff Station	Staff Intercept	Vertical Angle	Measured Horizontal Distance (m)
P	Q	0·70	+ 7° 24′	68·49
	R	1·86	− 4° 42′	183·58

Find the distance between the object glass and anallatic lens. How far and in what direction must the latter be moved so that the multiplying constant of the instrument is to be 100 exactly. (*L.U., B.Sc.*)

Let the formula for the instrument be

$$D = Cs + K$$

For inclined sights,

$$H = Cs \cos^2 \theta + K \cos \theta$$

$\therefore$ $68 \cdot 49 = 0 \cdot 70 \ C \cos^2 7° \ 24' + K \cos 7° \ 24'$

$183 \cdot 58 = 1 \cdot 86 \ C \cos^2 4° \ 42' + K \cos 4° \ 42'$

Whence $C = 100 \cdot 90$

From equation (7.8),

$$C = \frac{ff_1}{i(f + f_1 - l)} = \frac{14{,}478}{2 \cdot 5(241 - l)}$$

$\therefore$ $l = 183 \cdot 6 \ \text{mm}$

For $C = 100;\ \ 100 = \dfrac{14{,}478}{2 \cdot 5(241 - l)}$

Therefore $l = 183 \cdot 1 \ \text{mm}$

Hence the anallatic lens must be moved 0·5 mm nearer to the object glass.

Note that consideration of the anallatic lens is now of mainly academic interest in that it represented an important step forward in

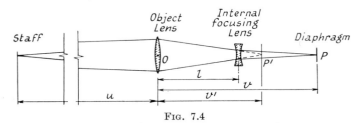

Fig. 7.4

the development of the tacheometer. There can be few external focusing instruments in general use to-day, so that of greater interest is the more detailed consideration of the internal focusing instrument in tacheometry.

Internal Focusing Instruments. It has already been mentioned that the concave internal focusing lens described in Chapter 2 (to which the student is referred for revision on internal focusing telescopes) is not an anallatic lens, since the latter is fixed relative to the object lens, whereas the former is moved to focus the telescope, i.e. l is variable in Fig. 7.4.

The focal length of the object lens is f. The focal length of the internal lens is f_1. Were the internal lens not present, the image from the objective would be formed at P'.

$$\frac{1}{u} + \frac{1}{v'} = \frac{1}{f}$$

$$v' = \frac{fu}{u - f} \qquad . \qquad . \qquad . \ (7.11)$$

Then, treating P' as object giving an image at P, we get

$$\frac{1}{v-l} - \frac{1}{v'-l} = -\frac{1}{f_1}$$

$$\therefore \qquad f_1 = \frac{(v-l)(v'-l)}{v-v'} \qquad . \qquad . \quad (7.12)$$

Substituting equation (7.11) in (7.12),

$$\left(v - \frac{fu}{u-f}\right)f_1 = \left(\frac{fu}{u-f} - l\right)(v-l)$$

$$\therefore \quad (u-f)l^2 - \{v(u-f) + uf\}l - vuf_1 + vuf + uff_1 + vff_1 = 0$$

This cumbersome expression is a quadratic in l and can be solved to give l in terms of f, f_1, v and u, of which the first three are constants. Thus, the tacheometric formula $D = \frac{f}{i} \cdot s + (f + c)$ does not apply directly, and in addition it would appear that no simple solution will enable the instrument to be made exactly anallatic. This might have proved to be a serious drawback to the use of the internal focusing telescope had it not been found* that a straight line formula of the same type could be applied with sufficient accuracy for normal work;

$$D = Cs + K$$

where C and K are constants.

In practice, K might be found to vary somewhat for short sights— e.g. less than 25 m, but unless great accuracy is required no correction is necessary. Most modern instruments are of the internal focusing type, and the makers aim at a low constant value for K. For example, all the standard instruments of Hilger & Watts Ltd. are now fitted with internal focusing telescopes, so designed that the focusing lens, when at infinite focus, is mid-way between the object glass and the diaphragm; it does not therefore upset the balance of the telescope, and the correction necessary when using the telescope for tacheometric work is so small as to be negligible, except for very short distances (which the makers give as up to 15 m). The Zeiss level tested by Major Henrici also had the concave lens mid-way between the object glass and the diaphragm, and had an additive correction of 6 mm at 35 m, rising to 76 mm at 4·5 m distance. These corrections are often ignored and the telescope is to all intents anallatic.

Measurement of Tacheometric Constants. Although these are given by the makers, it is sometimes necessary to measure them on old instruments, or an instrument on which the web or glass diaphragm has been changed, or instruments in which wear and tear may have caused some alteration. The simplest way, both for external- and internal-focusing instruments, is to regard the basic formula as being a linear one of the form

$$D = Cs + K \qquad . \qquad . \qquad . \qquad . \quad (7.13)$$

* Trans. Optical Society, Vol. XXII, 1920–21. Major E. O. Henrici, R.E. *The Use of Telescope with Internal Focusing for Stadia Surveying.*

On a fairly level site chain out a a line 100 to 120 m long, setting pegs at 25 to 30 m intervals (the exact distances being measured). Set up the tacheometer over one end, level on the plate bubbles, sight the telescope along the line and then level accurately using the altitude bubble. Since there are two unknowns, two readings of s, say at approximately 35 m and 100 m, are sufficient to give the required two simultaneous equations, though it is preferable to take readings at each point and solve selected pairs of equations, thus enabling mean values for C and K to be derived. If only two readings are used, they should cover the working range of the instrument so as to give a representative value for K.

With glass diaphragms, the distance between the stadia lines is usually guaranteed by the makers to a high degree of accuracy, but as a check the staff readings of the outer lines and of the centre cross-hair should be read. Then, by subtraction, it can be found whether the two parts which make up the total stadia interval are equal. If they are not —as may be the case where webs are used—then values of C for the two parts can be calculated. This knowledge is useiul in sights where part of the field of vision is obscured. Where the two parts are equal, the constant for each half interval is, of course, twice the value for the whole.

EXAMPLE. The following results were obtained using a Watts microptic theodolite to determine the constants.

Distance (m)	Readings			Interval		
	Upper Stadia	Centre	Lower Stadia	Upper	Lower	Total
30·00	1·433	1·283	1·133	0·150	0·150	0·300
55·00	1·710	1·435	1·160	0·275	0·275	0·550
90·00	2·352	1·902	1·452	0·450	0·450	0·900

$$D = Cs + K$$

$\therefore$
$$30{\cdot}00 = C\,0{\cdot}300 + K$$
$$90{\cdot}00 = C\,0{\cdot}900 + K$$

Therefore $C = 100$ and $K = 0$

Any other combination of equations gives the same result, showing that the telescope is anallatic over this range, to all intents and purposes.

Refraction and Curvature. In carrying out such an investigation, and indeed in carrying out any stadia survey, care must be taken that the reading of the lower hair is not too low on hot days. The problem of negative temperature gradient due to decrease in air temperature from ground level upwards has been discussed in Chapter 3. Differential refraction of the light rays occurs, and this can have a serious effect on the stadia readings, in particular the lower one. The remedy is to keep the lower reading wherever possible above 1·00 m and to restrict sighting lengths to 60–80 m as a maximum. Outside the region of differential refraction both stadia readings are equally affected, and there is no effect on the value of the intercept. With long inclined sights the middle hair will read too high, and the curvature and refraction correction should be applied. However, long sights are not usual in this type of surveying, and in normal civil-engineering work sight lengths are not likely to exceed those just quoted.

Inclined Sights. Although a stadia survey could be carried out with the telescope level, work would be tedious in broken and hilly terrain, and since it is on such ground that the tacheometer comes into its own,

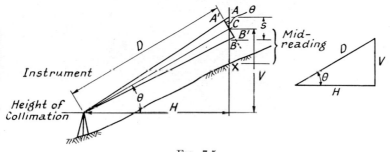

FIG. 7.5

we see that the basic formula $D = Cs + K$ must be modified to cover the general case when the line of sight is inclined to the horizontal. Two cases are to be considered: (1) sights taken on to a vertical staff; and (2) sights taken on to a staff inclined so as to be normal to the line of sight. The former method is more generally used in this country.

(1) STAFF VERTICAL

From Fig. 7.5, where A, C and B are the readings given by the three lines, and A', C' and B' are those which would be given if the staff were normal to the line of collimation,

$$D = C(A'B') + K$$

$$A'B' = AB \cos \theta \text{ (assuming } C\hat{A}'A = C\hat{B}'B = 90°)$$

$$= s \cos \theta$$

$$\therefore \quad D = Cs \cos \theta + K$$

$$\therefore \quad H = D \cos \theta$$

$$= Cs \cos^2 \theta + K \cos \theta \quad . \qquad . \qquad . \qquad . \quad (7.14)$$

$$V = D \sin \theta$$

$$= Cs \cos \theta \sin \theta + K \sin \theta$$

$$= \tfrac{1}{2} Cs \sin 2\theta + K \sin \theta \qquad . \qquad . \qquad . \qquad . \quad (7.15)$$

The importance of the anallatic condition, i.e. $K = 0$, in simplifying the reduction of readings is readily seen, but in most modern instruments where K is very small, if not actually zero, the following approximations are justified:

$$H = (Cs + K) \cos^2 \theta \qquad . \qquad . \qquad . \quad (7.16)$$

and
$$V = (Cs + K) \frac{\sin 2\theta}{2} \qquad . \qquad . \qquad . \quad (7.17)$$

When booking the vertical angle θ, the following convention is used—
Elevation—i.e. sight uphill—θ positive;
depression—i.e. sight downhill—θ negative.

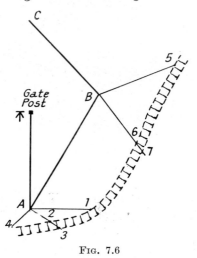

Fig. 7.6

The level at staff station X is given by—
(i) Sight uphill;
Level at X = Height of inst. $+ V -$ mid-reading (CX)
(ii) Sight downhill;
Level at X = Height of inst. $- V -$ mid-reading (CX)

The method of using the above formulae and booking readings is illustrated by the excerpt from the tacheometric field book (Table I).

The points surveyed are plotted schematically in Fig. 7.6. In this case the initial sight on to the bench mark has been taken as the meridian to which the bearings of all the other rays are related, but any other convenient datum, e.g. magnetic north (or south in the U.S.A.) can be used. When the instrument is moved, the technique adopted is similar to the "fast needle" method of traversing, i.e. the last sight with the instrument at A is to the proposed position B. With the plates locked at this bearing, set up at B and sight back on A using the *lower* clamp and tangent screw only, then transit the telescope. If the upper plate is now freed and brought to zero, the telescope will be pointing in a direction parallel to the meridian.

Where a contour plan is required, the points are plotted as shown, and the contour lines interpolated as described in Chapter 2.

Stadia Survey—Peel Park, Salford

Inst. Stn. and Ht. of Inst. Axis	Staff Stn.	Horizontal Angle	Vertical Angle θ	Stadia readings	$D = Cs + k = 100s$ (m)	Mid-reading	Horizontal distance $H = 100s\cos^2\theta$ (m)	Vertical distance $V = \dfrac{100s\sin 2\theta}{2}$ (m)	$\pm V -$ Mid-reading (m)	Ht. of Inst.	R.L. at Staff	Remarks
A 1·57 m	↑	0° 0′ 0″	0° 0′ 0″	$\dfrac{1·700}{0·475}$	122·5	1·085	122·5	—	—	27·38	26·30	B.M. on gatepost
	1	90° 0′ 0″	+ 3° 0′ 0″	$\dfrac{1·780}{1·030}$	75·0	1·405	74·8	3·92	2·51		29·89	Bottom of bank
	2	125° 0′ 0″	+ 6° 0′ 0″	$\dfrac{1·410}{1·135}$	27·5	1·270	27·2	2·86	1·59		28·97	Bottom of bank
	3	125° 0′ 0″	+ 11° 0′ 0″	$\dfrac{1·455}{1·000}$	45·5	1·225	43·8	8·52	7·30		34·68	Top of bank
	4	235° 30′ 0″	+ 5° 0′ 0″	$\dfrac{0·635}{0·260}$	37·5	0·450	37·2	3·26	2·81		30·19	Bottom of bank
B 1·57 m	B	59° 28′ 40″	+ 0° 20′ 0″	$\dfrac{2·425}{1·185}$	124·0	2·055	174·0	1·01	−1·04		26·34	
	A	239° 29′ 0″	− 0° 00′ 0″	$\dfrac{2·965}{1·205}$	176·0	2·085	176·0	—	−2·09	27·91	25·82	
	5	68° 16′ 0″	+ 4° 0′ 0″	$\dfrac{2·250}{1·240}$	100·0	1·745	100·5	6·96	5·24		33·15	Bottom of bank
	6	147° 0′ 0″	0° 0′ 0″	$\dfrac{1·200}{0·625}$	57·5	0·910	57·5	—	—		27·00	Bottom of bank
	7	147° 0′ 0″	+ 8° 40′ 0″	$\dfrac{1·465}{0·700}$	76·5	1·080	74·8	11·40	10·32		38·23	Top of bank
	C	320° 37′ 40″	+ 0° 40′ 0″	$\dfrac{3·195}{1·660}$	153·5	2·425	153·5	1·78	−0·64		27·27	
C 1·49 m	B	140° 37′ 40″	− 0° 40′ 0″	$\dfrac{2·525}{1·000}$	152·5	1·760	152·5	1·77	−3·53	28·76	25·23	

Note that since the last sight before moving the instrument is taken with the staff on the *next* instrument station, when the instrument is set up over this station, it is necessary to measure only the height of the collimation axis above the station to enable the height of the instrument above datum to be found by addition.

The first sight now taken is with the staff on the previous instrument station, and the R.L. of this is deduced, e.g. for $A = 25 \cdot 82$ A.O.D. This must be checked against the R.L. deduced directly, e.g. for

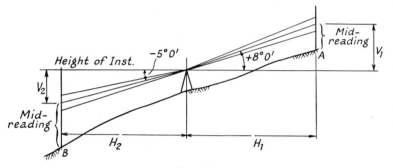

FIG. 7.7

$A = 27 \cdot 38 - 1 \cdot 57 = 25 \cdot 81$ A.O.D., which is sufficiently good agreement. In the case of bigger discrepancies the mean of the two values would be taken.

EXAMPLE. The following readings were taken on a vertical staff with a tacheometer fitted with an anallatic lens and having a constant of 100.

Staff Station	Bearing	Stadia Readings			Vertical Angle
A	27° 30′	1·000	1·515	2·025	+ 8° 00′
B	207° 30′	1·000	2·055	3·110	− 5° 00′

Calculate the relative levels of the ground at A and B, and the mean slope between A and B. (*L.U., B.Sc. (Ext.), Pt. II*, 1949.)

Refer to Fig. 7.7.

Staff Station A

$$\text{Staff intercept} = s = 2 \cdot 025 - 1 \cdot 000$$
$$= 1 \cdot 025 \text{ m}$$
$$\text{mid-reading} = 1 \cdot 515 \text{ m}$$

For this instrument,

$$H = 100 \, s \cos^2 \theta$$
$$V = 100 \, s \, \tfrac{1}{2} \sin 2\theta$$
$$H_1 = 100 \cdot 1 \cdot 025 \cos^2 8°$$
$$= 100 \cdot 5 \text{ m}$$

$$V_1 = 100 \cdot 1{\cdot}025 \, \frac{\sin 16°}{2}$$
$$= 14{\cdot}12 \text{ m}$$

Staff Station B

$$s = 3{\cdot}110 - 1{\cdot}000$$
$$= 2{\cdot}110 \text{ m}$$
$$\text{mid-reading} = 2{\cdot}055 \text{ m}$$
$$H_2 = 100 \cdot 2{\cdot}110 \cos^2 5°$$
$$= 209{\cdot}4 \text{ m}$$
$$V_2 = 100 \cdot 2{\cdot}110 \, \frac{\sin 10°}{2}$$
$$= 18{\cdot}32 \text{ m}$$

Let

$$X - \text{Ht. of inst. above datum.}$$

$$\text{Level at } A = X + 14{\cdot}21 - 1{\cdot}51$$
$$= X + 12{\cdot}61$$
$$\text{Level at } B = X - 18{\cdot}32 - 2{\cdot}06$$
$$= X - 20{\cdot}38$$

∴

$$\text{Fall from } A \text{ to } B = 12{\cdot}61 + 20{\cdot}38$$
$$= 32{\cdot}99 \text{ m}$$

From a consideration of the bearings, it will be seen that A, B, and the instrument lie on a straight line ($207° \, 30' - 27° \, 30' = 180°$), so that the mean slope is given by

$$\frac{\text{Diff. in Level}}{H_1 + H_2} = \frac{32{\cdot}99}{309{\cdot}9}$$
$$= \frac{1}{9{\cdot}39} = 1 \text{ in } 9{\cdot}39$$

Note: In reasonable sighting conditions a modern theodolite will allow estimation of staff readings to 5 mm at distances of 200 m. At 50 m or so estimation to 1 mm should be possible.

(2) STAFF INCLINED NORMAL TO COLLIMATION AXIS

In Fig. 7.8, A, C and B are the readings given by the three lines

$$D = C(AB) + K$$
$$= Cs + K$$
$$H = D \cos \theta + CE$$
$$= D \cos \theta + CX \sin \theta.$$

But

$$\text{mid-reading} = h$$

∴

$$H = (Cs + K) \cos \theta + h \sin \theta \quad . \qquad . \qquad (7.18)$$

Note that for a downhill sight, i.e. negative values of θ, when the staff leans the other way, the term $h \sin \theta$ is subtracted and equation (7.18) becomes

$$H = (Cs + K) \cos \theta - h \sin \theta . \qquad . \qquad . (7.18a)$$

$$V = D \sin \theta$$
$$= (Cs + K) \sin \theta \qquad . \qquad . \qquad . \qquad . (7.19)$$

For a sight uphill, the height of X *above* the collimation axis

$$= XF$$
$$= CF - EX$$
$$= V - h \cos \theta$$

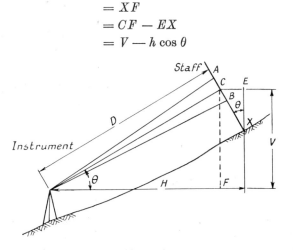

Fig. 7.8

For a sight downhill, the height of X *below* the collimation axis
$$= V + h \cos \theta.$$

Thus the level of ground at the staff station is given by—

(i) Sight uphill = Height of inst. $+ V - h \cos \theta$. $\qquad$. (7.20)

(ii) Sight downhill = Height of inst. $- V - h \cos \theta$. $\qquad$.(7.20a)

In practice, since θ is usually small, i.e. less than 10°, the approximations are often made that

$$h \sin \theta = 0$$
$$h \cos \theta = h$$

To reduce errors from this source, h is kept as small as possible, bearing in mind refraction effects close to the ground. Then equations (7.18) and (7.18a) simplify to

$$H = (Cs + K) \cos \theta$$

and equations (7.20) and (7.20a) to

$$\text{Level at staff} = \text{Height of instrument} \pm V - h$$

General procedure and booking are as for the staff vertical method, the only difference being that the staff man inclines the staff by sighting on to the tacheometer through a telescope, or a special sighting rule fixed to the staff for the purpose. Since the instrument man can also see this sight, he can check that the staff is normal to the line of sight. Also he has the check that if the staff is rocked slightly, the

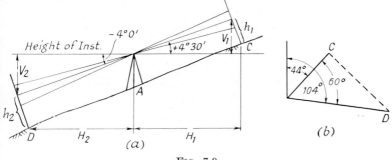

FIG. 7.9

minimum staff interval is given when the staff is truly normal to the line of sight.

EXAMPLE. In a tacheometric survey made with an instrument whose constants are $\frac{f}{i} = 100$ and $(f + d) = 0.5$, the staff was inclined so as to be normal to the line of sight for each reading. How is the correct inclination ensured in the field? Two sets of readings were as given below. Calculate the gradient between the staff stations C and D and the reduced level of each if that at A is 38.22 m.

Inst. Station	Ht. of Inst. Axis	Staff Station	Azimuth	Vertical Angle	Stadia Reading		
A	1·46	C	44°	+ 4° 30′	1·000	1·382	1·765
		D	104°	− 4° 00′	1·000	1·605	2·210

(*L.U.*, *B.Sc.* (*Ext.*) *Pt. II*, 1948)

Refer to Fig. 7.9.

Put
$$\frac{f}{i} = C, \ (f + d) = K$$

$$D = Cs + K$$
$$H = (Cs + K) \cos \theta + h \sin \theta$$
$$V = (Cs + K) \sin \theta$$

Staff Station C

Staff interval $= s_1 = 1.765 - 1.000$
$$= 0.765 \text{ m}$$

mid-reading

$$h_1 = 1 \cdot 382 \text{ m}$$

∴ $$D_1 = 100 \cdot 0 \cdot 765 + 0 \cdot 50$$
$$= 77 \cdot 00 \text{ m}$$

∴ $$H_1 = 77 \cdot 00 \cos 4° 30' + 1 \cdot 382 \sin 4° 30'$$
$$= 76 \cdot 87 \text{ m}$$

$$V_1 = 76 \cdot 87 \sin 4° 30'$$
$$= 6 \cdot 04 \text{ m}$$

Therefore

Ground level at $C = 38 \cdot 22 + 1 \cdot 46 + 6 \cdot 04 - 1 \cdot 382 \cos 4° 30'$
$$= 44 \cdot 34 \text{ m above datum.}$$

Staff Station D

$$s_2 = 2 \cdot 210 - 1 \cdot 000$$
$$= 1 \cdot 210 \text{ m}$$

$$h_2 = 1 \cdot 605 \text{ m}$$

∴ $$D_2 = 100 \cdot 1 \cdot 210 + 0 \cdot 50$$
$$= 121 \cdot 50 \text{ m}$$

∴ $$H_2 = 121 \cdot 50 \cos 4° 00' - 1 \cdot 605 \sin 4° 00'$$
$$= 121 \cdot 1 \text{ m}$$

$$V_2 = 121 \cdot 10 \sin 4° 00'$$
$$= 8 \cdot 45 \text{ m}$$

Therefore

Ground level at $D = 38 \cdot 22 + 1 \cdot 46 - 8 \cdot 45 - 1 \cdot 605 \cos 4° 00'$
$$= 29 \cdot 63 \text{ m above datum.}$$

Using the Cosine Formula,

$$CD^2 = 76 \cdot 87^2 + 121 \cdot 1^2 - 2 \cdot 76 \cdot 87 \cdot 121 \cdot 1 \cos 60°$$
$$= 5,909 \cdot 0 + 14,665 \cdot 2 - 9,309 \cdot 0$$
$$= 11,265 \cdot 3$$

∴ $$CD = 106 \cdot 1 \text{ m}$$

Therefore the gradient between C and D

$$= \frac{44 \cdot 34 - 29 \cdot 63}{106 \cdot 1}$$

$$= \frac{14 \cdot 71}{106 \cdot 1}$$

$$= 1 \text{ in } 7 \cdot 21$$

Comparison of the Two Methods of Staff Holding. It has already been remarked that the vertical staff method is normally favoured, the reason being that, provided the staff is fitted with a small spirit level, it is generally easier to train a staff man to use this than to use a telescopic sighting device. On the other hand, the surveyor has to trust the staff man to hold the staff truly vertical, whereas with the other method, by having the staff swung as in ordinary levelling, a check is theoretically available; it is not so easy to read a minimum intercept, however, as it is to note a minimum reading.

The reduction of the vertical staff formulae is somewhat easier than the formulae for normal holding, if the $L \sin \theta$ and $h \cos \theta$ terms are included in the latter, but otherwise there is little difference.

Errors due to tilt on the staff are more serious in vertical holding, and the following investigation into the relative effect of tilt in the two methods shows quite clearly the importance of using a spirit level to ensure true verticality when this method is used.

Effect of Tilt on the Staff in Stadia Tacheometry

1. STAFF VERTICAL. There are two cases to consider—

(*a*) tilt increasing the intercept so giving a positive error, i.e. telescope level, or pointing uphill—staff tilted away from the telescope (Fig. 7.10*a*); telescope level, or pointing downhill—staff tilted towards the telescope :

(*b*) tilt decreasing the intercept so giving a negative error, i.e. telescope pointing uphill—staff tilted towards the telescope (Fig. 7.10*b*); telescope pointing downhill—staff tilted away from the telescope.

The errors involved in these cases can be evaluated simply but within tacheometric accuracy if the following approximations are made—

(i) That the vertical angle θ is not greater than $\pm 30°$, and is always greater than δ, the angle of tilt.
(ii) That δ is small, i.e. $\not> 3°$.
(iii) That the telescope is anallatic.

Making the usual assumption that

$$C\hat{A}'A = R\hat{P}'P$$
$$\simeq 90°$$

then
$$A'B' = AB \cos \theta$$
$$= s \cos \theta$$
$$P'Q' = PQ \cos (\theta + \delta)$$
$$= s_1 \cos (\theta + \delta)$$

Now $P'Q'$ approximately equals $A'B'$, since they are very near, so that

$$s \cos \theta \simeq s_1 \cos (\theta + \delta)$$

$$\therefore \qquad s_1 = \frac{s \cos \theta}{\cos (\theta + \delta)} \qquad . \qquad . \qquad . \quad (7.21)$$

Similarly from Fig. 7.10*b*,

$$s_i = \frac{s \cos \theta}{\cos (\theta - \delta)} \qquad . \qquad . \qquad . \qquad . \qquad (7.22)$$

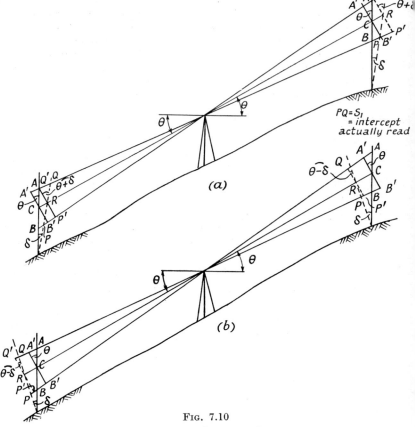

$PQ = S_1$
= intercept
actually read

(a)

(b)

Fig. 7.10

The errors are then best expressed as a ratio,

$$H_{true} = Cs \cos^2 \theta$$
$$H_{apparent} = Cs_1 \cos^2 \theta$$

$$\therefore \quad \text{Ratio of error } E = \frac{H_{app.} - H_{true}}{H_{app}} = \frac{Cs_1 \cos^2 \theta - Cs \cos^2 \theta}{Cs_1 \cos^2 \theta}$$

$$= \frac{s_1 - s}{s_1}$$

$$= 1 - \frac{\cos (\theta \pm \delta)}{\cos \theta} \qquad . \qquad . \qquad . \qquad . \qquad (7.23)$$

E has been evaluated by Professor F. A. Redmond* for values of θ ranging from $3°$ to $30°$, and three values of δ, i.e. $10'$, $1°$ and $2°$, corresponding to a 4 metre staff out of plumb by 0·01 m, 0·07 m and 0·14 m respectively, and the following table gives his results. E_1 represents

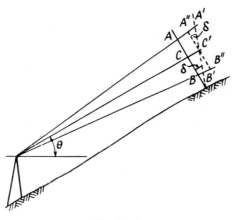

FIG. 7.11

relative error for the condition shown in Fig. 7.10a, i.e. positive errors, and E_2 for the condition shown in Fig. 7.10b, i.e. negative errors.

TABLE II

θ	E_1			E_2		
	$\delta = 10'$	$\delta = 1°$	$\delta = 2°$	$\delta = 10'$	$\delta = 1°$	$\delta = 2°$
$3°$	1/6,240	1/930	1/410	1/6,630	1/1,320	1/820
$5°$	1/3,840	1/595	1/270	1/4,000	1/730	1/410
$10°$	1/1,920	1/310	1/150	1/1,960	1/340	1/180
$15°$	1/1,280	1/205	1/100	1/1,300	1/220	1/115
$20°$	1/940	1/155	1/75	1/950	1/160	1/85
$25°$	1/735	1/120	1/60	1/740	1/125	1/65
$30°$	1/590	1/98	1/47	1/600	1/100	1/52

2. STAFF NORMAL. There is only one case to consider here; when the staff is truly normal the intercept AB is a minimum, and tilt of the staff away from this position must give an increased value, i.e a positive error.

In Fig. 7.11, δ is very small so that CC can be ignored. Also,

$$AB \simeq A''B''$$
$$= s$$

* *Tacheometry for Students and Surveyors*, Technical Press, London.

But $$A''B'' = A'B' \cos \delta$$
$$= s_1 \cos \delta$$

where s_1 is the intercept actually read.

∴ $$s = s_1 \cos \delta$$

∴ Ratio of error $E = \dfrac{Cs_1 - Cs}{Cs_1}$

$$= (1 - \cos \delta) \qquad . \qquad . \quad (7.24)$$

The error is thus independent of θ.

TABLE III

δ	E
1°	1/6,500
2°	1/1,640
3°	1/730

Comparing Tables II and III, it is clear that for small values of δ, say 1°, which does not after all represent much tilt—the staff normal method is preferable if θ is much more than 10°.

EXAMPLE. In a tacheometric survey an intercept of 0·755 m was recorded on a staff which was believed to be vertical, and the vertical angle measured on the theodolite was 15°. Actually the staff which was 4 metres long was 0·14 m out of plumb and leaning backwards away from the instrument position. Assuming it was an anallatic instrument with a multiplying constant of 100, what would have been the error in the computed horizontal distance?

In what conditions will the effect of not holding the staff vertical but at the same time assuming it to be vertical be most serious? What alternative procedure can be adopted in such conditions? (*Inst. C.E., Final Pt. II*, 1952.)

$$\text{Angle of tilt } \delta = \tan^{-1} \frac{0 \cdot 14}{4}$$

$$= 2°, \text{ say}$$

$$\text{Angle of elevation } \theta = 15°$$

By equation (7.23), ratio of error caused by tilt

$$= 1 - \frac{\cos (\theta + \delta)}{\cos \theta}$$

$$= 1 - \frac{\cos 17°}{\cos 15°}$$

$$= 1 - 0 \cdot 9902$$

$$\simeq 0 \cdot 01 \text{ or } 1 \text{ part in } 100.$$

Horizontal distance would have been computed as
$$H = 100 \cdot 0\cdot755 \cos^2 15°$$
$$= 70\cdot44 \text{ m}$$

Therefore error would have been $+ \dfrac{70\cdot44}{100}$, i.e. $+ 0\cdot70$ m, giving a true distance of 69·74 m.

The student may deduce the answer to the second part of the question from Tables II and III.

(B) Movable-hair or Subtense Tacheometers

Formula (7.3) still applies to this type of instrument,

$$D = \frac{f}{i} s + (f + c),$$

but i is now variable, and is measured by a micrometer. The staff intercept may also vary, but in practice this type of instrument is invariably used with a staff of known length with clearly marked targets at each end, on which to sight the movable hairs.

If p = pitch of the screw on the micrometer, and x = number of turns and part of a turn (measured on the micrometer drum) to bring hairs on to the targets, then $i = px$

Therefore
$$D = \frac{fs}{p} \cdot \frac{1}{x} + (f + c)$$

$$= C_1 \cdot \frac{1}{x} + K_1 \qquad . \qquad . \qquad . \qquad . \quad (7.25)$$

The inclined sight formulae are derived as before.

Methods to Simplify Reduction of Readings in Stadia Surveying, and Auto-reduction. Although the field work in stadia surveying is rapid, the office work involved in reducing the readings can be tedious, and many methods have been evolved which attempt to simplify reduction. They fall into two broad groups: those which involve special methods of reading or reduction, and those which involve modifications to the tacheometer.

Of the former group, the even-angle method of reading is typical. Only certain values of vertical angle are used; a preliminary sight is made on to the target, the most convenient even value of θ then being re-set, and the mid-reading noted. The stadia readings may also be noted, but it is usual in this method to bring the lower cross-hair to the nearest metre mark by means of the tangent screw, so simplifying the subtraction. Reduction tables have been prepared for use with this method, and since only a relatively few values of θ are used (the lowest subdivision used is 20'), the tables are not bulky or cumbersome to use. As well as these tables, special charts, diagrams, and slide rules are available.*

* *Tacheometry*, by Prof. F. A. Redmond (Technical Press).

There are many methods of reduction involving modified tacheo-meters and specially made instruments, and we will note only three of them here. A. L. Higgins,† himself an authority on self-reducing instruments, considers that there is no great future for automatic reduction, and that the Beaman arc will meet most requirements.

This was written before the Wild, Ewing and other reduction devices were on the market, however, and the authors consider that these two modern instruments, together with the Beaman arc, are worthy of consideration where large amounts of tacheometric surveying are to be undertaken. All three (*a*) are simple in construction, (*b*) do not add much to the weight of the tacheometer, and (*c*) do not reduce the efficiency of the instrument when it is used as a straightforward theodolite. The student should note also the new auto-reduction instruments under *Optical Wedge System*.

Beaman's Stadia Arc. These are two scales, known as the *H*-scale and the *V*-scale, mounted concentric with the vertical circle. On the

Fɪɢ. 7.12

V-scale are engraved values for 100 sin θ cos θ of 1, 2, 3, 4, etc., repre-senting values of sin θ cos θ of 0·01, 0·02, etc. The index mark is set to the zero which is often the 50 graduation, when the index on the vernier will be at 0°. For inclined sights the *V*-scale index is set against a convenient graduation on the scale, and *V* can be then quickly com-puted by multiplying the readings by the observed intercept *s*. If the corresponding value of θ is greater than $\pm$ 3°, a correction to *H* is re-quired (i.e. for $\theta < 3°$, $H \simeq D$) and this is given by the *H*-scale which is graduated to read 100 sin² θ, i.e. the appropriate reading on the *H*-scale multiplied by *s* gives the correction to the short range *D*, since

$$D = 100\ s \text{ (instrument anallatic)}$$
$$H = 100\ s \cos^2 \theta$$
$$D - H = 100\ s\ (1 - \cos^2 \theta)$$
$$= 100\ s \sin^2 \theta$$

As an example, consider the readings (Fig. 7.12) shown in the viewing eyepiece of a microptic telescopic alidade, to which reference ha

† *Higher Surveying* (Macmillan).

already been made in Chapter 6 (*see* Plane Table equipment). In this field, i.e. plane table surveying at scales of 1:500 to 1:10,000, the Beaman arc is especially useful as it enables a large number of readings to be made which can be quickly reduced.

$$\text{Observed stadia reading} = 1 \cdot 58 \text{ m}$$
$$\text{Staff reading at mid-hair} = 2 \cdot 55 \text{ m}$$
$$V\text{-scale reading} = 22$$
$$\therefore \quad \text{Vertical component} = 1 \cdot 58 \times 22$$
$$= 34 \cdot 76 \text{ m}$$

$\therefore$ Change in level relative to ht. of collimation

$$= 34 \cdot 76 - 2 \cdot 55$$
$$= 32 \cdot 21 \text{ m, say } 32 \cdot 2 \text{ m}$$

$$H\text{-scale reading} = 5$$

$\therefore$ True horizontal distance $= 1 \cdot 58 \times 100 - (1 \cdot 58 \times 5) = 150 \cdot 1$ m

The Ewing Stadi-altimeter. This device, the invention of Mr. Alistair Ewing, an Australian surveyor, converts a normal theodolite into a direct reading tacheometer giving distances and levels directly without vertical circle readings. Construction is in two parts, the cylindrical scale unit, which is mounted on one of the theodolite uprights, and the optical reader, mounted on the telescope axis above the trunnion Fig. 7.13). The index of the reader is a bright pin point of light which appears superimposed on the scale of the drum, on which there are two sets of curves. These represent the reduction equations

$$V = 100 \, s \cos \theta \sin \theta$$
$$D - H = 100 \, s \, (1 - \cos^2 \theta)$$

As the telescope is tilted the spot of light moves on the surface of the drum. When a reading of s has been taken the drum is rotated until the light spot is in coincidence with $100 \, s$ interpolated on the curves printed in black (in terms of $10 \, s$ for convenience). The correction $D - H$) is now read directly from the second set of curves, printed in red; V is then read on the circular curve at the eyepiece end. Fig. 7.14 shows a view of the cylindrical scale viewed through the optical reader. The instrument is made by Hilger & Watts, Ltd.

Wild Reduction Tacheometer R.D.S. In this instrument, the stadia lines are engraved as flat curves on an auxiliary glass circle, the image of the curves being seen in the field of vision in the plane of the diaphragm. One of the curves forms a zero line and is set against any convenient whole reading on the staff (an ordinary levelling staff or tacheometry staff being used). The intercept between this curve and the upper curve, multiplied by 100, gives the *horizontal* distance from the staff to the instrument; the curves having been so engraved and the circle so connected to the telescope trunnion axis that the effect of tilt on the telescope is automatically allowed for (i.e. between the working limits of elevation and depression, the stadia interval is infinitely variable).

Similarly, the intercept between the zero and middle or elevation curves, multiplied by the constant which is seen in the field of vision,

FIG. 7.13.　No. 1 MICROPTIC THEODOLITE WITH EWING STADI-ALTIMETER
(*Courtesy of Hilger & Watts, Ltd.*)

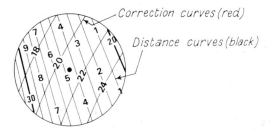

FIG. 7.14.　TYPICAL VIEW OF CYLINDRICAL SCALE
THROUGH OPTICAL READER

gives the *difference in elevation* between the trunnion axis and the zer line intersection on the staff.　Fig. 7.15 shows two typical readings one with a sight uphill, the other downhill.　Estimation of staff reading to 1 mm is stated by the manufacturer to give distance accuracies c the order of 0·1 m up to 100 m.

Further information on accuracies attainable in the field by certain tacheometers and reduction devices may be obtained from a paper by Bannister and Schofield* in the *Surveyor*; an indication of general performances is also given therein.

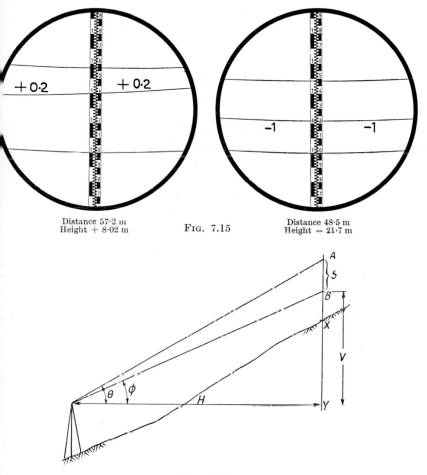

Distance 57·2 m
Height + 8·02 m

Fig. 7.15

Distance 48·5 m
Height — 21·7 m

Fig. 7.16

The Tangential System

In this system, in which the angle subtended by a known length is measured directly, two pointings of the instrument are required on to *A* and *B* respectively on a staff held vertically as in Fig. 7.16.

* "Reduction Tacheometers," Bannister and Schofield, *Surveyor*, 23 Dec. 1967.

$$AY = H \tan \theta$$
$$BY = H \tan \phi$$
$$AY - BY = s = H (\tan \theta - \tan \phi)$$

$$\therefore \qquad H = \frac{s}{\tan \theta - \tan \phi} \qquad . \qquad . \qquad . \quad (7.26)$$

for sights downhill

$$H = \frac{s}{\tan \phi - \tan \theta} \qquad . \qquad . \quad (7.26a)$$

$$V = H \tan \phi$$

Therefore level of ground at X = Height of instrument $\pm V - BX$, according to whether sight is uphill or downhill.

Staff tilt must be guarded against, as in stadia tacheometry, such tilt introducing serious errors when steeply-inclined sights are used.

It is possible to use those values of θ and ϕ whose tangents differ by 0·01, so that $\qquad H = 100 \ s$

where s is the intercept corresponding to values of θ and ϕ,

e.g. setting first $\qquad \theta = 4° \ 0' \ 15'' \ \tan \theta = 0·07$

then $\qquad \qquad \phi = 3° \ 26'1'' \ \tan \phi = 0·06$

It is doubtful however whether these angles can be set off with enough accuracy, bearing in mind that tacheometry is supposed to be rapid, and the method is not much used.

EXAMPLE. (1) The following readings were taken on a vertically held staff. Calculate the horizontal distance between the theodolite and the staff, and the elevation at the staff station, if the height of the instrument axis is 37·36 m above datum.

Vertical Angle	Staff Reading
+ 4° 13′ 30″	1·00
+ 5° 58′ 20″	3·00

(2) If the probable error in angle measurement is ± 5″, what is the probable error in horizontal distance, assuming that the staff is held vertical and that reading errors are negligible?

(1) $\qquad \theta = 5° \ 58' \ 20'' \quad \tan \theta = 0·10461$

$\qquad \qquad \phi = 4° \ 13' \ 30'' \quad \tan \phi = 0·07385$

$\therefore \qquad \qquad \tan \theta - \tan \phi = 0·03076$

$\qquad \qquad$ Staff intercept $s = 2·00$ m

$$\therefore \qquad H = \frac{s}{\tan \theta - \tan \phi} = \frac{2·00}{0·03076}$$

$$= 65·02 \text{ m}$$

$$V = H \tan \phi = 65·02 \times 0·07385$$

$$= 4·80$$

$\therefore \qquad$ R.L. at staff station $= 37·36 + 4·80 - 1·00$

$$= 41·16 \text{ m above datum}$$

(2)
$$H = \frac{s}{\tan \theta - \tan \phi} \simeq \frac{s}{\theta - \phi} \text{ in radians}$$

If $d\theta$ and $d\phi$ are the probable errors in angular measurement and dH the probable error in H

Then
$$dH = \sqrt{\left(\frac{\partial H}{\partial \theta}\right)^2 \cdot (d\theta)^2 + \left(\frac{\partial H}{\partial \phi}\right)^2 \cdot (d\phi)^2},$$

and, since $d\theta = d\phi$,

$$= \sqrt{2\left(\frac{\partial H}{\partial \theta}\right)^2 \cdot (d\theta)^2} = \sqrt{2} \cdot \frac{s}{(\theta - \phi)^2} \cdot d\theta$$

$$= \sqrt{2\frac{H^2}{s}} \, d\theta = \sqrt{2} \cdot \frac{65 \cdot 02^2}{2} \cdot \frac{5}{206,265} \text{ (note } 1'' = \frac{1}{206,265} \text{ radian)}$$

$$= \pm 0 \cdot 08 \text{ m}$$

FIG. 7.17. SUBTENSE BAR
(*Courtesy of Wild Heerbrugg, Switzerland*)

The Subtense System

A theodolite reading to one second of arc is required, and this is used to measure the angle subtended by a horizontal *subtense* bar, of accurately known length, stationed at the target. Theodolites capable of this accuracy were described in Chapter 4; several patterns of subtense bar are available, including the Watts 2-metre bar and the Wild 2-metre bar. The last-named two are of invar with their lengths guaranteed to the order of 0·0001 m. Fig. 7·17 shows the Wild subtense bar with levelling head and tripod.

The steel casing is hinged at the middle and contains invar wires anchored there and tensioned by springs at the target ends. The target holders themselves are of brass.

The horizontal angle subtended at A by the targets on the bar at B (the bar being set normal to AB, using the directing telescope) is measured. The principle of the subtense system is shown in Fig. 7.18; it will be appreciated that since the subtense bar is mounted horizontally : (1) refraction has an equal effect on both readings, and (2) the angle subtended (α) is measured in the horizontal plane so that the horizontal distance H obtains directly from

$$H = \frac{b}{2 \tan \dfrac{\alpha}{2}} \text{ where } b = \text{length of bar} \quad . \quad . \quad (7.27)$$

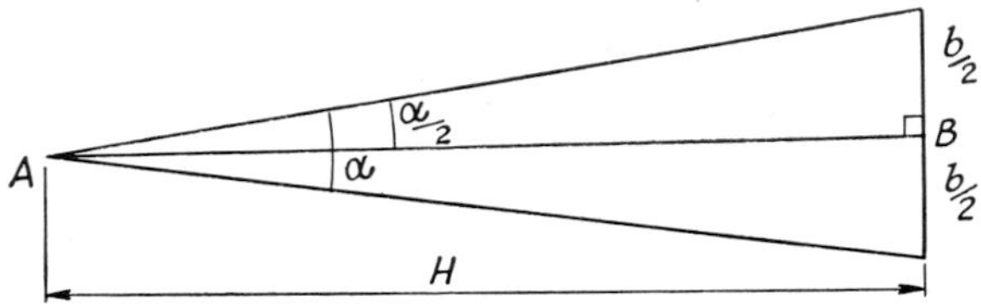

Fig. 7.18

Thus, if the vertical circle angle on the theodolite is θ, the difference in level is given by—

$$V = H \tan \theta \quad . \quad . \quad . \quad . \quad (7.28)$$

Accuracy of Subtense Tacheometry. The accuracy with which the horizontal distance H may be computed depends on three factors, (a) the accuracy of the angle measurement, (b) the accuracy of the length of the subtense bar, and (c) the accuracy of the right angle at B.

(a) *Influence of angle measurement.* This is the most important factor.

$$H = \frac{b}{2 \tan \dfrac{\alpha}{2}} \simeq \frac{b}{\alpha} \text{ for small values of } \alpha$$

$$\therefore \qquad dH = -\frac{b}{\alpha^2} . \, d\alpha = -\frac{H^2}{b} . \, d\alpha . \quad . \quad . \quad (7.29)$$

the negative sign indicating that if α is measured larger than it should be, the distance H is reduced. Thus if the error in angle measurement is $\pm 1''$, and the bar length is 2 m,

$$dH = \frac{\pm H^2}{2 \cdot 000} . \frac{1}{206,265} \text{ m} \quad . \quad . \quad (7.29a)$$

since $1'' = 1/206,265$ radian. The error in distance is proportional to the square of the distance, and so the best results are obtained at short distances, with a limiting minimum distance of 40 m say. At shorter distances than this, the error in angle measurement increases.

H (m)	40	50	75	100	125	150
lH (mm)	3·9	6·3	14·2	25·2	39·4	56·8
Fractional Error	1/10,200	1/7,940	1/5,290	1/3,970	1/3,170	1/2,640

At 40 m the limiting error from this source is about 1/10,000, and by keeping the distance down and by repetition measurement of α, it is possible to improve the relative accuracy to better than these figures, so long as the influences of (*b*) and (*c*) are small.

(*b*) Influence of length of subtense bar. The invar 2-metre bars are guaranteed to ± 0·05 mm, and since a temperature change of 20°C

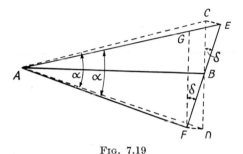

Fɪɢ. 7.19

will affect the bar length only up to these limits, the length can be considered constant.*

(*c*) Influence of orientation of the bar. In Fig. 7.19 the bar is set in the position EF instead of CD so as to give an error in the right angle at B of δ, then, ignoring the slight effect of such an error on α, i.e. taking $C\hat{A}D = E\hat{A}F$ and assuming $F\hat{G}E = 90°$, since FD and CE are small, the effective length of the bar is $GF = b \cos \delta$.

$$H = \frac{b}{\alpha}$$

$$H_{app.} = \frac{b}{\alpha} \cos \delta$$

$$\therefore \quad H - H_{app.} = dH = \frac{b}{\alpha}(1 - \cos \delta) = \frac{2b}{\alpha} \sin^2 \frac{\delta}{2}$$

$$= 2H \sin^2 \frac{\delta}{2}$$

Thus, for

$$\frac{dH}{H} \not> \frac{1}{20,000}$$

$$\delta \not> 36'$$

* *Optical Distance Measuring*, by E. Berchtold. (Wild Co., Heerbrugg.)

The directing telescope on the bar allows orientation to much finer limits than this, so that this error can also be ignored. Thus it is necessary to consider the accuracy of the angle measurement only.

Improvement of the Accuracy. For a given length the accuracy of distance measurement may be improved by repeating the angle measurement by the process already described (*see* Chapter 4), thus reducing the probable error in the angle measurement. For long lines, the length may be subdivided into smaller sections or measured by use of an auxiliary base, either of which will give improvement in accuracy;

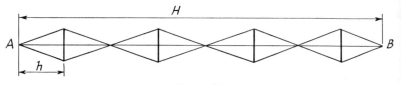

Fig. 7.20

this may be still further improved, of course, by repetition reading of the angles.

(*a*) Repetition of angle measurement. The following readings were obtained when sighting on a 2-metre subtense bar.

Measured angle	Residual error d (= measured − mean angle)	d^2
02° 04′ 46·5″	+ 0·1	0·01
48·1	+ 1·7	2·89
44·5	− 1·9	3·61
44·0	− 2·4	5·76
49·1	+ 2·7	7·29
46·2	− 0·2	0·04
45·5	− 0·9	0·81
47·6	+ 1·2	1·44
Mean 02° 04′ 46·4″		$\Sigma d^2 = 21\cdot85$

$$H = b/2 \tan 01° 02′ 23\cdot2″$$
$$= 55\cdot10 \text{ m.}$$

Bird* has suggested a minimum number of eight angular observations when using the subtense bar. It is also good practice to advance the horizontal circle after each individual angle has been observed: in this case a $22\frac{1}{2}°$ increase was given to enhance the coverage of the horizontal circle.

* "The Accuracy of Subtense Bar Measurements in Relation to the Number of Observations of the Subtended Angle," R. G. Bird, *Survey Review* No. 148.

The standard or mean square error of the mean (s) [Refer to Chapter 13].

$$= \sqrt{\frac{\Sigma d^2}{n\,(n-1)}} \quad \text{where } n \text{ is the number of observations}$$

$$= \sqrt{\frac{21 \cdot 85}{56}}$$

$$= \pm\, 0 \cdot 6''$$

This gives a measure of the improvement in accuracy. If the probable error of the mean is required, this is given by

$$0 \cdot 67 \sqrt{\frac{\Sigma d^2}{n\,(n-1)}} = \pm\, 0 \cdot 4''$$

(b) Subdivisions of long lengths. Referring to Fig. 7.20, if AB is measured with the theodolite at one end and the subtense bar at the other, the mean error in horizontal distance will be, from (7.29a)

$$m_H = \pm\, \frac{H^2}{412,530}\, m_\alpha = \pm\, \frac{H^2}{K_1}\, m_\alpha, \text{ say.}$$

where m_α is the mean error in angle measurement.

If H is divided into n sections of length h where $H = n \cdot h$

$$m_h = \pm\, \frac{h^2}{K_1}\, m_\alpha$$

The mean error on the entire length of n bays is then

$$= \pm\, \sqrt{n m_h^2}$$

i.e.

$$m_{nh} = \pm\, \sqrt{n}\, \frac{h^2}{K_1}\, m_\alpha$$

$$m_H = \pm\, \frac{H^2}{K_1}\, m_\alpha = \pm\, \frac{n^2 h^2}{K_1}\, m_\alpha$$

$$\therefore \qquad m_{nh} = \frac{m_H}{\sqrt{n^3}} \qquad . \qquad . \qquad . \qquad . \qquad . \qquad . \qquad (7.30)$$

EXAMPLE (i) What do you understand by systematic and accidental errors in linear measurement, and how do they affect the assessment of the probable error? Does the error in the measurement of a particular distance vary in proportion to the distance or to the square root of the distance?

(ii) Assume you have a subtense bar the length of which is known to be exactly 2 m and a theodolite with which horizontal angles can be measured to within a second of arc. In measuring a length of 600 metres, what error in distance would you get from an angular error of 1 sec?

(iii) With the same equipment, how would you measure the distance of 600 metres in order to achieve an accuracy of about 1/5,000?

(Aide-memoire: 1 second of arc = 1/206,265 radians.) (*I.C.E.*)

(i) This part of the question is dealt with in the earlier chapters.

(ii)

$$m_{600} = \pm\, \frac{600^2}{412,530} \cdot 1''$$

$$= \pm\, 0 \cdot 87 \text{ metre}$$

This corresponds to a fractional error of 1/690.

(iii) For a fractional error of say 1/5,000

$$m_{nh} = 600/5,000 = 0.87/\sqrt{n^3}$$

∴ $$n^{3/2} = 7.25$$

$$n = 4, \text{ say.}$$

∴ $$m_{4.150} = \pm 0.87/8 = \pm 0.11 \text{ metre}$$

i.e.   Fractional error = 1/5,520

(c) *Long lines using auxiliary base.* In the above example, a short *auxiliary base AB* could have been measured at right angles to the main

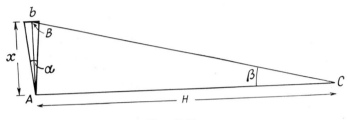

Fig. 7.21

line with the subtense bar at B, and, by measuring the angle β at C, the length $AC = H$ computed (Fig. 7.21).

The length $AB(= x)$ should be about $\sqrt{2H}$, so that $\alpha = \beta$ or thereabouts; this gives the most favourable conditions for extending to AC, which may be up to 1,000 metres long.

Errors are introduced in both the angle measurements, and if m_α and m_β are the mean errors in angle measurement, then

for small angles $H = x/\beta$

$$\text{Error in } H = \pm \sqrt{\left(\frac{\partial H}{\partial x}\right)^2 (dx)^2 + \left(\frac{\partial H}{\partial \beta}\right)^2 (d\beta)^2}$$

$$= \pm \sqrt{\frac{1}{\beta^2}(dx)^2 + \left(\frac{H^2}{x}\right)^2 (d\beta)^2}$$

Due to errors m_α and m_β in angular measurement

$$dx = \pm \frac{x^2}{bK} m_\alpha$$

$$d\beta = \pm \frac{m_3}{K}$$

$$\text{Error in } H = \pm \sqrt{\frac{H^2}{x^2} \frac{x^4}{b^2 K^2} m_\alpha{}^2 + \frac{H^4}{x^2 K^2} m_\beta{}^2}$$

$$= \pm \frac{H}{K} \sqrt{\frac{x^2}{b^2} m_\alpha{}^2 + \frac{H^2}{x^2} m_\beta{}^2}$$

Very long lines can be measured as in Fig. 7.20 but replacing the subtense bar by a series of auxiliary bases.

EXAMPLE. If $m_\alpha = m_\beta = \pm 1''$, $b = 2$ metres, and $H = 600$ m in the above expression, compute the fractional error in the measurement of H.

For $\alpha = \beta$, $x = \sqrt{2H} = 35$ metres, say.

Then
$$m_H = \pm \frac{600}{206,265} \sqrt{\left(\frac{35^2}{2^2} + \frac{600^2}{35^2}\right)}$$

$$= \pm 0\cdot071 \text{ metre}$$

The fractional error is thus 1/8,450.

Uses of Subtense Tacheometry.

It has been shown that if an accurate subtense bar is used in conjunction with a theodolite capable of reading to 1 sec of arc, distance measurement is possible whose accuracy is considerably higher than that of all but the best direct lineal measurement. Also this accuracy is not affected by the nature of the ground, which may indeed be so bad as to preclude any direct measurement with the steel band.

The method can be used to: (1) measure legs of traverses inside triangulation points; (2) measure base lines for small independent triangulations (*see* Chapter 9); (3) fix control points in aerial photogrammetry.

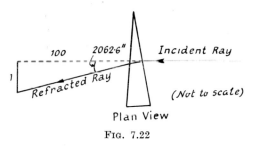

Plan View

FIG. 7.22

Optical Wedge System

The methods discussed so far have involved the formation of an angle known as the parallactic angle. In the case of the tacheometer this angle is given by stadia readings on an external base (the staff). It is constant, being equal to $\tan^{-1} i/f$, and for convenience in reduction is formed at the trunnion axis. In the case of the subtense bar a variable angle is now obtained at the instrument, since the bar forming the external base is of fixed length.

Most of the important advances in tacheometry of resent years take up a third alternative which involves the displacement of part of the field of view relative to the other part. This again can involve the use of constant or variable parallactic angles, depending upon the device itself. The simplest of these is the *Richards' wedge*, a glass wedge which can be inserted in front of the telescope. Thus, if the cross-hairs are sighted on to the zero of a horizontal staff (a vertical staff can be used, but is less convenient), and the wedge then placed in position, the cross-hairs will be shifted giving a different reading, the amount of the shift depending on the deflection factor of the glass wedge. If a wedge with a total deflection angle between incident and refracted (emergent) ray of 34′ 22·6″ (Fig. 7.22), i.e. the angle whose tangent is exactly 1/100, be used, then the amount of the shift multiplied by a hundred gives the distance D between the telescope lens and the staff.

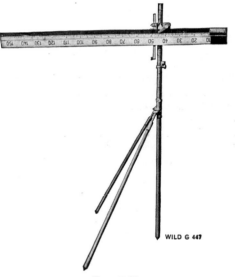

Fig. 7.23

The Telemeter. If the wedge covers only half the lens, only half the image will be subject to this shift. Thus if the main scale is viewed directly, while a vernier, whose index is coincident with the zero of the main scale, is viewed through the wedge, only the image of the vernier is shifted, and the amount of the shift can thus be read with vernier accuracy. Further accuracy can be obtained by the use of a parallel plate micrometer. Figs. 7.23, 7.24 and 7.25 show the staff, the Wild D.M.1 precision telemeter and a typical view of a staff as seen through the telescope with the micrometer reading 0·08. The distance of the staff from the instrument is 61·58 m.

The precision of this attachment (which is fitted in a few seconds to the theodolite) is given as 1–2 cm per 100 m, or 5 to 10 times better than stadia tacheometry.

It can be mounted in front of the object glass of either the T16 or T2 instruments, a counterweight being fitted at the eyepiece end. The given working range is up to 100 m and the slope distance is obtained on inclined sights. Watts manufacture a similar system.

The Wild R.D.H. Reduction Tacheometer. An extension of the wedge tacheometry principle to give automatic reduction was patented by the Swiss surveyor R. Bosshardt in 1923, and various instruments have been manufactured using his idea, e.g. Wild R.D.H., Kern DK-RT and Zeiss Redta 002. The first-named not only gives horizontal distance directly in lieu of the slant distance, but also can give the corresponding vertical intercept. For this reason, although it is now obsolescent, its construction is still discussed herein. The other instruments utilize a tangent scale reading or vertical circle reading to deduce the vertical intercept value. The single prism shown in Fig. 7.22 gives a maximum deflection in the horizontal plane when it is mounted as shown, with its parallel triangular end faces parallel to the horizontal plane through the collimation axis of the telescope. If the prism shown in

FIG. 7.24. WILD D.M.1
PRECISION TELEMETER
(*Courtesy of Wild Heerbrugg, Switzerland*)

Fig. 7.26 were rotated in the vertical plane, the effective wedge angle (δ) contained in the horizontal plane and producing horizontal deflection in that plane is given by the expression—

$$\tan \delta = \tan \gamma \cos \beta = \text{constant} \times \cos \beta$$

where γ is the maximum wedge angle and β the angle of rotation from the horizontal.

FIG. 7.25

Thus the horizontal deflection produced by so rotating the wedge is proportional to $\cos \beta$. A vertical deflection is also produced when a single wedge is rotated so that the emergent ray from an incident ray perpendicular to the entrance surface of the wedge would in fact generate a cone. For this reason a single rotating wedge cannot be used for automatic reduction, since the images of the vernier (viewed through the telescope) and of the main scale (viewed through the telescope and wedge) would not remain in contact owing to the vertical shift of the latter image.

If two equal wedges each having a wedge angle half as big as that of the single wedge are used, the maximum horizontal deflection will be given when the wedges are mounted as in Fig. 7.27(*a*) and will,

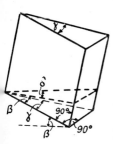

FIG. 7.26

for suitably designed wedges, be such that the total deflection angle between incident and emergent rays is $\tan^{-1} \frac{1}{100}$. If these two wedges are rotated in opposite directions in the vertical plane, however, the vertical components of the deflection, being equal and opposite, will cancel each other out, whilst the horizontal components, being additive, sum up to the same total horizontal deflection as would be given by a single wedge having twice the wedge angle.

Since δ and γ are small angles, the deflections due to the wedges can be written as $\alpha_0 = (\mu - 1)\gamma$ when the telescope is horizontal, and as $\alpha = (\mu - 1)\gamma \cos \beta$ when the line of sight is inclined. These give rise to staff image deflections of l_0 and l say.

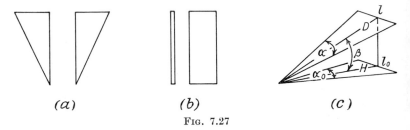

(a) (b) (c)

Fig. 7.27

Reduction to Give Horizontal Component of D. With the two wedges mounted as in Fig. 7.27a and the telescope horizontal, the distance from the instrument to the staff is given by

$$H = \frac{l_0}{\alpha_0} = 100 \times \text{deflection of the image}$$

Let the same horizontal distance be measured with the line of sight now inclined. The wedges are connected to the trunnion axis by a gear train in such a manner that as the telescope is elevated or depressed, each is rotated, one clockwise and the other anti-clockwise, through an angle β (*see* Fig. 7·26), and the lateral deflection decreases as β increases. The lateral deflection is now recorded as l for a slope distance $D = H/\cos \beta$. Thus we have $H = l_0/\alpha_0$ and $D = l/\alpha$, and therefore $H = l \cos \beta/\alpha$. But $\alpha = \alpha_0 \cos \beta$, so that $H = l/\alpha_0 = 100 \, l$, and the horizontal distance is given directly from l. But to obtain the horizontal distance now from the staff reading, using the R.D.H. instrument, a small negative correction, obtained directly from an auxiliary scale, is applied when β is large, since the apex of the angle of lateral deflection does not lie on the transverse axis of the telescope.

Reduction to Give Vertical Component of D. With the initial position of the wedges as shown in the plan view Fig. 7.27(b) (i.e. corresponding to the elevational view in Fig. 7.28) β equals 90°, the telescope is horizontal and the lateral deflection is zero. If, as the telescope is tilted, the wedges are rotated in opposing directions through $(90° - \beta)$, then the lateral deflection of the main scale relative to the vernier will increase proportionally to the sine of the angle of tilt as β decreases. A correction having the same sign, positive or negative, as the angle of elevation and obtained from the auxiliary scale, must be applied.

Fig. 7.28 shows the optical parts of the Wild R.D.H. Reduction Tacheometer, which incorporates twin wedges. The instrument has a knob which enables the wedges to be thrown into the positions of maximum and minimum deflection, and two readings are required at each pointing, one for *H* and the other for *V*. The rhombic prism shown to the left of, and above, the object lens, produces a parallel shift of the rays entering the upper half of the lens, thus leaving room for the rotating wedges, which are placed below the prism. This prism also serves as a parallel-plate micrometer, the drum for which is shown above the object lens.

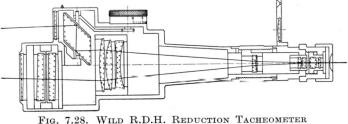

FIG. 7.28. WILD R.D.H. REDUCTION TACHEOMETER
(Courtesy of Wild Heerbrugg, Switzerland)

Rays deflected by the achromatic wedges pass through the lower part of the objective, whilst non-deflected rays pass through the rhombic prism and the upper part of the objective. The rays then pass successively through the internal focusing lens, a prism with a horizontal middle edge and a slot diaphragm which eliminates those rays which would give interfering images. When taking observations, the horizontal middle prism edge is set on the middle of a horizontal staff whose main divisions deflected by the wedges are seen above that line, whilst its undeflected vernier divisions are below. The micrometer knob allows coincidence between main and vernier divisions to be obtained.

The accuracy of the instrument is given as 1 to 2 cm for horizontal distance measurement, and 4 to 5 cm for measurement in the vertical plane. Tests made in the University of Durham, King's College, Mining Department,* show that these claims are justified over the *moderate* distances (30–150 m) over which the instrument is intended to be used. Some of the results are given in the following table in which distances measured by tape and corrected for standard temperature and slope, are compared with the same distances measured with a Wild R.D.H. Tacheometer, three observations being made on each face in each case.

---

* *The Application of the Wild R.D.H. Tacheometer to Mining Problems,* University of Durham, King's College, Mining Dept., Vol. 7, Bulletin No. 5.

It will be good practice to take the mean of several coincidences for one distance measurement when using this, and other, wedge tacheometers. In addition they should be calibrated at frequent intervals of time against known distances.

Corrected Horizontal Distance (tape) in (m)	Differential in level in (m)	Observation with RDH		Error in H (mm)	Error in V (mm)	Gradient
		H (m)	V (m)			
30·352	2·777	30·356	2·789	4	12	+ 1 in 11
30·436	1·535	30·434	1·558	2	23	− 1 in 20
30·480	0·067	30·477	0·050	3	17	+ 1 in 440
91·314	4·693	91·297	4·688	17	5	+ 1 in 20

Zeiss Jena BRT006 Telemeter. This instrument also measures reduced horizontal distances, utilizing a pair of five-sided prisms to form a parallactic angle. Essentially a right-angled triangle is obtained since one of the prisms ensures an effective displacement of the rays from the target through 90° whilst the other is so positioned with

FIG. 7.29. ZEISS JENA BRT006 TELEMETER
(*Courtesy of C. Z. Scientific Instruments, Ltd.*)

respect to the first that an equivalent angle of $\tan^{-1} 1/200$ is subtended at the target when the sighting is horizontal. This angle is caused to change to cater for $\cos \beta$ as the line of sight is elevated, as with the Wild RDH instrument, and the first prism is moved along a base so that the images of the target produced by the two prisms can be lined up, one above the other, in the field of view of the eyepiece.

The advantage of these two instruments is that α_0 does not change, and since $H = l/\alpha_0$ distance errors are directly proportional to l, whereas with the subtense bar α varies, thereby inducing errors proportional to the square of distance.

Distance is read off directly from the base rail of the BRT006 (see Fig. 7.29) and a graduated vertical circle is provided for the measurement of vertical angles. This instrument is very useful in urban surveying and a discussion of its use therein can be seen in *Surveying News* No. 20, published by Messrs. Zeiss. One point of importance is that the prisms must be shielded from direct rays of the sun, otherwise the standard of accuracy may fall below that of conventional tacheometry.

For ranges up to 60 m a target is required for optimum accuracy, although suitable natural features may be viewed directly providing loss of accuracy below that of the 0·06 per cent implied for the device can be accepted. From 60 to 180 metres target staves must be used to achieve that accuracy.

EXERCISES 7

1. In a tacheometric traverse, of which the first three stations are A, B, and C, at station B it is found that the instrument height had not been measured above the peg A. Find from the tabulated readings the missing instrument height and the reduced level of the ground at C, if that at A is 83·44 m. The instrument constants are 100 and 0.

Station	Point	Vertical Circle Reading	Stadia Readings	Height of Instrument (ft)
A	B	+ 5° 42′	2·43; 2·07; 1·71	?
B	A	− 5° 24′	1·68; 1·34; 1·00	1·28
B	C	− 5° 24′	1·68; 1·44; 1·20	1·28

Answer: 1·38 m; 85·2 m.

2. Explain with diagrams and deduce the basic tacheometric formulae (using an anallatic lens) for determining the horizontal distance H, and the difference in level V, between two points, X and Y. Let the angle of elevation from X to Y be θ; h the height of the instrument axis above the ground level at station X, and m the middle-hair reading on the staff at Y. (*I.C.E.*)

3. The following readings were taken for one boundary of a building site by a tacheometer fitted with an anallatic lens giving a multiplying constant for the instrument of 100.

Instrument Station	Staff Station	Stadia Hair Readings			Bearing E of N	Vertical Angle
A	B	1·00	2·05	3·10	0°	+ 2°
B	C	1·00	2·12	3·24	30°	+ 3°
C	D	1·00	2·38	3·76	60°	+ 1°
D	E	1·00	2·09	3·18	120°	− 5°

Find the length of AE. If the reduced level of A is 36·58 m, and the instrument height at A is 1·28 m, determine the reduced level of B.
Answer: 690·3 m; 43·1 m.

4. Discuss the two methods of staff holding normally associated with fixed-hair stadia tacheometry, giving an approximate analysis of the effect of staff tilt on the results. Under what conditions are errors from this source most serious?

In a tacheometric survey an intercept of 0·76 m was recorded on a staff which was believed to be vertical, and the elevation angle was recorded as 15°. Actually, the staff, which was 4 m long, was 150 mm out of plumb leaning towards the instrument. The constant for the instrument was 100. What would have been the error in the computed horizontal distance?

Answer: Error = −0·66 m.

5. Assume you are planning a tacheometric traverse through mixed types of country. Write down what instruments and apparatus you would consider, and what order of accuracy you would expect, together with the factors controlling the accuracy. Give the basic formula to be used, and also the formulae for curvature and refraction, stating in what conditions the latter would be applicable, if at all. (*I.C.E.*)

6. In making a reconnaissance survey for a proposed reservoir, a plane table and telescopic alidade fitted with a Beaman arc were used for contouring. When the alidade was aligned on a staff held at a point *A*, the telescope was elevated until the *vertical scale* of the arc read 25 and the horizontal scale read 6·7. The staff readings were 1·94, 1·47, and 1·00.

What do the *vertical* and *horizontal* scales of the Beaman arc mean, and in this case what is the difference in height and the horizontal distance between the point *A* and the plane table? (*I.C.E.*)

Answer: 22·03 m; 87·7 m.

7. Describe a method of tacheometry using a horizontal 2-metre subtense bar and an accurate theodolite, giving (*a*) the merits of the system; (*b*) the conditions under which the system would be used; (*c*) the factors affecting the accuracy of the system; and (*d*) two methods whereby accuracy may be improved when measuring long lines.

How can a distance of 600 m be measured with an accuracy of about 1/10,000 when the error of angular measurement is ± 1″?

Answer: Subdivide into 6 bays.

8. Describe recent developments in tacheometry, dealing with stadia-type and non-stadia-type instruments, and give clearly the principles involved in the various developments and their main aims.

CHAPTER 8

CURVE RANGING

INCLUDED IN THE DUTIES of an engineer in charge of constructional schemes is the setting out of works, an operation which is in some sense the reverse of surveying, in that measurements and other data are taken from the plan and transferred to the ground.

Where the works are completely defined by straight lines, nothing need be said, as the setting-out operations are simple. The setting out, or ranging of curves, however, will be dealt with at some length, as, in many types of construction, curves will be required. For example, in road, railway or pipeline construction, two straights will normally be connected by a curve whenever there is a change in direction.

The types of curve to be dealt with in this chapter are the circular curve, the transition curve, and the vertical curve. In the last type, the setting-out operations consist of fixing level pegs, and since this has already been dealt with, only the design method will be described.

Circular Curves

Two straights meet at the point of intersection I, and a circular arc is run between the straights, meeting them tangentially at the tangent points T and U.

The radius of the curve is R, and the angle of deflection is as shown in Fig. 8.1 (this angle is also sometimes referred to as the angle of deviation, or the angle of intersection).

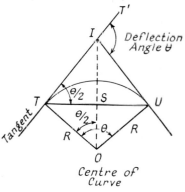

FIG. 8.1

Then, since $\quad I\hat{T}O = I\hat{U}O = 90°$

$$T\hat{I}U + T\hat{O}U = 180°$$

but $\quad T\hat{I}U + T'\hat{I}U = 180°$

Therefore $\quad T'\hat{I}U = \theta = T\hat{O}U.$

Thus the long chord TU subtends an angle equal to the deflection angle at the centre of the curve.

LENGTH OF TANGENTS. Triangles IOT, IOU, are congruent and so,
(i) $\quad IT = IU$, i.e. the tangent lengths are equal.

$\therefore \qquad\qquad ITU$ is isosceles

but $\qquad\qquad I\hat{T}U + I\hat{U}T = T'\hat{I}U = \theta$

Therefore $\qquad\qquad I\hat{T}U = \dfrac{\theta}{2}$

(ii) Also, $\qquad\qquad T\hat{O}S = U\hat{O}S$

$$= \dfrac{\theta}{2}$$

Therefore $\qquad\qquad IT = \text{tangent length}$

$$= R\tan\dfrac{\theta}{2}$$

Thus, if the location of the intersection point I be known, then by chaining a distance $R\tan\dfrac{\theta}{2}$ back along the straight from I, the tangent point T can be located. Similarly, chaining the same distance along the other straight will locate the second tangent point U.

LENGTH OF CURVE. Circumference of a circle of radius $R = 2\pi R$. This subtends an angle of 360° at the centre of the circle,

$\therefore\qquad$ the length L of the arc TU which subtends θ at the centre

$$= 2\pi R\,\dfrac{\theta}{360}$$

$$= \dfrac{\pi}{180}\,R\theta$$

CHAINAGE OF TANGENTS. Knowing the chainage of I,

then $\qquad\qquad$ chainage of T = chainage of $I - IT$,

and $\qquad\qquad$ chainage of U = chainage of $T + \dfrac{\pi}{180}\,R\theta$

(*Note:* the chainage of U is *not* the chainage of $I + IU$.)

DESIGNATION OF CURVES. Curves can be designated in two ways—

(i) by radius, this being the usual method in England, e.g. 600 m radius curve, or, possibly but less preferably,

(ii) by the number of degrees subtended at the centre by a chord 100 links long, e.g. 2° curve.

In Fig. 8.2, LN is a chord 100 links long. OM is the perpendicular bisector of chord LN. Let LN subtend an angle of D at O. Then

$$L\hat{O}M = N\hat{O}M = \dfrac{D}{2}$$

$$LM = R\sin\dfrac{D}{2}$$

$\therefore\qquad\qquad 50 = R\sin\dfrac{D}{2}$

For small angles $\quad \sin \dfrac{D}{2} = \dfrac{D}{2}$ radians nearly

$\therefore \qquad\qquad \sin \dfrac{D}{2} = \dfrac{D}{2} \cdot \dfrac{\pi}{180}$

$\therefore \qquad\qquad 50 = R \cdot \dfrac{D}{2} \cdot \dfrac{\pi}{180}$

whence $\qquad\qquad RD = 5{,}729 \cdot 6$

i.e. $\qquad\qquad R = \dfrac{5{,}729 \cdot 6}{D}$ or $D = \dfrac{5{,}729 \cdot 6}{R}$

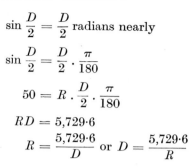

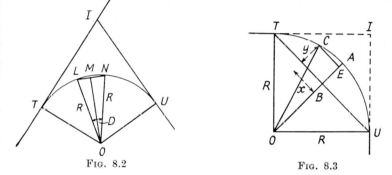

FIG. 8.2   FIG. 8.3

Therefore if $\qquad\qquad R = 600 \text{ m} = 3{,}000 \text{ links}$

$$D = \frac{5{,}729 \cdot 6}{3{,}000} = 1 \cdot 9099^\circ = 1 \cdot 91^\circ, \text{ say}$$

Hence a curve of radius 600 m is equivalent to a $1 \cdot 91^\circ$ curve on this basis.

Setting Out

Circular curves may be set out with the following equipment—(1) Chain and linen tape, (2) single theodolite and chain or steel tape, (3) two theodolites.

1. Methods Using Chain and Tape. OFFSETS FROM THE LONG CHORD. This method is suitable for curves of small radius such as kerb lines at road intersections, and boundary walls. TU is the long chord of length L, T and U being the tangent points. AB is the versed sine of the curve, as shown in Fig. 8.3.

$$AB = AO - OB$$
$$= AO - \sqrt{OU^2 - UB^2}$$
$$= R - \sqrt{R^2 - \left(\frac{L}{2}\right)^2}$$

To set out the curve, points such as C must be located. The offsets y are calculated for corresponding distances x from origin B, which is the mid point of TU, and are set off at right angles, using any of the methods already described.

Draw CE parallel to TU

Then
$$y = EB = EO - BO$$

$$EO^2 = CO^2 - CE^2 \quad \therefore \ EO = \sqrt{R^2 - x^2}$$

whence
$$y = \sqrt{R^2 - x^2} - \sqrt{R^2 - \left(\frac{L}{2}\right)^2}$$

Or, since $\quad OB = R - AB = R - \text{versed sine}$,

$$y = \sqrt{R^2 - x^2} - (R - \text{versed sine})$$

EXAMPLE. Derive data for setting out the kerb line shown in Fig. 8.3 if the radius be 12 m, and $T\hat{O}U = 90°$. Offsets are required at 2 m intervals.

Then
$$TU^2 = TO^2 + OU^2 = 12^2 + 12^2 = 288$$

$$\therefore \qquad TU = 16\cdot97 \text{ m}$$

Versed sine $AB = 12 - \sqrt{12^2 - \left(\frac{16\cdot97}{2}\right)^2}$

$$= 3\cdot51 \text{ m}$$

Therefore $\qquad R - \text{versed sine} = 8\cdot49 \text{ m}$

x (m)	x^2	$(R^2 - x^2)$	$\sqrt{R^2 - x^2}$	$-(R - \text{V.S.})$	Offset (y) (m)
0	0	144	12·00	− 8·49	3·51
2	4	140	11·83	− 8·49	3·34
4	16	128	11·31	− 8·49	2·82
6	36	108	10·39	− 8·49	1·90
8	64	80	8·94	− 8·49	0·45

Points T and U would be located by measuring $IT(= IU)$ from the intersection point I.

OFFSETS FROM THE TANGENT. This method is also suitable for short curves, and, as in the previous method, no attempt is made to keep the chords of equal lengths. In Fig. 8.4

$$R^2 = AB^2 + AO^2 \text{ (Pythagoras' theorem)},$$

$$= y^2 + (R - x)^2$$

$$\therefore \qquad x = R - \sqrt{(R^2 - y^2)}$$

$$= R - R\left(1 - \frac{y^2}{R^2}\right)^{\frac{1}{2}}$$

Expand $\left(1 - \dfrac{y^2}{R^2}\right)^{\frac{1}{2}}$ using Binomial theorem,

then
$$x = R - R\left(1 - \tfrac{1}{2}\dfrac{y^2}{R^2} + \ldots\right)$$

$$= \dfrac{y^2}{2R} \text{ approximately}$$

The curve is set out in two parts, starting from each tangent.

OFFSETS FROM CHORDS PRODUCED. For longer curves of larger radius, the curve can be set out by offsets from chords using a chain and tape. This method has the advantage that not all the land between the two tangents T and U need be accessible, and can be used when the accuracy attainable with the theodolite is not required, e.g. giving the line for soil stripping and excavation in road works. In order that the assumptions made in the derivations of formulae are not invalidated, namely, that the length of the arc is very nearly equal to the length of the corresponding chord, the length of chord chosen should not exceed $\dfrac{R}{20}$. It is usual to peg the curve at regular intervals with cross sections at per-

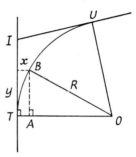

FIG. 8.4

haps 100 m intervals. Thus a convenient interval for the pegs is 20 m thereby allowing a curve of radius 400 m to be so treated. If the chainage of T be, say, 22,186 then the first chord can be made 14 m long (assumed to agree with the $R/20$ concept) in order that the next peg will be placed at the even chainage of 22,200 m.

Procedure. Taking chainage of T as 22,186 (referring to Fig. 8.5), (a) Hold 6 link at T with the chain arranged on the line TI. The end of the chain at A, is swung round through a calculated offset $A_1 A$, with T as centre, thus locating peg A (at chainage 22,200) on the curve. (Note, however, that chainages are often not kept continuous in this method, i.e. chainage of T is assumed to be zero, especially if cross sections are not required.)

(b) Pull chain forward along TA produced until forward end is at B_1, and rear end is at A, if a length of 10 m be required for AB, then the 50 link will be held at A. The end B is now swung round through a calculated offset $B_1 B$ to locate B.

(c) Repeat for all other points and the final offset should be such that U is located on the tangent (*see* Fig. 8.6). As a check, U can be located by chaining tangent length IU from I and the final offset should then agree with this position of U. If not, then it will be necessary to check all points such as A, B, adjusting their positions until coincidence at U is obtained.

Calculations. Referring to Fig. 8.5, let

$$A_1 \hat{T} A = \alpha$$

Then for small angles

$$\text{chord } AA_1 = \text{arc } AA_1$$

$$\therefore \qquad AA_1 = TA_1\alpha = TA \cdot \alpha$$

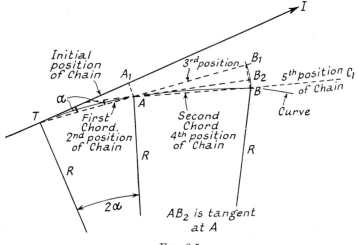

Fig. 8.5

If $A_1 \hat{T} A = \alpha$, then the angle subtended at the centre of the curve will be $2\,\alpha$,

$$\text{Arc } TA \simeq \text{chord } TA \left(\text{so long as } TA \not> \frac{R}{20}\right)$$

$$= R \cdot 2\,\alpha$$

Thus

$$\alpha = \frac{TA}{2R}$$

$$\therefore \qquad AA_1 = TA \cdot \alpha = \frac{TA}{2R} \cdot TA = \frac{TA^2}{2R}$$

which is the first offset.

$$\textit{Also} \quad B_1 B_2 = AB_1 \cdot \alpha$$

$$= AB_1 \cdot \frac{TA}{2R} = AB\frac{TA}{2R}, \text{ since } AB_1 = AB$$

Since B_2B is the offset from the tangent AB_2, then exactly as for AA_1 which was offset from TI we may derive B_2B to be $\dfrac{AB^2}{2R}$

$$\therefore \quad B_1B = B_1B_2 + B_2B = AB \cdot \frac{TA}{2R} + \frac{AB^2}{2R}$$

$$= \frac{AB}{2R}(TA + AB)$$

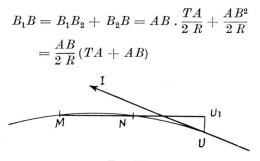

FIG. 8.6

For all subsequent offsets but the last, the chord lengths will now be equal and so

$$C_1C = \frac{BC}{2R}(AB + BC) = \frac{BC^2}{R}$$

It may be shown that the final offset U_1U will be $\frac{NU}{2R}(MN + NU)$ (*see* Fig. 8.6).

EXAMPLE. Tabulate data needed to set out, using chain and tape, a circular curve of radius 600 m to connect two straights having a deflection angle of 18° 24'. The chainage of the intersection of the tangents is 2,140 m.

$$IT = 600 \tan \frac{18° 24'}{2} = 97 \cdot 20 \text{ m}$$

T may be located by measuring 97·20 m back from I

$$\therefore \qquad \text{Chainage of tangent point } T = 2042 \cdot 8 \text{ m}$$

$$\text{Length of curve} = 600 \times 18 \cdot 4 \times \frac{\pi}{180} = 192 \cdot 7 \text{ m}$$

$$\therefore \qquad \text{Chainage of tangent point } U = 2{,}042 \cdot 8 + 192 \cdot 7$$

$$= 2{,}235 \cdot 5 \text{ m}$$

Now $\frac{R}{20} = 30$ m, and this interval would suffice for the peg spacings, but for convenience the chord length of 20 m will be adopted. Thus the first chord would be made 17·2 m followed by 8 chords each 20 m long, together with a final chord of 15·5 m length (*see* Table p. 270).

2. Setting Out Using one Theodolite and a Steel Tape or Chain by the Method of Deflection Angles. Again it is recommended that the chords do not exceed $\frac{R}{20}$ in length, so that the length of the chord may be approximated to the length of the corresponding arc.

Field Procedure. (i) Set up theodolite at T, having calculated angles, etc., for their particular chord lengths, set vernier I to 360°,

Chainage	Chord Length (m)	Offset (m)
2,042·8	0	—
2,060·0	17·2	$\dfrac{17 \cdot 2^2}{2 \times 600} = 0 \cdot 25$
2,080·0	20·0	$\dfrac{20(20 \cdot 0 + 17 \cdot 2)}{1,200} = 0 \cdot 62$
2,100·0	20·0	$\dfrac{20^2}{600} = 0 \cdot 67$
2,120·0	20·0	$\dfrac{20^2}{600} = 0 \cdot 67$
2,140·0	20·0	$\dfrac{20^2}{600} = 0 \cdot 67$
2,160·0	20·0	$\dfrac{20^2}{600} = 0 \cdot 67$
2,180·0	20·0	$\dfrac{20^2}{600} = 0 \cdot 67$
2,200·0	20·0	$\dfrac{20^2}{600} = 0 \cdot 67$
2,220·0	20·0	$\dfrac{20^2}{600} = 0 \cdot 67$
2,235·5	15·5	$\dfrac{15 \cdot 5(15 \cdot 5 + 20 \cdot 0)}{2 \times 600} = 0 \cdot 46$

clamp the vernier and scale plates together, direct the telescope to I, obtaining coincidence with the lower clamp and tangent screw.

(ii) Unclamp the vernier plate (but ensure that the scale plate remains clamped and stationary throughout the operation), and set vernier I to read deflection angle α, thus directing the telescope to A. Clamp the scales together, hold the chain at T, so that chord TA of correct length will be obtained. As in the previous method with chain and tape, the first chord length TA will generally be some residual length calculated to bring the chainage of A to the next whole number of chains. A chaining arrow is held at the 100 link end of the chain, which is now swung taut about T until exact coincidence is obtained between the instrument cross-hairs and chaining arrow. This fixes A.

(iii) Set vernier I to read β and then reclamp. Hold chain at A at that link which will give chord AB of correct length and again swing the taut chain, obtaining coincidence on the cross-hairs.

(iv) Repeat for all points on the curve, and coincidence should be obtained at U located by chaining the tangent distance IU along the tangent from the intersection point. The final deflection angle ITU will of course equal $\dfrac{\theta}{2}$, and this could be read immediately on setting up at T to serve as a check that the tangent lengths have been correctly set out.

Calculation. Referring to Fig. 8.7

$$I\hat{T}A = \alpha$$

∴ angle subtended at centre of curve by $TA = 2\alpha$

$$\text{arc } TA = 2R\alpha \qquad (\alpha \text{ in radians})$$

$$= 2R\alpha \times \frac{\pi}{180} \qquad (\alpha \text{ in degrees})$$

$$= \text{chord } TA \qquad \left(\text{if chord} \not> \frac{R}{20}\right)$$

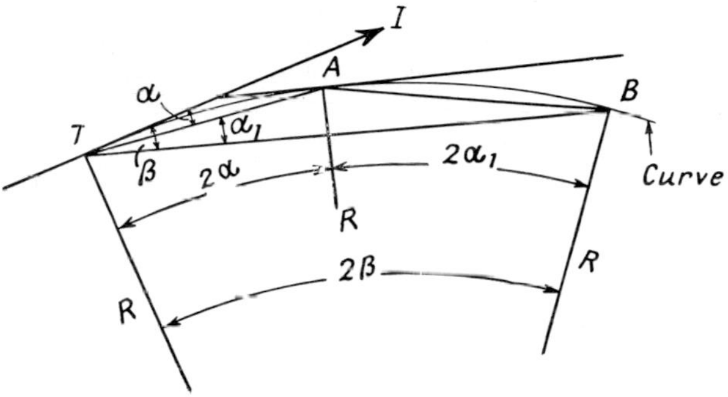

Fig. 8.7

∴

$$\alpha = \frac{TA \times 180}{2\pi R} \text{ deg}$$

$$= \frac{TA \times 180 \times 60}{2\pi R} \text{ min}$$

$$= 1718 \cdot 9 \times \frac{TA}{R} \text{ min}$$

or, if

$$TA = O_1 = \text{length of first chord}$$

$$\alpha = 1718 \cdot 9 \times \frac{O_1}{R} \text{ min}$$

If α_1 equals the angle between the tangent at A and chord AB, then, the angle subtended by AB at the centre of the curve is $2\alpha_1$, and since an angle subtended by a chord at the circumference is half that subtended by the chord at the centre, $A\hat{T}B = \alpha_1$.

∴

$$\text{arc } AB = 2R\alpha_1 \times \frac{\pi}{180}$$

and

$$\alpha_1 = 1718 \cdot 9 \times \frac{AB}{R} \text{ min}$$

or, if
$$AB = O_2$$

$$\alpha_1 = 1{,}718{\cdot}9 . \frac{O_2}{R} \text{ min}$$

$\therefore$
$$I\hat{T}B = \beta = \alpha + \alpha_1$$

= angle set out from T while the chord AB is being set out.

Note that AB is the chord, not TB. It should also be noted that α_1 is the angle which would have to be set off from the tangent at A if we were setting out B with the instrument at A.

It will be seen that by repeating the above calculations a series of deflection angles α, $\alpha + \alpha_1$, $\alpha + \alpha_1 + \alpha_2 + \ldots$, etc., will be obtained corresponding to arcs TA, AB, BC, etc. The actual calculations are, of course, simplified because all the chords except the first and the last are equal, i.e. $\alpha_1 = \alpha_2 = \alpha_3 =$, etc.

EXAMPLE. Tabulate data needed to set out by theodolite and chain a circular curve of radius 600 m to connect two straights having a deflection angle 18° 24′ the chainage of the intersection point being 2,140·0 m.

$$IT = 600 \tan \frac{18° \ 24'}{2} = 97{\cdot}20$$

Chainage of tangent point T is $2{,}140{\cdot}0 - 97{\cdot}2 = 2{,}042{\cdot}8$ m

$$\text{Length of curve} = 600 \times 18{\cdot}4 \times \frac{\pi}{180} = 192{\cdot}7 \text{ m}$$

Chainage of tangent point $U = 2{,}235{\cdot}5$ m

Chainage (m)	Chord (m)	Deflection Angle	Total Deflection Angle	Total Deflection Angle set on 20″ inst.
2,042·8	0	0	0	0
2,060·0	17·2	$1{,}718{\cdot}9 \times \dfrac{17{\cdot}2}{600} = 49{\cdot}27'$	00° 49·27′	00° 49′ 20″
2,080·0	20·0	$1{,}718{\cdot}9 \times \dfrac{20{\cdot}0}{600} = 57{\cdot}30'$	01° 46·57′	01° 46′ 40″
2,100·0	20·0	$1{,}718{\cdot}9 \times \dfrac{20{\cdot}0}{600} = 57{\cdot}30'$	02° 43·87′	02° 44′ 00″
2,120·0	20·0	$1{,}718{\cdot}9 \times \dfrac{20{\cdot}0}{600} = 57{\cdot}30'$	03° 41·17′	03° 41′ 20″

Chainage (m)	Chord (m)	Deflection Angle	Total Deflection Angle	Total Deflection Angle set on 20″ inst.
2,140·0	20·0	$1{,}718{\cdot}9 \times \dfrac{20{\cdot}0}{600} = 57{\cdot}30'$	04° 38·47′	04° 38′ 20″
2,160·0	20·0	$1{,}718{\cdot}9 \times \dfrac{20{\cdot}0}{600} = 57{\cdot}30'$	05° 35·77′	05° 35′ 40″
2,180·0	20·0	$1{,}718{\cdot}9 \times \dfrac{20{\cdot}0}{600} = 57{\cdot}30'$	06° 33·07′	06° 33′ 00″
2,200·0	20·0	$1{,}718{\cdot}9 \times \dfrac{20{\cdot}0}{600} = 57{\cdot}30'$	07° 30·37′	07° 30′ 20″
2,220·0	20·0	$1{,}718{\cdot}9 \times \dfrac{20{\cdot}0}{600} = 57{\cdot}30'$	08° 27·67′	08° 27′ 40″
2,235·5	15·5	$1{,}718{\cdot}9 \times \dfrac{15{\cdot}5}{600} = 44{\cdot}40'$	09° 12·07′	$\underline{9° \; 12' \; 00''} = \dfrac{\theta}{2}$

WHEN THE CURVE DEFLECTS TO THE LEFT. It will be appreciated that the above procedure only applies directly when the curve deflects to the right of the tangent when proceeding in the direction of the chainage (as in Fig. 8.5) since only then will the telescope be traversing in a clockwise direction, and vernier *I* reading the actual deflection angles α, $\alpha + \alpha_1$, etc., during the setting-out operation. Where the curve deflects to the left when approaching in the direction of the chainage it is set out either,

(*a*) from the first tangent point as a left-hand curve,

(*b*) from the second tangent point as a right-hand curve.

In the former case, if the scale plate is graduated in a clockwise direction and since the telescope is traversing anti-clockwise, the deflection angles to be set out are $(360 - \alpha)$, $(360 - (\alpha + \alpha_1))$, $(360 - (\alpha + \alpha_1 + \alpha_2))$, etc., where α, α_1, α_2, etc., are computed as outlined above. In the latter case, procedure is as above though the chainages must be kept running as from the first tangent.

Possible Difficulties in Setting Out Simple Curves

1. IF ENTIRE CURVE CANNOT BE SET OUT FROM ONE TANGENT POINT (Fig. 8.8).

Deflection angle from *T* to $A = \alpha_A$.

Deflection angle from *T* to *B* would be α_B.

(i) Set out as far as possible in normal manner, say to peg *A*, and position a ranging rod at *S* on *TA* produced.

(ii) Move instrument to *A*, set vernier to 360°, and direct telescope on to ranging rod *S*, obtaining coincidence with lower clamp and tangent screw (or, sight back on to tangent point *T* and then transit the telescope to locate *S*).

(iii) Free upper plate and set off α_B on the vernier. The telescope is now directed along *AB* and *B* is located in the normal manner, and the remainder of the curve set out.

If T_1AT_2 is a tangent at A, the triangle TT_1A must be isosceles

and
$$T_1\hat{T}A = T_1\hat{A}T = \alpha_A$$

But
$$T_1\hat{A}T = S\hat{A}T_2$$

Also
$$T_2\hat{A}B = A\hat{T}B$$

$$T_1\hat{T}B = S\hat{A}B = \alpha_B$$

Thus, although the instrument has been moved to peg A, the same total deflection angles apply to all points subsequent to A as if the instrument were still set up at the tangent point, T.

Therefore, should another move be required in the general case where the instrument is moved from peg N of total deflection angle δN to peg X of total deflection angle δX,

(i) sight back to peg N with the vernier set at δN,

(ii) transit and set off the next peg $(X+1)$ with total deflection angle $\delta(X+1)$.

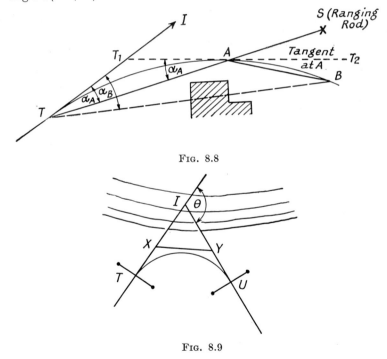

Fig. 8.8

Fig. 8.9

2. Intersection Point Inaccessible (Fig. 8.9). To locate T and U on the ground,

(i) Measure angles $T\hat{X}Y$ and $X\hat{Y}U$ by theodolite. Measure XY by chain or tape, XY being any pair of suitable points on the tangents

(ii) Deduce $I\hat{X}Y = 180 - T\hat{X}Y$, and $I\hat{Y}X = 180 - X\hat{Y}U$, and thence evaluate $\theta = I\hat{X}Y + I\hat{Y}X$, and $X\hat{I}Y = 180 - \theta$.

(iii) Evaluate $IX = XY\dfrac{\sin I\hat{Y}X}{\sin X\hat{I}Y}$, and $IY = XY\dfrac{\sin I\hat{X}Y}{\sin X\hat{I}Y}$.

(iv) Knowing θ (from (ii)) and the desired radius of the curve, determine $IT = R\tan\theta/2 = IU$.

(v) Determine $XT = IT - IX$, and $YU = IU - IY$, and chain these distances back from X and Y respectively to obtain the tangent points.

3. WHEN THE INSTRUMENT CANNOT BE SET UP AT THE FIRST TANGENT POINT. There are two cases to consider: (i) when the curve deflects to the right in the direction of the chainage; (ii) when the curve deflects to the left.

(i) *Curve deflects to right.* Although it may be possible to peg out tangent point T, it may not be convenient to set up the instrument here because the ground is too soft or for some other reason. The procedure is then to set out the curve from the second tangent point U, but using the total deflection angles computed from the first tangent (to keep the chainages running).

Referring to Fig. 8.10, with the instrument at U the first chord is, say UE.

Angle subtended at T by UE

$$= E\hat{T}U$$
$$= \frac{\theta}{2} - \alpha_E$$
$$\therefore \qquad I\hat{U}E = \frac{\theta}{2} - \alpha_E$$

where α_E is the total deflection angle for setting out point E with the instrument at T.

Since the scale graduations read clockwise, the external angle will be set off on the vernier

$$\text{External } I\hat{U}E = 360° - \frac{\theta}{2} + \alpha_E$$

$$- \text{ first deflection angle}$$

Similarly, the subsequent deflection angles setting out from U are $\left(360° - \dfrac{\theta}{2} + \alpha_D\right)$, $\left(360° - \dfrac{\theta}{2} + \alpha_C\right)$, etc., with a final value of $\left(360° - \dfrac{\theta}{2}\right)$ which should close on T.

(ii) *Curve deflects to the left.* If such a curve has to be set out from the second tangent point, the telescope will be swinging in a clockwise direction as the points are set out, and the student will readily appreciate from the above reasoning that the deflection angles to be set out will be $\left(\dfrac{\theta}{2} - \alpha_n\right)$, $\left(\dfrac{\theta}{2} - \alpha_{n-1}\right)$, $\left(\dfrac{\theta}{2} - \alpha_{n-2}\right)$, etc., where α_n is the total deflection angle, computed from the first tangent point of the last peg before reaching the second tangent point, α_{n-1} is the total deflection angle of the last but one peg and so on.

3. Setting Out Using Two Theodolites. This method is suitable when the ground between the tangent points T and U is of such a

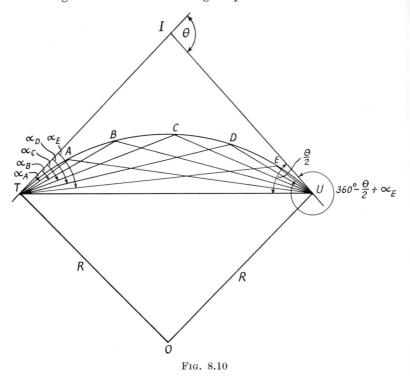

Fig. 8.10

character that chaining or taping proves difficult, e.g. very steep slopes, or if the curve is wholly or partly over water or marsh. Two theodolites are used, one being set at each tangent point and the main disadvantage of the method is that it requires two surveyors and two instruments as well as the assistants who locate the pegs.

Referring to Fig. 8.10 and to the previous section, it will be seen that if the theodolite at T is set so that it deflects angle α_E from the tangent IT, and the theodolite at U set so that it deflects $(360° - \dfrac{\theta}{2} + \alpha_E)$ from tangent IU, the two lines of sight will intersect at point E. To locate E, therefore, the assistant is directed until the signal is seen at the intersections of the cross-hairs of both theodolites. Good liaison between the three groups is, of course essential at this stage. This procedure is repeated for the location of the other points, and as an example the curve already tabulated for setting out with one theodolite and chain is given below for setting out with two theodolites.

Point	Chainage	Chord	Deflection Angle	Deflection Angle from *T*	Deflection Angle from *U*
T	2,042·8	0	0	0	350° 48′ 00″
	2,060·0	17·2	49·27′	00° 49′ 20″	351° 37′ 20″
	2,080·0	20·0	57·30′	01° 46′ 40″	352° 34′ 40″
	2,100·0	20·0	57·30′	02° 44′ 00″	353° 32′ 00″
	2,120·0	20·0	57·30′	03° 41′ 20″	354° 29′ 20″
	2,140·0	20·0	57·30′	04° 38′ 20″	355° 26′ 20″
	2,160·0	20·0	57·30′	05° 35′ 40″	356° 23′ 40″
	2,180·0	20·0	57·30′	06° 33′ 00″	357° 21′ 00″
	2,200·0	20·0	57·30′	07° 30′ 20″	358° 18′ 20″
	2,220·0	20·0	57·30′	08° 27′ 40″	359° 15′ 40″
U	2,235·5	15·5	44·40′	09° 12′ 00″	360° 00′ 00″

Transition Curves

The forces acting on a vehicle change when it moves from the tangent on to the circular curve.

The centrifugal force P acting on the vehicle as it traverses the curve is given by $P = \dfrac{WV^2}{gR}$, or $P \propto \dfrac{V^2}{gR}$, where W = weight of vehicle, V = velocity, R = radius of curve and g = acceleration due to gravity.

Also the centrifugal ratio = $\dfrac{P}{W} = \dfrac{V^2}{gR}$.

It will be seen from both formulae that the centrifugal force and ratio increase as R decreases, and thus, since the radius of the straight is infinity, the centrifugal force would increase instantaneously from zero to its maximum value (assuming no change in V), as the vehicle moved from straight to curve. Passengers in the vehicle would thus experience a lateral shock as the tangent point was passed. A curve of variable radius is inserted between the straight and the circular curve in order that the centrifugal force may build up in a gradual and uniform manner, and in this way the lateral shock is minimized. This curve is called a transition curve.

The radius of a transition curve varies from infinity at its tangent point with the straight, to a minimum value at its tangent point with the circular curve. It will readily be seen that in any case where a single circular curve is to be replaced, two transition curves are required, joined together by the circular curve. In some cases the circular curve joining the two transitions is of zero length so that the single circular curve is replaced by two transition curves having one common tangent point.

Super-elevation. A consideration of Fig. 8.11 will show that whereas a car or train going round a flat curve is subject to a side thrust equa to the centrifugal force, by lifting the outer edge of the road or rails i.e. by applying cant or super-elevation to the curve, the resultant M can be made to lie along the normal to the road surface or rails for a given speed.

Now
$$P = \frac{WV^2}{gR}$$

and from the triangle of forces
$$\tan \alpha = \frac{WV^2}{gR/W}$$
$$= \frac{V^2}{gR}$$

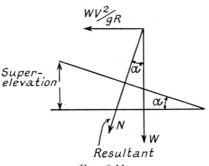

FIG. 8.11

The amount of cant or super-elevation is thus a constant for given values of V and R, and is given by

Super-elevation $= G \sin \alpha = G \tan \alpha$ for small angles

where $G =$ gauge of railway track or width of road.

The transition curve allows this super-elevation to be introduced i a gradual manner so that it varies from zero on the straight to it maximum value at the tangent point where the transition curve meet the circular curve.

Before proceeding with the design of transition curves, it is appro priate to say a word here about the actual amounts of super-elevatio used in practice. The formula
$$\tan \alpha = \frac{V^2}{gR}$$

contains three variables so that values of two of them will have t be used in the formula to enable the third term to be evaluated Generally V is known, in railways being the maximum probable spee at which trains will use the track, and in roads the design averag speed for the road, while the radius R is generally fixed by the land available. Thus tan α and hence the cant can be worked out.

In *railways*, a maximum of, say, 150 mm lift on the outer rail is allowed, this figure being determined by considering the stability of lightly loaded vans moving slowly with high cross-wind. If the theoretical cant is more than this, then either

    (*a*) a larger radius curve is used, or if this is impossible
    (*b*) the value of 150 mm is used and velocity limited.

In *roads*, standards are fixed by the Department of the Environment, having regard to the radius of the road and its design speed: For urban areas the following table is relevant.

Design Speed	Normal Radius (m) for Super-elevation of:		Minimum Radius (m) for Super-elevation of:	
(km/h)	4 per cent	7 per cent	4 per cent	7 per cent
80	500	300	260	230
60	275	170	150	130
50	200	120	90	80
30	75	50	35	30

Normally super-elevation can be determined from the expression

$$\text{Super-elevation} = 1 \text{ in } 314 \frac{R}{V^2}$$

where V is the design speed in km/h and R is the radius in metres; the minimum values include an allowance for side friction. Super-elevation should not be steeper than 7 per cent (1 in 14·5, say) nor should it be flatter than 1 in 48 for effective drainage.

Derivation of the Transition Curve Equations. Summarizing the points which have been made so far, we have

1. P increases uniformly with distance l from the beginning of the transition curve, i.e. $P \propto l$.

2. At this point where the radius of the transition curve $= r$

$$P = \frac{WV^2}{gr}$$

$\therefore$ $P \propto \dfrac{1}{r}$ for constant velocity

and so $l \propto \dfrac{1}{r}$ since $P \propto l$

3. The amount of super-elevation provided increases uniformly with distance from beginning of curve and so is directly proportional to

l. Thus $lr = K$ (where K is some constant) and if the transition curve
has a length L and the radius of the circular curve entered is R,

$$LR = K$$

or
$$lr = LR = K$$

TT_1 is the transition curve along which the radius varies from infinity
at T to R at T_1. The tangent at T_1 is common to the circular curve and
transition curve (*see* Fig. 8.12).

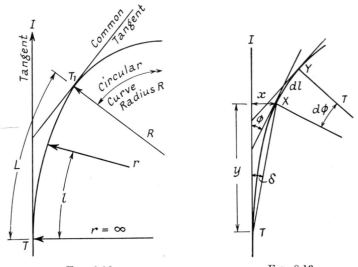

FIG. 8.12   FIG. 8.13

Length of Transition Curve (Fig. 8.12).   The length of transition
curve may be taken

> (i) as an arbitrary value from past experience, say 50 m,
> (ii) such that the super-elevation is applied at a uniform rate,
> say 0·1 m in 100 m,
> (iii) such that the rate of change of radial acceleration equals a
> certain chosen value.

Let L = total length of transition curve, R = radius of circular
curve entered and V = velocity of vehicle.

Then radial acceleration at circular curve tangent point T_1 is $\dfrac{V^2}{R} = f$,
say.

There is no radial acceleration just as vehicle is about to leave the
tangent straight.

∴ If time taken to travel along transition curve be t sec,

$$t = \frac{L}{V}$$

*Rate of change of radial acceleration $= \dfrac{f}{t} = \alpha$

then
$$\alpha = \frac{V^2}{R} \Big/ \frac{L}{V} = V^3/LR$$

and
$$L = V^3/R\alpha$$

4. Consider near points X and Y on the transition curve (Fig. 8.13),
$$dl = rd\phi$$
$$d\phi = -\frac{1}{r}\, dl$$

but since
$$rl = K, \text{ then } \frac{1}{r} = \frac{l}{K}$$

$$d\phi = \frac{l}{K}\, dl$$

integrating
$$\phi = \frac{l^2}{2\,K} + A$$

where A is a constant of integration and will be zero, since
$$\phi = 0$$

when
$$l = 0$$

Then
$$\phi = \frac{l^2}{2\,K} = \frac{l^2}{2\,LR} = \frac{l^2}{2\,lr}$$

and
$$\phi = \frac{l^2}{2\,LR} \text{ is the intrinsic equation of the } ideal$$

transition spiral.

Now
$$dy = dl \cos\phi \text{ and } dx = dl \sin\phi$$

Then
$$dy = \left(1 - \frac{\phi^2}{2!} + \frac{\phi^4}{4!} - \cdots\right)dl$$

$$dy = \left\{1 - \left(\frac{l^2}{2\,K}\right)^2 \frac{1}{2!} - \left(\frac{l^2}{2\,K}\right)^4 \frac{1}{4!} + \cdots\right\}dl$$

$$y = l - \frac{l^5}{40\,K^2} + \cdots$$

Since $K = LR$, the second term is small compared to l.

$\therefore$
$$y = l \text{ nearly}$$

---

* *See* paper by W. H. Shortt in *Journ. Inst. C.E.* (Vol. CLXXVI), "A Practical Method for the Improvement of Existing Railway Curves." Shortt gives a rate of change of radial acceleration of about $\frac{1}{3}$ m/s³ as a "comfort limit," above which side-throw will be noticed.

and
$$dx = \left(\phi - \frac{\phi^3}{3!} + \frac{\phi^5}{5!} - \ldots\right)dl$$

$$= \left\{\frac{l^2}{2\,K} - \left(\frac{l^2}{2\,K}\right)^3 \frac{1}{6} + \left(\frac{l^2}{2\,K}\right)^5 \frac{1}{120} - \ldots\right\}dl$$

$$\therefore \qquad x = \frac{l^3}{6\,K} - \frac{l^7}{336\,K^3} - \ldots$$

or
$$x = \frac{l^3}{6\,K} \text{ nearly}$$

In neither integration will there be a constant of integration, since
$$\phi = 0, \text{ when } l = 0$$

Thus
$$x = \frac{l^3}{6\,K} = \frac{l^3}{6\,LR}$$

This is the equation for the *cubic spiral.*

Writing $y = l$, then $x = \dfrac{y^3}{6\,LR}$, which is the *cubic parabola.*

Relation Between δ and ϕ

$$\tan \delta = \frac{x}{y}$$

$$= \frac{l^3}{6\,LR} \Big/ l$$

$$= \frac{l^2}{6\,LR}$$

$$\therefore \qquad \text{For small angles } \delta = \frac{l^2}{6\,LR} \text{ radians}$$

But
$$\phi = \frac{l^2}{2\,LR}$$

$$\therefore \qquad\qquad \delta = \phi/3 \text{ radians}$$

This expression is important in the setting out of the curve.

The maximum value of ϕ, which is denoted by ϕ_1, is equal to $L^2/2LR = L/2R$. It is the angle between the common tangent at T_1 and the tangent TI.

SHIFT. Where transition curves are introduced between the tangents and a circular curve of radius R (*see* Fig. 8.14), the circular curve is "shifted" inwards from its original position by an amount $BP = S$ (the shift). This is equivalent to having a circular curve of radius $(R + S)$ connecting the tangents replaced by two transition curves and a circular curve of radius R, although the tangent points are not the same, being B and T respectively in Fig. 8.13.

Referring to Fig. 8.15, TBN is tangential to the circular curve of radius $(R + S)$ and RT_1 is the common tangent at T_1

$$N\hat{B}O = R\hat{T}_1O$$
$$= 90°$$

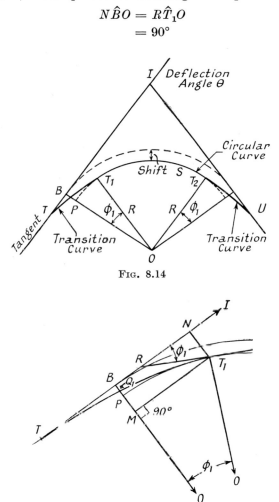

Fig. 8.14

Fig. 8.15

i.e. R, B, O and T_1 are concyclic.

Therefore ϕ_1 is subtended at O by arc PT_1. Draw MT_1 perpendicular to OB as shown.

$$BM = NT_1$$
$$= \text{maximum offset on transition curve}$$

also $\qquad$ shift $S = BP$

$$= BM - PM$$

$$= NT_1 - PM$$

$$= NT_1 - (PO - MO)$$

$$= \frac{L^3}{6\,LR} - (R - R\cos\phi_1)$$

$$= \frac{L^3}{6\,LR} - \left\{ R - R\left(1 - \frac{\phi_1{}^2}{2!} + \frac{\phi_1{}^4}{4!} - \cdots\right)\right\}$$

Ignoring higher powers than the second,

$$S = \frac{L^3}{6\,LR} - \frac{R\phi_1{}^2}{2}$$

$$= \frac{L^3}{6\,LR} - \frac{R}{2}\left(\frac{L^2}{2\,LR}\right)^2$$

$$= \frac{L^2}{6\,R} - \frac{L^2}{8\,R}$$

$$= \frac{L^2}{24\,R}$$

also $\qquad$ $QT_1 \simeq PT_1$

$$\simeq R \cdot \phi_1$$

$$= \frac{RL^2}{2\,LR}$$

$$= \frac{L}{2}$$

Therefore Q is the mid-point of the transition curve and since the deviation of this from the tangent is small, then

$$TQ \simeq TB \left(= \frac{L}{2}\right)$$

Setting Out Transition Curves. To locate the tangent point T—

(i) Calculate the shift S from the expression

$$S = \frac{L^2}{24\,R}$$

(ii) Calculate $\qquad$ $IB = (R + S)\tan\frac{\theta}{2}$

(iii) Since $\qquad$ $BT = \frac{L}{2}$

Then
$$IT = (R + S) \tan \frac{\theta}{2} + \frac{L}{2}$$

Chain this length back from I and so fix point T.

SETTING OUT THE TRANSITION CURVE. The next step depends on whether it is intended to set out the transition with chain and tape using the cubic spiral or cubic parabola, or by the theodolite using the cubic spiral. Often both methods are employed on one job, the chain and tape being used for preliminary work and the theodolite for final location. Usually the chords adopted are about half the length of those used for setting out the circular curve.

(iv) *Either* calculate offsets from

$$x = \frac{l^3}{6\,LR} \text{ or } x = \frac{y^3}{6\,LR}$$

Each peg is located by swinging a chord length from the preceding peg. *Note:* l is the progressive chainage from T.

Or calculate the deflection angles δ for particular distances l from T using the fact that $\delta = \phi/3$ (*see* Fig. 8.13), and so since

$$\phi = \frac{l^2}{2\,RL}$$

then
$$\delta = \frac{l^2}{6\,RL} \text{ radians}$$

$$= \frac{180}{\pi}\,\frac{l^2}{6\,RL} \text{ degrees}$$

$$= \frac{1,800}{\pi}\,\frac{l^2}{RL} \text{ min}$$

The final deflection angle which locates T_1 is $\dfrac{\phi_1}{3}$ where

$$\phi_1 = \frac{L^2}{2\,RL} \text{ radians}$$

$$\therefore \qquad \delta \text{ max} = \frac{\phi_1}{3} \text{ radians}$$

$$= \frac{1,800\,L}{\pi R} \text{ min}$$

SETTING OUT THE CIRCULAR CURVE (Fig. 8.16)

(v) Calculate the deflection angles for the circular curve from

$$\delta \text{ circular arc} = 1,718\cdot9 \times \frac{c}{R} \text{ min}$$

where c is the chord length for the circular arc.

The angle subtended by the circular arc T_1T_2 at the centre of that arc is $(\theta - 2\phi_1)$ as shown in Fig. 8.14

$\therefore \qquad$ length of arc $T_1T_2 = R(\theta - 2\phi_1)\dfrac{\pi}{180}$

(vi) Set up the theodolite at T_1, and sight back on T; then, transit the telescope and locate tangent T_1T_3 by setting off an angle $\frac{2}{3}\phi_1$ (Fig. 8.16). Set out the circular curve by the deflection angle method, from this tangent.

Alternatively when T_1 has been located as in (iv) set a ranging rod at T_4 on TT_1 produced and sight T_4 after setting up at T_1.

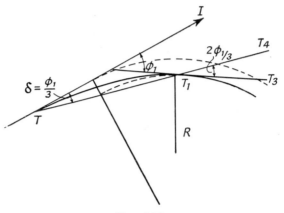

FIG. 8.16

SETTING OUT THE OTHER TRANSITION

(vii) Point U is located from I using the relationship $IU = IT$. The transition is then set out from tangent point U and tangent UI by either of the methods given in (iv).

EXAMPLE. Three straights AB, BC, and CD have whole circle bearings of 30°, 90°, and 45° respectively. AB is to be connected to CD by a continuous reverse curve formed of two circular curves of equal radius together with four transition curves. BC, which has a length of 800 m is to be the common tangent to the two inner transition curves. Determine the radius of the circular curves if the maximum speed is to be restricted to 80 km/h and a rate of change of radial acceleration is 0·3 m/s³ obtains. Give (i) the offset, and (ii) the deflection angle, with respect to BC, to locate the intersection of the third transition curve with its circular curve.

$$\alpha = 0{\cdot}3 = \frac{V^3}{LR}$$

$$V = \frac{80 \times 1{,}000}{3{,}600} = \frac{1{,}000}{45} \ \text{m/s}$$

$\therefore \qquad$ $$L = \frac{1{,}000 \times 1{,}000 \times 1{,}000}{45 \times 45 \times 13{\cdot}5R}$$

$$= \frac{36{,}580 \text{ m}}{R}$$

$$\text{Shift} = \frac{L^2}{24R}$$

$$= \frac{55{,}760{,}000}{R^3}$$

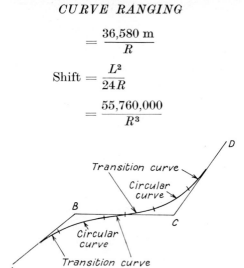

Fig. 8.17

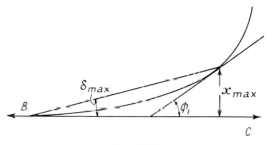

Fig. 8.18

$$800 = \left(R + \frac{55{,}760{,}000}{R^3} \right) \tan 30^\circ + \frac{36{,}580}{R}$$

$$+ \left(R + \frac{55{,}760{,}000}{R^3} \right) \tan 22 \cdot 5^\circ$$

Whence $\quad R^4 - 806 \cdot 7 R^3 + 36{,}580 R^2 + 55{,}760{,}000 = 0$

$\therefore \qquad\qquad R = 758 \cdot 5 \text{ m}$

and $\qquad\qquad L = \dfrac{36{,}580}{R} = 48 \cdot 23 \text{ m}$

Offset to locate circular curve (*see* Fig. 8.18)

$$= \frac{L^3}{6LR} = \frac{48 \cdot 23^3}{6 \times 36{,}580}$$

$$= 0 \cdot 51 \text{ m}$$

$$Deflection\ Angle = \frac{1,800\ L}{\Pi\ R} = \frac{1,800 \times 48\cdot23}{\Pi \times 758\cdot5}$$

$$= 36\cdot42'$$

$$= 00°\ 36'\ 20''$$

EXAMPLE. *Cubic Spiral Transition Curves and Circular Curve* (Fig. 8.19). It is required to join two straights having a total deflection angle 18° 36′ right by a

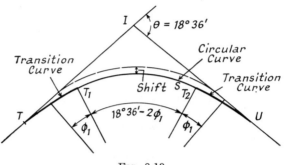

FIG. 8.19

circular curve of 450 m radius, having cubic spiral transition curves at each end. The design velocity is 70 km/h and the rate of change of radial acceleration along the transition curve is not to exceed 0·3 m/s³. Chainage of *I* is 2,524·20 m.

(i) *Determine Length of Required Transition Curve*

$$Design\ velocity = 70\ km/h$$

$$= \frac{700}{36}\ m/s$$

$\alpha =$ Rate of change of radial acceleration $= 0\cdot3$ m/s³

$$\alpha\ L = \frac{V^3}{R}$$

$$\therefore \qquad L = \frac{\left(\dfrac{700}{36}\right)^3}{450 \times 0\cdot3}$$

$$= 54\cdot44^2$$

(ii) $$Shift = \frac{L^2}{24\ R}$$

$$= \frac{54\cdot44\ m}{24 \times 450}$$

$$= 0\cdot27\ m$$

(iii) *Tangent length* $IT = (R + S)\tan\dfrac{\theta}{2} + \dfrac{L}{2}$

$$= 450{\cdot}27 \tan 9° \; 18' + \frac{54{\cdot}44}{2} \, \text{m}$$

$$= 100{\cdot}97 \; \text{m}$$

$$\therefore \quad \text{Chainage of } T = 2{,}524{\cdot}20 - 100{\cdot}97$$
$$= 2{,}423{\cdot}23 \; \text{m}$$

$$\therefore \quad \text{Chainage of } T_1 = 2{,}423{\cdot}23 + 54{\cdot}44$$
$$= 2{,}477{\cdot}67 \; \text{m}$$

(iv) *Cubic Spiral Transition Curve with 10 m Chord Lengths.* Deflection angle to locate a point on the transition distant l ft from T

$$= \frac{1{,}800}{\pi R L} \, . \, l^2 \; \text{min}$$

$$= 0{\cdot}02339 \; l^2 \; \text{min}$$

Chord (m)	l (m)	Chainage	Deflection Angle	Angle set on 20″ instrument
0	0	2,423·23	0	0
6·77	6·77	2,430·00	$0{\cdot}02339 \times 6{\cdot}77^2 = 01{\cdot}07'$	00° 01′ 00″
10·00	16·77	2,440·00	$\times 16{\cdot}77^2 = 06{\cdot}58'$	00° 06′ 40″
10·00	26·77	2,450·00	$\times 26{\cdot}77^2 = 16{\cdot}76'$	00° 16′ 40″
10·00	36·77	2,460·00	$\times 36{\cdot}77^2 = 31{\cdot}62'$	00° 31′ 40″
10·00	46·77	2,470·00	$\times 46{\cdot}77^2 = 51{\cdot}17'$	00° 51′ 20″
7·67	54·44	2,477·67	$\times 54{\cdot}44^2 = 69{\cdot}31'$	01° 09′ 20″

(v) *Circular Arc $T_1 T_2$*

$$\phi_1 = 3 \; \delta_1$$
$$= 3 \times 69{\cdot}31'$$
$$= 207{\cdot}93'$$
$$= 3° \; 27{\cdot}93'$$

Angle subtended by the circular arc

$$= 18° \; 36' - 2 \, \phi_1$$
$$= 18° \; 36' - 6° \; 55{\cdot}86'$$
$$= 11° \; 30{\cdot}14'$$
$$= 11{\cdot}502°$$

Length of circular arc $= \dfrac{\pi R}{180} \, (\theta - 2 \, \phi_1)$

$$= \frac{450}{180} \times \pi \times 11{\cdot}502$$

$$= 90{\cdot}34 \; \text{m}$$

∴ Chainage at end of circular curve

$$= 2{,}477{\cdot}67 + 90{\cdot}34$$
$$= 2{,}568{\cdot}01 \text{ m}$$

∴ Chainage of tangent point $U = 2{,}568{\cdot}01 + L$
$$= 2{,}568{\cdot}01 + 54{\cdot}44$$
$$= 2{,}622{\cdot}45 \text{ m}$$

Chord length for circular arc is to be less than $\dfrac{R}{20}$, i.e. say 20 m normally.

Chord (m)	Chainage (m)	Deflection Angle	Total Deflection Angle from tangent at T_1	Angle set on 20″ instrument
0	2,477·67	0	0	0
22·83	2,500·00	$1{,}718{\cdot}9 \times \dfrac{22{\cdot}23}{450}$ $= 85{\cdot}29'$	01° 25·29′	01° 25′ 20″
20·00	2,520·00	$1{,}718{\cdot}9 \times \dfrac{20}{450}$ $= 76{\cdot}40'$	02° 41·69′	02° 41′ 40″
20·00	2,540·00	$1{,}718{\cdot}9 \times \dfrac{20}{450}$ $= 76{\cdot}40'$	03° 58·09′	03° 58′ 00″
20·00	2,560·00	$1{,}718{\cdot}9 \times \dfrac{20}{450}$ $= 76{\cdot}40'$	05° 14·49′	05° 14′ 20″
8·01	2,568·01	$1{,}718{\cdot}9 \times \dfrac{8{\cdot}01}{450}$ $= 30{\cdot}60'$	05° 45·09′	05° 45′ 00″

Note that the first chord length just exceeds $\dfrac{R}{20}$ in length. A chord length of 2·83 m could have been adopted and then followed by one of 20 m length providing that the shortest focusing distance of the available/instrument allowed this. Alternatively a combination of say 12·83 m and 10·00 m could have been selected.

(vi) *The Second Transition Curve.* The final portion of the work is to set out the second transition curve from U. For running chainages pegs will be placed as shown in the following Table.

Note: When locating this transition curve it is a *left*-hand curve from UI and if the theodolite reads clockwise and the telescope is swung anti-clockwise the angles will be as given.

Chainage (m)	l m from U	Chord (m)	Deflection Angle $0.02339\,l^2$	Angle set on 20" instrument	Offsets from UI $x = \dfrac{l^3}{6\,LR}$ (m)
2,568·01	54·44	2·01	$0.02339 \times 54.44^2 = 69.31'$	358° 50' 40"	1·10
2,570·00	52·45	10·00	$0.02339 \times 52.45^2 = 64.35'$	358° 55' 40"	0·98
2,580·00	42·45	10·00	$0.02339 \times 42.45^2 = 42.15'$	359° 18' 00"	0·52
2,590·00	32·45	10·00	$0.02339 \times 32.45^2 = 24.63'$	359° 35' 40"	0·23
2,600·00	22·45	10·00	$0.02339 \times 22.45^2 = 11.77'$	359° 48' 20"	0·08
2,610·00	12·45	10·00	$0.02339 \times 12.45^2 = 3.63'$	359° 56' 20"	0·01
(U) 2,622·45	0	0	0	360° 00' 00"	0

Vertical Curves

Curves will be required at the intersection of gradients, such curves being known as vertical curves. Two cases occur, gradients meeting at *summits,* and gradients meeting at *sags,* as shown in Fig. 8.20.

The gradients themselves are conveniently expressed as percentages, thus a 1 in 20 gradient is a 5 per cent gradient, and depending on its

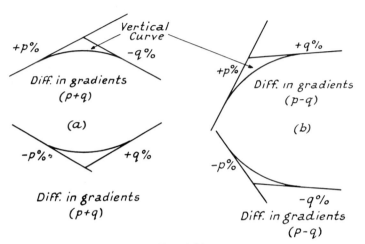

Fig. 8.20

direction, is positive or negative. The following convention is used in this section:

Gradients rising to the right—positive
Gradients falling to the right—negative
Left-hand gradient—p per cent,

$$\text{i.e. } 1 \text{ in } \frac{100}{p} \text{ gradient}$$

and the gradient makes an angle of $\frac{p}{100}$ radians with the horizontal

Right-hand gradient—q per cent,

$$\text{i.e. } 1 \text{ in } \frac{100}{q} \text{ gradient}$$

Algebraic difference of gradients—$(p - q)$ per cent.

It will be seen that every 100 m forward the level along the gradient changes by p or q m.

SHAPE OF CURVE. Where the ratio of length of curve to radius is less than 1 to 10, there is no practical difference between the shapes of a circle, a parabola, and an ellipse, and since this condition can be shown to apply in the cases normally met with, the parabola will be used.

LENGTH OF CURVE. Factors affecting the length of a vertical curve are: (a) centrifugal effect, (b) visibility.

At sags, and at summits formed by flat gradients, centrifugal effect is the chief factor, but at summits where the algebraic change of gradient is large, visibility is the ruling factor.

Curve Design. *Note:* In the following analysis the conditions shown

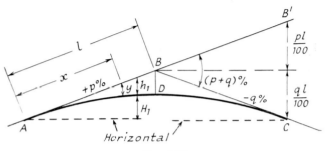

FIG. 8.21

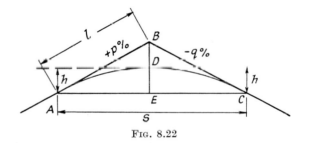

FIG. 8.22

in Figs. 8.21 and 8.22 are used, and in any particular case the appropriate signs would have to be applied to $(p + q)$.

The properties of the parabola give—

(a) The vertical through the intersection of the tangents B is a diameter and bisects AC.

(b) $AB = BC$, and $BD = DE$. D is the vertex of the parabola.

(c) Offsets from the tangent AB are proportional to the square (Fig. 8.21) of the distance from A. (These offsets should be at right angles to the tangent, but in the case of the flat gradients usually involved it is sufficiently accurate to take them vertically.)

(d) Offsets from the tangent FDG are proportional to the square of the distance from D (Fig. 8.24).

If
$$y = Kx^2$$

$$\frac{dy}{dx} = 2\,Kx$$

$$\frac{d^2y}{dx^2} = 2\,K = \text{constant}$$

i.e. the parabola gives an even rate of change of gradient.

When
$$x = 2l$$

$$y = B'C$$

$$= (p + q)\frac{l}{100} \text{ in Fig. 8.21.}$$

$$\therefore \qquad (p + q)\frac{l}{100} = K\,4l^2$$

$$K = \frac{p + q}{400\,l}$$

$$y = \frac{p + q}{400\,l}\,x^2$$

when
$$x = l$$

$$y = BD$$

$$= \frac{p + q}{400\,l}\,l^2$$

$$= (p + q)\frac{l}{400} = DE$$

In setting out the curve it is necessary to compute the offsets at various chainages and apply these to the known levels on the gradients; thus the final levels for the vertical curve may be determined. The procedure is shown in the examples given. For flat gradients it is accurate enough to treat the length along the tangents (i.e. $2l$) as equal to (a) the length of the curve, (b) the chord AC and (c) the horizontal projection of AC.

HIGHEST POINT. Taking A as the datum point, the highest point of the curve will be at a height of $\left(\dfrac{xp}{100} - h_1\right)$ above A, where h_1 is the offset from the tangent AB at this point, whose distance from A is x.

$$\text{Say} \quad H_1 = \frac{xp}{100} - \frac{(p + q)}{400\,l}\,x^2$$

for maximum value of H_1

$$\frac{dH_1}{dx} = 0$$

i.e. $\qquad \dfrac{p}{100} - \dfrac{2(p+q)}{400\,l}\,x = 0$

$\therefore \qquad\qquad x = \left(\dfrac{2p}{p+q}\right) l$

CENTRIFUGAL EFFECT. It has already been pointed out that the parabolas in vertical curves can be approximated to circular curves.

Let $\qquad\qquad\qquad R = $ radius

$$\dfrac{d^2y}{dx^2} = 2\,K$$

$$= \dfrac{1}{R}$$

$\therefore \qquad\qquad \dfrac{1}{R} = 2\cdot\dfrac{p+q}{400\,l}$

$$= \dfrac{p+q}{200\,l}$$

$\therefore \qquad\qquad R = \dfrac{200\,l}{p+q} = \dfrac{v^2}{f}$

where $f = $ allowable centrifugal acceleration for velocity v.

Therefore length of curve

$$= (p+q)\,\dfrac{v^2}{100\,f}$$

$$= K(p+q)$$

At summits on roads where speeds of 100 km/h are contemplated, the requirements of visibility, i.e. the sight line, can lead to longer curves than are given by the formula above.

SIGHT DISTANCES. Let two points on the curve at a height h from the ground be intervisible, and let the distance between them be S. The sight line is taken to pass tangentially through D parallel to AC Fig. 8.22) and the sight distance thus represents the length of road over which an observer whose eye level is h above the road surface can just see an object h above the surface on the other side of the crest to the observer. A value of 1·05 m is usually taken as the eye level height above the road surface for an observer sitting in a motor car. Sight distances are laid down in the interests of road safety and the choice of any particular distance depends on the nature of the road and the speed of the traffic using it. The following K values are suggested in *The Layout of Roads in Rural Areas*, issued by the Department of the Environment, in respect of single carriageways—

Speed	Carriageway Width	Stopping Sight Distance		Overtaking Sight Distance on Summit
		Summit	Sag	
100 km/h	10·0 m	50	50	240
100 km/h	7·3 m	50	50	240
80 km/h	7·3 m	25	30	150

There are three cases to consider—

(*a*) Sight distance equal to length of curve. Then in Fig. 8.22,

$$S = 2\,l, \text{ and } DE = h = \left(\frac{p+q}{400}\right)l$$

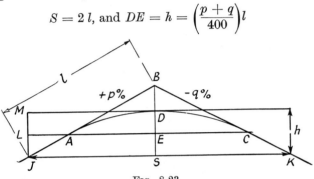

FIG. 8.23

Given h, p and q, then l may be determined and the offsets for the vertical curve computed from

$$y = \left(\frac{p+q}{400\,l}\right)x^2$$

(*b*) Sight distance longer than the curve (Fig. 8.23).

$$h = JL + ML$$
$$= JL + DE$$
$$= JL + BD$$

Ht. of B above $A = \dfrac{pl}{100}$

Ht. of C above $A = \dfrac{pl}{100} - \dfrac{ql}{100}$

$$= \frac{l}{100}\,(p - q)$$

Then the angle between AC and the horizontal

$$= \frac{l}{100}\frac{p-q}{2\,l} = \frac{p-q}{200} \text{ radians}$$

The reader can check, by considering the rate of change of gradient, that AC and MD are parallel.

Thus
$$L\hat{A}J = \frac{p}{100} - \frac{p-q}{200}$$

$$= \frac{p+q}{200} \text{ radians}$$

But
$$AL = \frac{S}{2} - l$$

$\therefore$
$$JL = \left(\frac{p+q}{200}\right)\left(\frac{S}{2} - l\right)$$

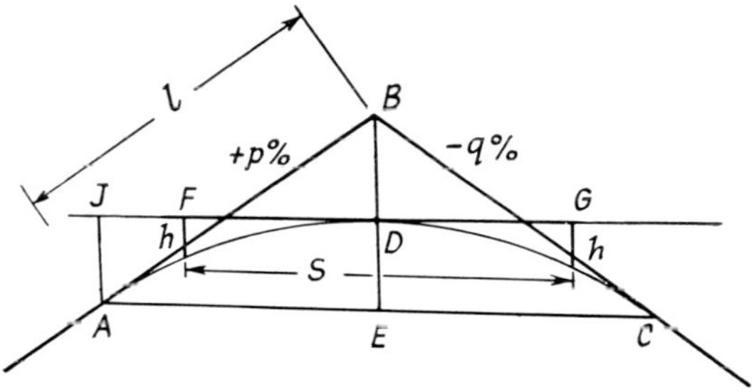

FIG. 8.24

now
$$h = JL + BD$$

$$- \left(\frac{p+q}{200}\right)\left(\frac{S}{2} - l\right) + \frac{l}{400}(p+q)$$

$$= \left(\frac{S-l}{400}\right)(p+q)$$

EXAMPLE. If $S - 4l$, $p - 1$ per cent, $q = -1$ per cent, and $h - 1·05$ m,

hen
$$1·05 = \left(\frac{4l-l}{400}\right)(1+1)$$

$$6l = 1·05 \times 400$$

$$l = 70 \text{ m}$$

Therefore equation for offsets from tangent measuring from A is
$$y = \left(\frac{p+q}{400\,l}\right)x^2$$

$$= \frac{(1+1)}{400\,.\,70}\,.\,x^2$$

$$= 0·0000714\,x^2$$

(c) Sight distance less than length of curve (Fig. 8.24). The equation or offsets from the tangent FDG with the origin at D is

$$y_1 = Kx^2$$

Hence $\qquad\qquad h = K\left(\dfrac{S}{2}\right)^2$ since $y_1 = h$ when $x = \dfrac{S}{2}$

$\therefore \qquad\qquad\qquad K = \dfrac{4\,h}{S^2}$

at point J, $\qquad\qquad x_1 = l$

$\therefore \qquad\qquad\qquad JA = \dfrac{4\,h}{S^2}\cdot l^2$

But $\qquad\qquad\qquad JA = DE = BD$

$\qquad\qquad\qquad\qquad = \dfrac{l}{400}\,(p+q)$

$\therefore \qquad\qquad \dfrac{4\,h}{S^2}\cdot l^2 = \dfrac{l}{400}\cdot(p+q)$

$\therefore \qquad\qquad\qquad l = \dfrac{(p+q)}{1,600}\cdot\dfrac{S^2}{h}$

EXAMPLE.

If $\qquad\qquad S = $ half the total length of the curve

$\qquad\qquad\qquad = l$

and $\qquad\qquad p = 2$ per cent, $q = -2$ per cent, and $h = 1\cdot05$ m,

then $\qquad\qquad l = \dfrac{4}{1,600}\cdot\dfrac{l^2}{1\cdot05}$

$\therefore \qquad\qquad\qquad l = 420$ m

The offsets are computed in the usual way.

EXAMPLE. Design a vertical curve 200 m long connecting a rising gradient of 1 in 50 with a falling gradient of 1 in 75 which meet in a summit of R.L. 30·35 A.O.D., and chainage 2,752 m. Give offsets at 20 m intervals. What is the rate of change of gradient?

Referring to Fig. 8.25, $AB = BC = l = 100$ m

$\qquad\qquad$ Chainage of $A = 2,752 - 100$

$\qquad\qquad\qquad\qquad = 2,652$ m

$\qquad\qquad$ Level of $B = 30\cdot35$ m

$\qquad\qquad$ Level of $A = \dfrac{lp}{100}$ below B

$\qquad\qquad\qquad\qquad = \dfrac{100 \times 2}{100}$

$\qquad\qquad\qquad\qquad = 2$ m below B, i.e. 28·35 A.O.D.

$\qquad\qquad$ Level of $C = \dfrac{lq}{100}$

$\qquad\qquad\qquad\qquad = \dfrac{100 \times 1\cdot33}{100}$

$\qquad\qquad\qquad\qquad = 1\cdot33$ m below, i.e. 29·02 A.O.D.

Measuring offsets from tangent AB throughout, the final offset

$$B_1C = \frac{p+q}{100} \cdot l$$

$$= \left(\frac{2 + 1\cdot33}{100}\right) \times 100$$

$$= 3\cdot33 \text{ m}$$

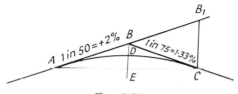

FIG. 8.25

Equation for offsets from tangent ABB_1 is

$$y = \left(\frac{p+q}{400\,l}\right)x^2$$

$$= \frac{3\cdot33}{400 \cdot 100} \cdot x^2$$

$$= \frac{1}{12,000} \cdot x^2$$

The grade levels are determined along ABB_1 from the formula

$$\text{Grade level} = \text{Level of } A + \frac{xp}{100}$$

$$\text{Height of curve above } A = \frac{px}{100} - Kx^2$$

$$= h$$

to locate highest point put $\dfrac{dh}{dx} = 0$

i.e.
$$\frac{p}{100} = 2\,Kx$$

Setting-out Table

Point	A						B
Chainage	2,652	2,660	2,680	2,700	2,720	2,740	2,752
x m	0	8	28	48	68	88	100
Grade Level (Level of A + $xp/100$)	28·35	28·51	28·91	29·31	29·71	30·11	30·35
Offset (m) $y = \dfrac{1}{12,000}x^2$	0	0·01	0·07	0·19	0·39	0·65	0·83
Curve Level	28·35	28·50	28·84	29·12	29·32	29·46	29·52

Point	B (vertex)						C
Chainage	2,752	2,760	2,780	2,800	2,820	2,840	2,952
x m	100	108	128	148	168	188	200
Grade Level (Level of A + $xp/100$)	30·35	30·51	30·91	31·31	31·71	32·11	32·35
Offset (m) $y = \dfrac{1}{36,000}x^2$	0·83	0·97	1·36	1·82	2·35	2·94	3·33
Curve Level	29·52	29·54	29·55	29·49	29·36	29·17	29·02

$$\therefore \qquad x = \frac{p}{200\,K}$$

$$= \frac{2}{200} \cdot \frac{12,000}{1}$$

$$= 120 \cdot 0 \text{ m}$$

$$\text{Grade Level} = 120 \times \frac{2}{100} = 2 \cdot 4 \text{ m above } A$$

$$= 30 \cdot 75 \text{ m A.O.D.}$$

$$\text{Offset} = \frac{1}{12,000} \times 120 \times 120 = 1 \cdot 20 \text{ m}$$

$$\therefore \qquad \text{Curve level} = 29 \cdot 55 \text{ A.O.D.}$$

Rate of Change of Gradient

Total change of gradient $= 3·33$ per cent in 200 m

$$\therefore \qquad \text{Rate of change} = \frac{3 \cdot 33}{200}$$

$$= 0 \cdot 0167 \text{ per cent per m}$$

It should be noted that offsets may be taken from tangent CB instead of AB, and that the final offset will still be

$$\frac{2 + 1 \cdot 33}{100} \times 100 \text{ to give } A$$

An alternative method is to take offsets from both tangents working from A to B and C to B. The formula $y = Kx^2$ will still obtain, and since $y = BD$ when $x = l$,

$$y = \frac{BD}{l^2} \cdot x^2$$

relates to offsets to the curve distant x from either A or C.

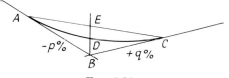

Fig. 8.26

EXAMPLE. Design a vertical sag curve 200 m long connecting a falling gradient of 1 in 50 with a rising gradient of 1 in 75. The R.L. of the intersection point of the gradients is 30·35 A.O.D. and the chainage is 2,750 m (*see* Fig. 8.26).

$$AB = BC = l = 100 \text{ m}$$

$$A \text{ is now } 100 \times \frac{2}{100} = 2 \cdot 00 \text{ m above } B$$

$$C \text{ is now } 100 \times \frac{1 \cdot 33}{100} = 1 \cdot 33 \text{ m above } B$$

As in previous example

$$y = \frac{p + q}{400\, l} \cdot x^2$$

$$= \frac{1}{12,000}\, x^2$$

$$\therefore \qquad BD = \frac{100^2}{12,000}$$

$$= 0 \cdot 83 \text{ m}$$

Therefore formula for offsets from either tangent AB or CB is

$$y = 0 \cdot 83 \left(\frac{x}{100}\right)^2$$

The lowest point on the curve is determined exactly as the highest point on the curve in the previous example. The rate of change of grade is the same as in the previous case.

Note: Offsets are added to grade levels since the curve is above the gradients. Offset 50 m from A = offset 50 m from C, and so on.

Setting-out Table

Point	A						B
Chainage	2,650	2,660	2,680	2,700	2,720	2,740	2,750
x (from A)	0	10	30	50	70	90	100
Grade Level (Level of A $- xp/100$)	32·35	32·15	31·75	31·35	30·95	30·55	30·35
Offset $y = \dfrac{x^2}{1^2} BD$	0	0·01	0·07	0·21	0·41	0·67	0·83
Curve Level	32·35	32·16	31·82	31·54	31·36	31·22	31·18

Point	B						C
Chainage	2,750	2,760	2,780	2,800	2,820	2,840	2,850
x (from C)	100	90	70	50	30	10	0
Grade Level (Level of C $- xq/100$)	30·35	30·48	30·74	31·01	31·28	31·55	31·68
Offset $y = \dfrac{x^2}{1^2} BD$	0·83	0·67	0·41	0·21	0·07	0·01	0
Curve Level	31·18	31·15	31·15	31·22	31·35	31·56	31·68

The second portion of the curve may be set off by offsets from AE and it is recommended that the reader completes the Table, locates the lowest point, and determines the rate of change of grade.

EXAMPLE. Two gradients of 1 in 50 and 1 in 75 meet at a summit (R.L. 30·35 chainage 2,752). What length of vertical curve is required if two points 1·05 m above road level and 200 m apart are to be intervisible? (*See* Fig. 8.27.)

$$\text{Gradient of } AC = \frac{l(p-q)}{100} \cdot \frac{1}{2\,l}$$

$$= \frac{p-q}{200}$$

$$JL = (100 - l)\left(\frac{p}{100} - \frac{(p-q)}{200}\right)$$

but $$JL + JM = 1·05$$

and $$JM = DE = DB$$

$$1·05 = (100 - l)\left(\frac{p+q}{200}\right) + BD$$

Now BD is the offset at l m from A measured from tangent AB using

$$y = \left(\frac{p+q}{400\,l}\right)x^2$$

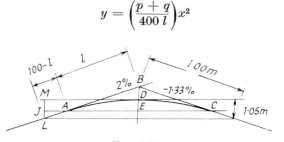

FIG. 8.27

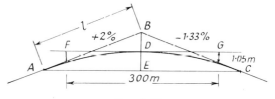

FIG. 8.28

Then
$$BD = \frac{p+q}{400}.l$$

$\therefore$
$$(100 - l)\left(\frac{2+1\cdot33}{200}\right) + (2+1\cdot33).\frac{l}{400}$$

$$= 1\cdot05 \text{ m}$$

$\therefore$
$$l = 74 \text{ m}$$

So that the curve will be 148 m long (this can be checked using Fig. 8.29).

The offsets will be determined from the formula

$$y = \left(\frac{p+q}{400\,l}\right)x^2$$

as shown in previous examples.

EXAMPLE. Design a vertical curve connecting two gradients of 1 in 50 to 1 in 75 at a summit (R.L. 30·35 m, chainage 2,752). The curve is to be such that two points 300 m apart and 1·05 m above the curve are intervisible (*see* Fig. 8.28).

Offsets from line FDG parallel to AEC are to obey the law

$$y = Kx^2$$

F and G being the intervisible points.

At a distance of 150 m from D, the offset must therefore be 1·05 m

$$\therefore \qquad 1 \cdot 05 = K \times 150^2$$

$$K = \frac{1 \cdot 05}{150^2}$$

$$\text{Offset at } A = \frac{1 \cdot 05}{150^2} \times l^2 = BD = \frac{p + q}{400} \, l$$

$$\therefore \qquad = \frac{3 \cdot 33}{400} \times \frac{150^2}{1 \cdot 05}$$

$$= 178 \cdot 57 \text{ m}$$

$$\therefore \qquad BD = \frac{3 \cdot 33}{400} \times 178 \cdot 57$$

$$= 1 \cdot 49 \text{ m}$$

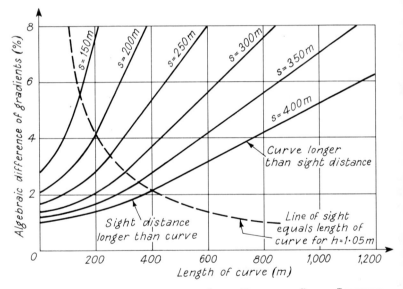

Fig. 8.29. Relationship between Sight Distance, Curve Lengths, and Algebraic Difference of Gradients

The setting-out data for the vertical curves designed previously have been given assuming the reduced level of the controlling feature is known to 0·01 m. Differences when offsets are calculated to 0·001 m and incorporated with grade levels taken to the same limits are shown below for this particular example.

Chainage	2,573·43	2,600·00	2,650·00	2,700·00	2,750·00	2,752·00
Grade Level	299·929	300·460	301·460	302·460	303·460	303·500
Offset	0	0·033	0·274	0·748	1·455	1·488
Curve Level	299·929	300·427	301·186	301·712	302·005	302·012

If working to 0·01 m the following values would have been derived—

Curve Level	299·93	300·43	301·19	301·71	032·01	302·01

A good quality level set up between A and B would allow the setting-out of these values expressed to 0·001 m, but for highway setting-out those to 0·01 m may be considered reasonable.

The second portion of the curve may be set off in the usual way and again the reader is recommended to complete the Table.

GRAPHICAL REPRESENTATION OF VERTICAL CURVE FORMULAE. The relationship between sight distance, curve length and algebraic difference of gradients for slopes meeting at a summit is shown graphically in Fig. 8.29. The student will readily appreciate, from a comparison of the graph and the formulae, that where the curve length is longer than the sight distance, the relation between algebraic difference of gradients and curve length is a linear one, so that the graph is easily extended to cover steeper slopes.

EXERCISES 8

1. Two roads having a deviation angle of 45° 50′ are to be joined by a 180 m radius curve. Calculate the necessary data and explain in detail how to set out the curve,

   (a) by chain and offsets only,
   (b) if a theodolite is available. (*L.U., B.Sc.*)

2. To locate the exact position of the tangent point T_2 of an existing 250 m radius circular curve in a built-up area, points a and d were selected on the straights close to the estimated positions of the two tangent points T_1 and T_2 respectively and a traverse *abcd* was run between them (*see* below).

Station	Length (m)	Deflection Angle
a		9° 54′ R
	89·0	
b		19° 36′ R
	115·5	
		30° 12′ R
	101·5	
d		5° 18′ R

The angles at a and d were relative to the straights. Find the distance T_2d. (*L.U., B.Sc.*)

   Answer: 16·98 m.

3. Two straights AI and BI meet at I on the far side of a river. On the near side of the river, a point E was selected on the straight AI, and a point F on the straight BI, and the distance from E to F measured and found to be 85 m.

   The angle, $A\hat{E}F$, was found to be 165° 36′ and the angle, $B\hat{F}E$, 168° 44′. If the radius of a circular curve joining the straights is 500 m, calculate the distance along the straights from E and F to the tangent points. (*I.C.E.*)

   Answer: 75·55 m; 65·10 m.

4. A transition curve is required for a circular road curve which has a centrifugal ratio of $\frac{1}{4}$ for a maximum speed of 100 km/h. If the rate of radial acceleration is not to exceed $\frac{1}{3}$ m/s³, what length of transition curve is required?

If the transition curve is to be in the form of a cubic spiral, what will be the deflection angles required in order to set it out ? Assume pegs are to be placed at 10 m intervals.

Answer: 204·3m; Final angle 6° 11·9′.

5. A 100 m long transition curve connects a straight motor road to a 600 m radius circular curve. Over this length the road is in cutting, all sections being level across. Depth of cutting varies uniformly from 2·50 m at the tangent point to 4·50 m at the junction point, width of road is 9·00 m, side slopes 1½ hor. to 1 vert. The road is to be widened gradually on the inner side of the curve from zero at the T.P. to 2·0 m extra at the J.P., the amount of the widening being proportional to distance from the T.P. along the surveyed centre line: new side slopes to be the same as the old.

Determine the volume of excavation, considering sections at 25 m intervals. (*L.U., B.Sc.*)

Answer: 289 m³.

6. An upgrade of 1 per cent meets a downgrade of 2 per cent at chainage 5,700 where the grade level is 130·43 m above datum. If the tangents are each 25 m long, design a suitable vertical curve giving all levels and offsets at 5 m intervals.

Answer: Offset at intersection = 0·19 m.

7. A parabolic vertical curve of length 100 m is formed at a summit between grades of 0·7 per cent up and 0·8 per cent down. The length of the curve is to be increased to 120 m, retaining as much as possible of the original curve and adjusting the gradients on both sides to be equal. Determine this gradient. (*L.U., B.Sc.*)

Answer: 0·9 per cent.

8. On a straight portion of a new road, an upward gradient of 1 in 100 was connected to a downward gradient of 1 in 150 by a vertical parabolic summit curve of length 150 m. A point *P*, at chainage 5,910·0 m, on the first gradient was found to have a reduced level of 45·12 m, and a point *Q*, at chainage 6,210·0 m on the second gradient of 44·95 m.

(*a*) Find the chainages and reduced levels of the tangent points to the curve.

(*b*) Tabulate the reduced levels of the points on the curve at intervals of 20 m from *P* and of its highest point.

Find the minimum sighting distance to the road surface for each of the following cases—

(*c*) the driver of a car whose eye is 1·05 m above the surface of the road;

(*d*) the driver of a lorry for whom the similar distance is 1·80 m.

(Take the sighting distance as the length of the tangent from the driver's eye to the road surface.) [*L.U., B.Sc.*]

Answer: (*a*) Chainages 5,944·8, 6,094·8; (*b*) R.L. 45·92; (*c*) 163·8 m; (*d*) 253·8 m.

CHAPTER 9

TRIANGULATION SURVEYS

WE HAVE ALREADY SEEN in chain surveying that stations may be located by building up a network of triangles measuring the lengths of the lines, but if this technique were extended to cover more than relatively small areas the errors inherent in the making of the linear measurements would become so great as seriously to reduce the accuracy of the plan. Thus, when a large area is to be surveyed a more rigorous approach is necessary and recourse is made to *triangulation surveys*.

This is the name given to surveys in which the area is divided into geometrical figures, the corners of which form a series of accurately located control stations from which more detailed surveys or location are carried out by the methods already described. The distinguishing features of triangulation surveying are shown in Fig. 9.1. If angles A, B, and C are measured in triangle ABC, and the length of side AB is measured, then the lengths of sides BC and CA can be calculated. In triangles ACD, DCE, etc., it is then only necessary

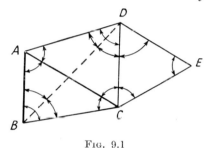

FIG. 9.1

to measure the angles, the lengths of all the other sides of the whole triangulation being worked out from these observed angles (after making necessary adjustments) and the length of the initial side, which is called the *base line*. Maximum use is thus made of the most accurate of surveying instruments, the theodolite, while linear measurement by taping, which is difficult and tedious when carried out to the same order of accuracy as good angular measurement, is reduced to the measurement of a single base line. (In practice, as will be discussed later, two or more bases are often measured to assist in distributing any errors which may occur.) This method of surveying was first introduced by the Dutchman, Snell, in 1615.

The introduction of electronic methods of distance measurement discussed later in the chapter not only allows the measurement of base lines, but also the measurement of selected sides of the triangulation network, which leads to further reliability.

In the geodetic triangulation survey carried out by the Ordnance survey, most of the sides of the primary triangles have lengths between 40 km and 60 km, while some, such as the three sides of the triangle whose corners are Scafell Pike, Snowdon, and Slieve Donard (Ireland), are over 150 km long. It is evident, therefore, that the greatest possible accuracy is required in such work. These large primary triangles are then broken down into smaller (secondary) triangles of somewhat lesser accuracy. There is a further breakdown to third and fourth order points which are used for detail surveys and for providing control in large-scale photogrammetric work. The fourth order points give closer spacing in towns—tertiary and higher order points cover almost the whole country at a density of 0·05 trig point per km², with a density of about 0·1 per km² near towns; fourth order points increase this density to 0·7 trig point per km² in towns. For less extensive surveys, i.e. topographical surveys and triangulations carried out to locate engineering works, less accuracy is required than for geodetic work, but since nowadays a similar type of apparatus for base line and angle measurement is often used in both classes of work, the principles involved are the same. In dealing with triangulation in more detail, the measurement of base lines will be considered first, followed by the principles involved in the actual triangulation itself.

APPLICATION OF TRIANGULATION SURVEYS. As will be apparent from the foregoing, triangulation is used for—

1. the establishment of accurately located control points for plane and geodetic surveys of large areas,

2. the establishment of accurately located control points in connexion with aerial surveying,

3. the accurate location of engineering works such as (i) centre lines, terminal points and shafts for long tunnels, and (ii) centre lines and abutments for bridges of long span.

Base Line Measurement

At one time base lines were measured by either (*a*) the short-length, or (*b*) the long-length system, but as it is probably true to say that the latter system is used almost exclusively nowadays, only the briefest mention will be made here of the short-length methods.*

(*a*) **Short-length Methods.** The earliest base lines were nearly all measured by short bars, and the following list gives some of the most famous types of apparatus.

COLBY'S APPARATUS. Two parallel bars, one of steel, the other of brass, are riveted together at their centres. They are joined at each end by tongues of metal which project on the same side as the steel bar, the distance between two points, one on the end of each tongue, being equivalent to 3·048 m. The tongues are pivoted at their connexions with the bars, and by making in each case the ratio between the lateral displacement of the point from the steel bar pivot, and its displacement from the brass bar pivot as 3:5, i.e. in the same ratio as the coefficients of linear expansion of steel and brass, the longitudinal distance between the points is maintained exactly over normal temperature ranges. The bars were used in measuring base lines in England, Ireland, and India.

BORDA'S RODS. These are compound bars of platinum and copper firmly riveted together at one end. Expansion or contraction of the

* For a full account see *Plane and Geodetic Surveying*, Vol. II, by Clark (Constable).

copper relative to the platinum is measured by a scale and vernier on the other ends, and gives a means of finding the average temperature. Borda's rods were used in the French Survey in 1805.

BESSELL'S APPARATUS. The same principle is employed here as in Borda's apparatus, except that the bimetal combination is iron and zinc. The apparatus was used on the Baltic Coast bases in 1836.

WOODWARD ICED BAR APPARATUS. The bar, which is of steel, has two points 5·02 metres apart on the neutral axis, and rests in a bath of crushed ice. This apparatus was used in the U.S. Coast and Geodetic Survey.

The bars in each case rested on accurately levelled and aligned trestles, micrometers being used to align the index marks of successive bars and, by the exercise of great care (which rendered such base line measurement costly in both time and money), accuracies of the order of ± 1 in 1×10^6 and higher were achieved. Apart from the difficulties involved, however, in the actual use of such short-length equipment, the need for long flat sites over which the trestle supports could be laid was a serious limitation, since such sites are not always available in the most convenient positions relative to the triangulation as a whole.

(*b*) **Long-length Methods.** Because of the limitations of short-length equipment, attention began to be paid to the use of long-length measuring devices, these consisting of tapes or wires of length varying from 24 m to some 90 m. They are hung clear of the ground, in catenary, and since corrections may be applied for slope this method does not impose the same limitation on choice of site. In addition, the apparatus is cheaper than the old short-bar equipment, and is quicker and easier to use. This combination of general convenience, versatility and economy has led, therefore, to the measurement of longer bases at more frequent intervals in recent surveys.

In 1889 Jáderin used 25-m tapes of steel and brass, and in 1890 Wheeler in the Missouri River Survey used steel tapes of length equivalent to 91·440 m draped at a known tension over a series of equidistant supporting pegs. Both achieved probable errors of one part of a million, but owing to the difficulty of measuring temperature effects over such long lengths, the method still lacked convenience. In 1896 the French scientist Dr. Guillaume discovered the alloy "invar" (a nickel-iron alloy containing approximately 36 per cent nickel) which has the lowest coefficient of expansion of any known metal or alloy, and its use in the manufacture of tapes, by considerably removing temperature effects, was a great step forward. Amongst base lines which have been measured with invar tapes are the Lossiemouth base, Elgin, 1909—·24 km long; the Ridge Way base, near Swindon, 1951—11·06 km long; the Caithness base, near Thurso, 1952—24·83 km long. All were measured with invar tapes of length equivalent to 30·480 m.

Principles of Base Measurement with the Tape

The base line should be located for preference on level ground, or on evenly-sloping ground. A limiting gradient of 1 in 12 may be specified, since at much steeper gradients the tape tends to run downhill over the pulleys which support it, but it may be noted that gradients of 1 in 3 have been encountered.* Such slopes modify the shape of the catenary, and correction must be made for the difference in tension from one end of the tape to the other (*see* O.S. Professional Papers, New Series, No. 1). The site should be free of obstructions, though it is not essential that the line be straight since a correction for departure from the straight can be applied. The ends of the base line should be intervisible and should be suitable for use as triangulation stations;† otherwise they should permit easy extension of the base as described later.

Tapes. For the measurement, invar tapes or wires are used, lengths of 24 m, 30 m and 50 m being available, with widths of about 6 mm and thicknesses of 0·5 mm. These can be graduated directly on the wide face over a short distance at each end in the region of the zero marks but wires need to have short graduated scales, called *reglettes*, attached. In addition, tapes wind on to the carrying drums more easily, and twists and kinks are more readily detected. In either case, however, careful attention will have to be paid to standardization when (*a*) newly put into service, and (*b*) during continuous service. The reason for (*a*) is that some sort of molecular creep takes place during the first months of use, and though this may be accelerated by special annealing by the maker,‡ a certain amount still occurs. Standardization to detect such creep must of course be done either against an already "aged" tape, or by the National Physical Laboratory. Routine checking (*b*) must always be carried out at regular intervals during measuring operations. Two invar tapes should be carried solely for such checking, which should be done say every two days.

Because of the greater stability, strength and kink-resistance of steel tapes (and because of their relative cheapness), attempts have been made to overcome the difficulties inherent in accurate correction for thermal expansion, thus enabling them to be brought back into use in base measurement. In one of these methods, the steel tape is hung in catenary in the usual way, and by means of electrical connexions at each end it is made to form part of a potentiometer or Wheatstone Bridge circuit. Current is passed through the tape causing its temperature to rise and its resistance to alter. To calibrate, resistance is measured simultaneously with length over a range of currents from 0·5 A to 5·0 A. In the field, it is only necessary to pass the current long enough for equilibrium conditions to be achieved, and then read the resistance. The calibration accuracy is given‖ as 1 part in 2 million, and the tape needs to be re-calibrated every 5 years. This method was used in 1938 for base measurement in Australia.

* *Maps and Surveys*, by A. R. Hinks (Camb. U.P.).
† "The East African Arc. IV Base Measurement," by Major (now Brigadier) M. Hotine, *Empire Survey Review*, Vol. III, No. 18, 1935.
‡ *Maps and Surveys*, op. cit.
‖ "Standardization of Steel Surveying Tapes," by J. S. Clark, A.R.C.S., B.Sc., D.I.C., and L. O. C. Johnson, *Empire Survey Review*, July 1951, Vol. XI, No. 81.

Straining Equipment, Measurement of Base. During measurement the tape or wire is, as already stated, hung in catenary, the actual support being provided by wires of small diameter which are attached to the brass rings at each end of the tape. These wires pass over pulleys in the straining trestles and are tensioned by weights attached to their other ends. The tensioning weight should be that at which the tape was calibrated; but in any case must be known, and not less than twenty times the weight of the tape. Should it not be possible to tension the tape by this amount, then it may be supported at mid-span or third points, the supports to be at constant gradient.

Fig. 9.2 shows a straining trestle, index head and optical plummet which form part of the standard equipment manufactured by Hilger & Watts, Ltd. The equipment, which is typical, utilizes tripod-mounted, standard centring bases which have been designed to take an index head, a plummet, target or theodolite; the index head has a central index line and roller support for the tape. Transfer to, and from, the ground is arranged by means of the plummet, a procedure which is required at the beginning and end of the line, and whenever work is stopped.

At the commencement a plummet is set up and centred over the ground point, or alternatively the point fixed on the ground using the plummet. This plummet is replaced by a theodolite which is used to align six targets, say, the positions of these having been located by invar tape or wire. The theodolite and targets can now be replaced by index heads, and the measuring tape is suspended in catenary between the two self-aligning pulleys of the straining trestles, so that its graduated portions are in position relative to the first two index heads. Adjustments are now made to the pulleys in the horizontal and vertical planes in order to bring the tape in line with the index heads. As may be seen in Fig. 9.2 the head is cut away so that the tape may be brought close to the reference mark, and a roller support is provided so that the catenary is not disturbed.

By means of removable reading glasses on the index heads, readings are now taken of the positions of the two reference marks relative to the graduated scales at each end of the tape. The booked distance between measuring heads is thus the known distance between the tape zero-marks with the correction given by the two scale readings. To avoid errors due to pulley friction, the tape is displaced by small amounts, and perhaps ten readings taken.

Unless protecting screens are erected, work is not possible on windy days, owing to tape movement. Temperature is measured at a number of points along the tape while the scale readings are being made, and the mean temperature obtained. In general, dull and cloudy days are preferable for this work.

The following equipment is suggested.

1. Invar tapes on metal reels, 635 mm dia.
2. Straining trestles, with weights to give adequate tension.
3. Tape index heads for measuring distance.
4. Tape standardizing heads for comparing field tape with a standard tape.

Fig. 9.2a

Fig. 9.2b

FIG. 9.2. BASE MEASURING EQUIPMENT
(Courtesy of Hilger & Watts, Ltd.)

Fig. 9.2c

Fig. 9.2d

(a) ST524 straining trestle; (b) ST504 tape indexing head; (c) ST542 target; (d) ST555 optical plummet.

The tape-supporting roller and index line are mounted on a slide whose movements can be measured by micrometer.

5. Levelling and centring bases (together with tripods). These may have a movement of 50 mm and the heads can be fitted into them. Spirit levels are fitted to the latter but the bases possess the footscrews for levelling up.

6. Optical plummet, which can be placed on the levelling base; it must be downward sighting for overground work, but can also include upward sighting for underground work.

7. Targets, which also must fit into the levelling bases.

For an authoritative (and pre-eminently readable) account of similar equipment and its use in base measurements the student is recommended to read *The Measurement of the Ridge Way and Caithness*

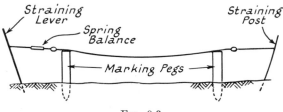

Fig. 9.3

Bases 1951–1952, by Major M. H. Cobb, M.A., A.R.I.C.S., R.E., O.S. Professional Papers, New Series, No. 18, (H.M.S.O.).

Less Precise Base Measurement. As an alternative to the above procedure, but with lower accuracy, wooden stakes about 1 m long and firmly driven into the ground will serve as measuring heads (*see* Fig. 9.3). These pegs are aligned by theodolite, and after being driven, zinc strips are tacked on. The levels of the tops are found and a transverse scratch mark is made on the first peg to serve as the beginning of the base line, or to act as an index mark similar to that on the previously-described measuring head. A longitudinal scratch may also be lined in. Lever-type straining arms and a spring balance may conveniently be used for tensioning and supporting the tape, which is adjusted so that the first zero is aligned with the scratched reference mark. A transverse scratch is then made on the second peg-head against the second zero, and this serves as reference mark for the second bay, the process being repeated. The tape itself is aligned by theodolite and temperatures are measured as before. Note that in this operation the tape should float just clear of the stakes. Instead of aligning the zero of the tape with the scratch mark made when taping the previous bay, it is also possible to make another scratch to mark the beginning of the new bay, the necessary correction to the bay length then being made by measuring the distance between the two scratches with a scale or optical measuring device. Alternatively the two scratches will act as index marks, across which the graduations at each end of the tape will fall parallel. The pegs may be driven in flush with the ground and the tape used in the flat instead of in catenary. For triangulation surveying, the ground over which the tape passes must be levelled off carefully, and thus it will be seen that work in catenary is to be preferred since such preparation will not normally be required.

The ends of the base line are transferred to the ground by means of two theodolites which are set up with their collimation lines at right angles to each other and sighted on to the mark on the measuring head. The telescopes may then be depressed and a mark, coincident with both sets of cross-hairs, made on a suitable permanent station.

Corrections to Taped Measurements. The measured bay lengths now require correction for—

(*a*) Elasticity, sag and thermal changes; in those cases where the field conditions differ from those at which the tape was standardized.

(*b*) Change in standard length; where this has occurred due to creep, or kinking.

(*c*) Slope and deviation from the straight.

(*d*) Height above mean sea level.

Let L = measured length of tape in the span

A = cross-sectional area of tape

E = Young's modulus for the material of the tape

P = pull on tape

W = weight of tape

[or w = weight of tape per unit length]

h = difference in level of measuring heads at each end of the bay

α = coefficient of linear expansion for the material of the tape

t = temperature of the tape during measurement

H = height of bay above mean sea level

R = mean radius of the earth

The standard conditions for the tape will be given by the manufacturers, i.e. L_s m at t_s°C under a tension of P_s N in catenary *or* on the flat. L_s could be 30 m under 89 N pull in catenary at 20°C. Standardization in catenary is to be preferred since (i) it reduces the amount of correction to be applied (correction for sag may not be required), and (ii) it is easier to check (tapes, standardized on the flat, require long flat stretches such as railway lines for checking).

1. CORRECTION FOR TENSION

$$\text{Correction} = (P - P_s)\frac{L}{AE} \qquad \left[\text{since } \frac{\text{stress}}{\text{strain}} = E\right]$$

It will be seen that if $P > P_s$ the correction will be positive. In this case the tape will have increased in length, although the graduations still give the nominal length, so that tape is actually longer than shown by the readings, and the line will be measured shorter than it is, i.e. the error is negative and the correction is positive. Where the tension is less than standard the correction is of course negative.

2. CORRECTION FOR SAG (Fig. 9.4). A freely suspended tape hangs in a catenary curve, but the required length is the chord distance. If the tape has been standardized in catenary, no correction for sag is required, so long as the field tension is the same as the tension applied during

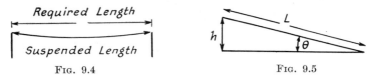

FIG. 9.4   FIG. 9.5

standardization, since the zero marks give the standard length along the *chord* under these conditions.

When the tape has been standardized on the flat, however, a negative correction must be applied, since the chord distance must always be less than that given by the graduations on the tape.

$$\text{Correction} = -\frac{L}{24}\left(\frac{W}{P}\right)^2$$

$$= -\frac{L^3}{24}\frac{w^2}{P^2}$$

P should be greater than 20 W.

3. CORRECTION FOR SLOPE (Fig. 9.5). Since the slope distance is always greater than the horizontal distance, this correction must always be negative.

In terms of the angle of slope θ, the horizontal length is $L \cos \theta$,

$$\text{Correction} = -L(1 - \cos \theta)$$

In terms of the difference in level between measuring heads,

$$\text{Correction} = -[L - (L^2 - h^2)^{\frac{1}{2}}]$$

$$= -\left\{\frac{h^2}{2\,L} + \frac{h^4}{8\,L^3} - \cdots\right\}$$

$$\simeq -\frac{h^2}{2\,L}$$

an approximation which is acceptable if h/L is less than 1 in 25, i.e. 1·2 m per bay.

4. CORRECTION FOR TEMPERATURE

$$\text{Correction} = \alpha L(t - t_s)$$

If the temperature during measurement is greater than the standard temperature, the tape will expand so that more ground is taped out than is shown by the graduations, i.e. the reading will be too low, indicating a negative error. A positive correction is thus required. If the temperature is lower than standard, a negative correction is required, and it will be noted that the correct sign for the correction is given by the formula.

5. CORRECTION FOR STANDARD. Tapes which have been standardized

in catenary will be checked at regular intervals, say daily, against a tape kept for the purpose, and also hung in catenary. The field tape and the standard tape are hung successively between two measuring heads, and after correcting for difference in tension and temperature from standard, the two results can be compared.

If the tape has been standardized on the flat, the same method may be used except that the additional correction for sag is required. Alternatively, if a long flat surface such as a railway line is available, the standard tape may be stretched along this, and two marks made on the rail opposite to the tape zeros. Temperature and tension corrections are applied to determine the true distance between the marks, and the field tape is then checked between the two marks.

The small difference between the two measurements gives the correction per bay, and is positive if the field tape is longer than the standard, negative if shorter.

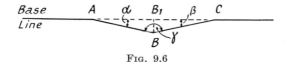

Fig. 9.6

6. CORRECTION FOR LACK OF ALIGNMENT (Fig. 9.6). Lengths AB and BC are measured in the normal way. Angles α, β, and γ are also measured.

Then
$$AB_1 = AB \cos \alpha$$
$$B_1C = BC \cos \beta$$

Correction
$$= -[AB(1 - \cos \alpha) + BC(1 - \cos \beta)]$$

Note that the correction is deducted. If γ be measured, $(\alpha + \beta + \gamma)$ should equal 180°, and the most probable values for α, β and γ may be computed. As a check, calculate AC from the cosine formula:

$$AC^2 = AB^2 + BC^2 - 2\,AB \cdot BC \cdot \cos \gamma$$

7. REDUCTION TO MEAN SEA LEVEL (Fig. 9.7). The length of the base line as measured is reduced to its equivalent length at mean sea level.

$$l = (R + H)\theta$$
$$l_1 = R\theta$$
$$\theta = \frac{l}{R + H} = \frac{l_1}{R}$$

$\therefore$
$$l_1 = l \cdot \frac{R}{R + H}$$

Therefore   correction $= l - l_1 = l\left(1 - \frac{R}{R + H}\right)$

$$= l \cdot \frac{H}{R + H} \text{ and is deducted.}$$

Since R may be taken to be 6,367 km, H can be neglected in the term $(R + H)$,

$$\therefore \qquad \text{Correction} = -\, l \cdot \frac{H}{R}$$

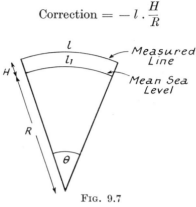

FIG. 9.7

EXAMPLE. During the measurement of a base line of four bays the following information was obtained—

Bay	Measured Length (m)	Temp. (°C)	Difference in Level (m)	Tension (N)
1	29·8986	18·0	+ 0·064	178
2	29·9012	18·0	+ 0·374	178
3	29·8824	18·1	− 0·232	178
4	29·9496	17·9	+ 0·238	178

The tape has a mass of 0·026 kg/m and a cross-sectional area of 3·24 mm². It was standardized on the flat at 20°C under a pull of 98 N. The coefficient of linear expansion for the material of the tape is 0·0000009/°C, and Young's modulus is $15·5 \times 10^4$ MN/m². The mean level of the base is 26·89 m above mean sea level. Determine the absolute length of the base line reduced to sea level.

Bay	L	L^3	h	h^2	$\dfrac{h^2}{2\,L}$
1	29·8986	26,727·09	+ 0·064	0·0036	0·0001
2	29·9012	26,734·12	+ 0·374	0·1399	0·0023
3	29·8824	26,683·72	− 0·232	0·0538	0·0008
4	29·9496	26,864·16	+ 0·238	0·0566	0·0008
	119·6318	107,009·09			0·0040

Pull Correction

$$(P - P_s)\frac{\Sigma L}{AE} = (178 - 89) \times \frac{119·6318}{3·24 \times 15·5 \times 10^4}$$

$$= + 0·0212 \text{ m}$$

Sag Correction

$$\frac{w^2 \Sigma (L)^3}{24\,P^2} = - \frac{(0·026 \times 9·806)^2 \times 107,009·1}{24 \times 178^2} = - 0·0091 \text{ m}$$

Temperature Correction. Based on average of 18°C, since there is little variation

$$\alpha \Sigma L(t - t_s) = 0 \cdot 0000009 \times 119 \cdot 6318(18 - 20)$$
$$= -0 \cdot 0002 \text{ m}$$

Slope Correction

$$\Sigma \frac{h^2}{2 L} = -0 \cdot 0040 \text{ m}$$

Reduction to Mean Sea Level

$$\Sigma L \cdot \frac{H}{R} = \frac{-119 \cdot 6318 \times 26 \cdot 89}{6,367,000} = -0 \cdot 0005 \text{ m}$$

$$\text{Absolute length} = 119 \cdot 6318 + 0 \cdot 0212 - 0 \cdot 0091 - 0 \cdot 0040$$
$$- 0 \cdot 0002 - 0 \cdot 0005$$
$$= 119 \cdot 6392 \text{ m}$$

EXAMPLE. A nominal distance of 30 m was set out with a steel tape from a mark on the top of one peg to a mark on the top of another, the tape being in catenary under a pull of 150 N and at a mean temperature of 25°C. The top of one peg was 0·442 m above the top of the other. Determine the horizontal distance between the marks on the two pegs reduced to mean sea level if the top of the higher peg is 195·57 m above mean sea level.

The tape which was standardized in catenary under a pull of 120 N and at a temperature of 20°C had a mass of 0·026 kg/m and had a cross-sectional area of 3·25 mm². The coefficient of linear expansion for the material of the tape may be taken as 0·0000009 per °C, and E as 15·5 × 10⁴ MN/m². The radius of the earth may be taken as 6,367 km.

Let L be the true length of the tape on the flat under a pull of 120 N at a temperature of 20°C. Since the tape has been standardized in catenary, if the sag correction be added to the catenary length then the length on the flat will be found.

Thus
$$L = 30 + \frac{30^3 \times (0 \cdot 026 \times 9 \cdot 806)^2}{24 \times 120^2} = 30 \cdot 0051 \text{ m}$$

To revert to the conditions of measurement with the tape now in catenary under a pull of 150 N

$$\textit{Pull Correction} \ = \ + \frac{(150 - 120) \times 30 \cdot 0051}{15 \cdot 5 \times 10^4 \times 3 \cdot 25} = + 0 \cdot 0018 \text{ m}$$

$$\textit{Sag Correction} \ = \ - \frac{30 \cdot 0051^3 \times (0 \cdot 026 \times 9 \cdot 806)^2}{24 \times 150^2} = - 0 \cdot 0033 \text{ m}$$

$$\textit{Slope Correction} = \ - \frac{0 \cdot 442^2}{2 \times 30 \cdot 0051} = - 0 \cdot 0033 \text{ m}$$

$$\textit{Temperature Correction} = + (25 - 20) \times 30 \cdot 0051 \times 0 \cdot 0000009$$
$$= + 0 \cdot 0001 \text{ m}$$

$$\text{Reduction to mean sea level} \quad = \frac{-\ 195{\cdot}35\ \times\ 30{\cdot}0051}{6{,}367\ \times\ 10^3} = -\ 0{\cdot}0009 \text{ m}$$

$$\text{Required Length} = 30{\cdot}0051 + 0{\cdot}0018 + 0{\cdot}0001 - 0{\cdot}0033 - 0{\cdot}0033$$
$$-\ 0{\cdot}0009$$
$$= 29{\cdot}999 \text{ m}$$

The corrections could have been based on a nominal length of 30·00 m as there would have been no differences in any of the values.

It will be seen that the procedure adopted reduced the standardization in catenary to standardization on the flat, then the normal corrections were applied to the field measurements. Had the tape been found to be say 3 mm longer than nominal, the length on the flat under a pull of 120 N at 20°C would have been 30·0051 + 0·003 = 30·0081 m and this value would have then been subject to correction.

Extension of Base Lines and Connexion to the Main Network. The length of the base line varies according to the size and nature of the triangulation, but should be as long as possible in comparison with the mean length of side of the triangles. A ratio of base length to mean side length of at least 0·5 has been suggested as desirable.* Thus in triangulations connected with engineering location work, where the mean length of triangle side will be from say $\frac{1}{2}$ to 1 km, a base $\frac{1}{2}$ km or so in length could be measured to form one side of a triangle.

The determining factor is the nature of the ground, for while flat ground is desirable for base measurement, high points are most suitable for stations as they may enable longer sights to be made without obstruction. In geodetic and other primary triangulations where the mean side length may be of the order of 50 km, a base length of 20 km desirable but often it is not practicable to measure more than 10 km or so. Thus the base must be extended by an initial triangulation and so connected to the main network. Fig. 9.8 shows how this may be done.

AB is the original base of, say 10 km length, and by measuring the angles marked in triangles ABC and ADB the length of line CD may be calculated after finding the most probable values of the angles (*see Advanced Surveying*). This new base can be further extended to give line EF, where the points E and F may be well suited for use as triangulation stations though not as terminal points in a directly measured base line. The accuracy of this "base net" is not necessarily diminished by extension with angular measurements; with a base measured on a good site, well located for extension and connexion to the main triangulation as shown in Fig. 9.8, the accuracy may well be higher than that of a longer base measured over poorer terrain. The proviso that the base be well located for extension is important, however. If the triangles are poorly conditioned or have weak rays, then the accuracy attained in the initial lineal measurement will be rapidly lost. (The Lossiemouth base extension lowered the accuracy from 1/900,000 to 1/70,000 in twelve triangles.†)

* *Plane and Geodetic Surveying*, Vol. II, Clark (Constable).
† O.S. Professional Papers, New Series, No. 18.

It has been mentioned that with modern measuring technique the tendency has been to measure longer bases, and it is interesting to note in this connexion that the two most recent bases in the U.K., Ridge Way and Caithness, terminate in points which form part of the primary triangulation. All four terminals stand well up from the surrounding country and have good, clear rays which are not grazing. No base extension is required and hence there is no loss of accuracy from this operation. Major M. H. Cobb (*op. cit.*) who carried out the work, however, considers that 10 to 12 km is sufficient length for a base.

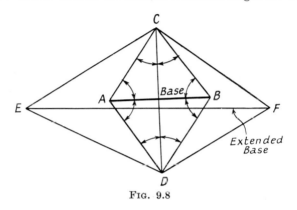

Fig. 9.8

Modern Developments: The Geodimeter

One of the most recent developments in the field of accurate linear measurement does not require a tape or measuring rod. Instead, the distance is obtained indirectly by measuring the time taken by a light beam to travel from one end of the line to the other and back. The velocity of light being known, it is then possible to calculate the distance. It will be recalled that Fizeau measured the velocity of light in 1849 by measuring the time taken for light to travel along a path of known length, and (after reflection by a mirror placed at the end of the path) back to the source. The time interval was measured mechanically, by passing the outgoing ray through the gaps in a rotating toothed wheel and then increasing the speed of the wheel to the critical speed at which the return ray hits the spurs, so cutting off all light from the distant mirror.

In recent investigations*† into the velocity of light by the Swedish physicist, Dr. E. Bergstrand, a similar principle was used but with much more complex electronic equipment for measuring the time interval between the emission of the light and its reception. The speed of light was determined with such accuracy that Dr. Bergstrand further

* *Measurement of Distances by High Frequency Light Signalling*, by E. Bergstrand. L'Activité de la Commission Géodésique, Baltique, 1944–47. Helsinki, 1948.
† "A Determination of the Velocity of Light," by E. Bergstrand, *Arkiv. für Fysik*, Vol. 2, No. 15, Stockholm, 1950.

investigated the possibility of measuring the time interval, and using this to measure the distance. As a result, he developed the *geodimeter,* manufactured by the Aga Company of Stockholm and shown in Fig. 9.9*a*.

The first model was introduced about 1950 and one current version — Model 6A—naturally incorporates several improvements. Information on the earlier model can be obtained by reference to a very comprehensive article by J. Clendinning in the *Survey Review* (Nos. 85/86, Vol. XI, 1952) while the *Proceedings of the Conference of Commonwealth Survey Officers,* 1967, may be consulted for some information on the Model 6.

A light beam is modulated by a Kerr cell in the Geodimeter and is projected outwards by the coaxial optics towards a prism reflector system set up at the other end of the line. The reflectors return the light back to the instrument where the light signals are converted into electrical signals by the photocell. The difference in phase between the outgoing and incoming signals is measured by the resolver whose knob is rotated until the null indicator reads zero, and the resolver setting is recorded on the digital readout. The student can compare the circuit shown (Fig. 9.9*b*) with that for the Geodimeter Model 8 shown in Chapter 14.

Three or four different frequencies are applied to the light beam during the modulation, three of them being used if the distance to be measured be known to within 1,000 m, and the fourth if the distance be known to 2,000 m. Phase settings are also changed during the observations and the distance between instrument and reflectors is calculated from the indicated readouts. Corrections are applied depending upon (*a*) atmospheric and meteorological conditions and (*b*) the heights of the instrument and reflectors above mean sea level. The light beam is unaffected by adverse ground conditions and working ranges of 5–10 km are suggested by the manufacturer in daylight, with 15–25 km in darkness, depending upon whether a standard lamp or mercury lamp is used: an accuracy of 5 mm $\pm$ 1 mm/km (mean square error) is quoted for the measurement of length. Corrections are applied for temperature, pressure and humidity, and in addition, corrections for slope, curvature and height above mean sea level as already mentioned.

Tests with the Geodimeter. Comprehensive tests in this country were carried out by Major Mackenzie* on the Ridge Way and Caithness Bases, using a geodimeter loaned by the U.S. Army Map Service, and the comparison with the base measurements made the previous year was good. For the Ridge Way base, the fractional error on a directly-measured length of 11,260·677 m was about 1/450,000, while the consistency of the geodimeter results was excellent. On the Caithness base, the error was about 1/360,000 on a measured length of

---

* *The Geodetic Measurement of the Ridge Way and Caithness Bases,* 1953, by Major I. C. C. Mackenzie, R.E., O.S. Professional Papers, New Series, No. 19 (H.M.S.O.)

FIG. 9.9a. AGA Geodimeter Model 6A
(*Courtesy of A.G.A. (U.K.), Ltd.*)

24,828·626 m. These results are very striking, for it must be borne in mind that geodimeter measurements can be made of all sides in a primary triangulation, not just the base lines. This should lead to a better balancing out of errors.

The Tellurometer System. This system was developed in South Africa for precise distance measurement. The travel time of radio micro-waves between two stations is determined, corrections applied

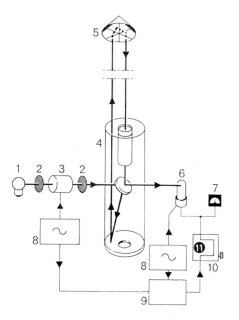

FIG. 9.9*b*. PRINCIPLES OF THE GEODIMETER 6A

1. Lamp
2. Polaroid filters
3. Kerr cell
4. Coaxial optics
5. Reflector
6. Photocell
7. Null indicator
8. Crystal-controlled generator
9. Mixer
10. Resolver
11. Digital readout

(Courtesy of A.G.A. (U.K.), Ltd.)

for meteorological conditions to determine the travel time in vacuum, and the distance between the stations then deduced from a velocity of 299,792·5 km/sec. The distance thus given is the slope distance and this must be reduced to the horizontal and to mean sea level from observations of heights by aneroid barometer or by more accurate levelling methods. The Tellurometer gives satisfactory observations under reasonably adverse conditions, though the best are attained in gentle sunshine with a light breeze and low relative humidity. A line relatively free of obstacles is required between the two stations, gently undulating terrain covered with medium vegetation probably being

FIG. 9.10a. TELLUROMETER MRA101 IN USE
(*Courtesy of Tellurometer (U.K.), Ltd.*)

the best; grazing rays should be avoided since they can lead to disturbed readings, and water surfaces and flat sandy surfaces may cause some inaccuracy. Comprehensive accounts of its development have

Fig. 9.10*b*. Tellurometer MRA101
Dimensions 381 mm × 368 mm × 184 mm.
(*Courtesy of Tellurometer (U.K.), Ltd.*)

been given by T. L. Wadley in *Survey Review*, Vol. XIV, Nos. 105/106, and by Sandover and Bill, *Proc. Instn. Civ. Engnrs.* (Jan. 1963).

As with the Geodimeter, improvements have occurred since the first models were produced, and a very sophisticated instrument, MRA4, has been introduced which can measure distances, in the range of 50 m to 50 km, with a resolution of 1 mm and with a stated accuracy of ± 3 mm ± 3 parts per million. The MRA3, which was designed to a military specification, and its civil counterpart MRA101 both work to the above maximum range, but resolve distance to 0·01 m. Naturally they are not so costly as the MRA4, and the quoted accuracy of

FIG. 9.10c. ANTENNA ASSEMBLY AND REFLECTOR DISH
OF TELLUROMETER MRA4
(Courtesy of Tellurometer (U.K.), Ltd.)

± 15 mm ± 3 parts per million is lower, but in very adverse weather conditions of heavy rain or fog their performance in so far as maximum range is concerned is not reduced as much.

In all cases a master station and a remote station are set up, one at each end of the line, observations being taken at the former by an operator who is linked by radio telephone to another operator at the latter: both instruments are interchangeable.

A continuous radio wave (wavelength 30 mm in the MRA101 and 8 mm in the MRA4) is radiated on to a Cassegrain reflector, thence back to the main parabolic reflector. This carrier wave, modulated by various frequencies, is received by the remote unit and returned to the master where the phase of the received modulation is compared with the transmitted modulation. The phase difference indicates the time taken for the wave to cover the outward and inward paths, i.e. double distance; this is converted electronically into a distance reading which is presented directly to the operator. In the MRA3 for instance the applied frequencies are—

A pattern 7·492377 MHz (Mc/s)
B ,,   7·490879 ,,   ,,
C ,,   7·477392 ,,   ,,
D ,,   7·342529 ,,   ,,
E ,,   5·993902 ,,   ,,

Thus,

$A - B$	=	1,498	Hz and equivalent wavelength			=	200,000 m
$A - C$	=	14,985	,, ,,	,,	,,	=	20,000 m
$A - D$	=	149,848	,, ,,	,,	,,	=	2,000 m
$A - E$	=	1,498,475	,, ,,	,,	,,	=	200 m
$A - (-A)$	=	14,984,754	,, ,,	,,	,,	=	20 m

It will be noted that the $A - (-A)$ frequency implies a wavelength of 20 m and so, allowing for double distance a 10 m scale reading is given, and this is duly resolved to single metres and hundredths; similarly the $A - B$ frequency implies a resolution of distance to 100 m. Considering the various frequencies in turn, together with the knowledge of phase differences, allows the length of the line to be determined. Some further discussion on this topic can be found in Chapter 14.

As mentioned, the MRA101 is compatible with this instrument, but is intended for civil use rather than military, and this has allowed the insertion of somewhat less expensive, but nevertheless high-grade, components.

To obtain a resolution of 1 mm, frequencies of about 75 Mc/s are applied in the MRA4 and, whereas the beam width of the MRA101 is 6°, that of the MRA4 is 2° and accurate alignment is essential. All these instruments can be used for base-line measurement and for direct measurement of the sides of the network: this latter operation is known as trilateration. Other uses include precise traversing and precise surveying in construction projects.

The instrument is handled by Tellurometer (U.K.) Ltd. in the United Kingdom.

It is essential to check all electronic distance-measuring instruments at frequent intervals of time to ensure that absolute values are being recorded. This is equivalent to calibrating the invar band as described previously.

Principles of Triangulation

Basic figures used in triangulation are: (i) simple triangles, (ii) quadrilaterals with diagonals, (iii) quadrilaterals with a central point, and some typical basic patterns are shown in Fig. 9.11.

Areas such as the British Isles may have a primary network consisting of intersecting quadrilaterals or polygons which covers the

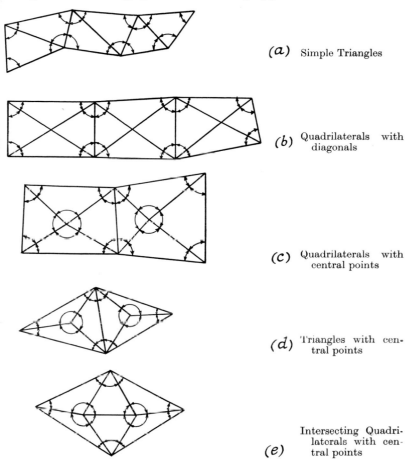

(a) Simple Triangles

(b) Quadrilaterals with diagonals

(c) Quadrilaterals with central points

(d) Triangles with central points

(e) Intersecting Quadrilaterals with central points

Fig. 9.11

whole country. If the country is large, however, such a procedure would be execssively lengthy and costly, and instead, on continental land masses, chains of figures may run along the meridians of longitude and the parallels of latitude. Triangulation networks may then be run between the various stations on the chains, and detail surveys put in hand as required. Fig. 9.12 shows the chain of figures used in an early Survey of India, and for comparison the main triangulation of the British Isles is given in Fig. 9.13.

Similarly, if the country is long and narrow, it is sufficient to have a backbone of quadrilaterals as in Fig. 9.14, which shows the triangulation of the Nile Valley from Cairo to Beba. Simple triangles are not desirable in such chains, due to the fact that the other figures enable better appreciations to be made of the most probable values of the angles.

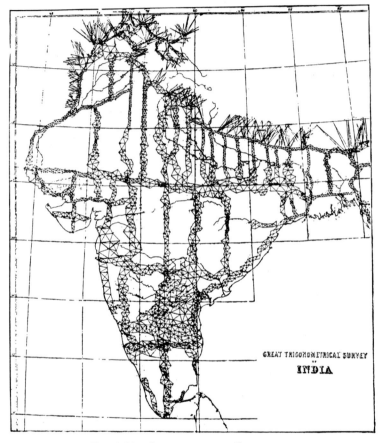

GREAT TRIGONOMETRICAL SURVEY
OF
INDIA

FIG. 9.12. INDIA, SHOWING TRIANGULATION

Note also in Fig. 9.14 the way in which the base lines are extended and connected with the main network, and the way in which the chain starts from one base and closes on another. The length of the second base as calculated through from the first base is compared with its length as measured, thus enabling control to be exercised over the accuracy of the whole chain.

Choice of Stations. In selecting the triangulation stations, certain

RETRIANGULATION OF GREAT BRITAIN

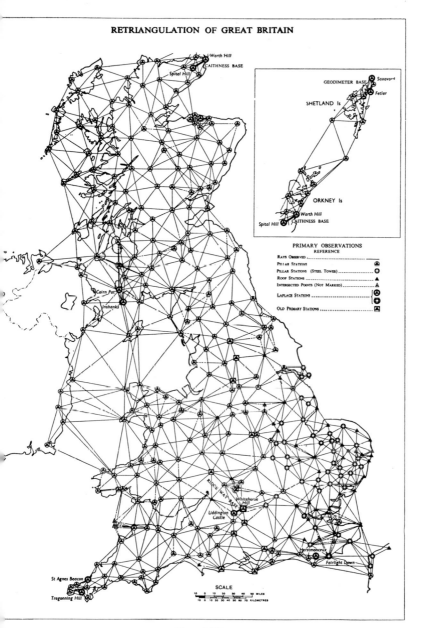

FIG. 9.13

intersected stations are fixed by means of intersecting sights from adjacent stations. Azimuth control is obtained at Laplace stations (*see* Chapter 12).

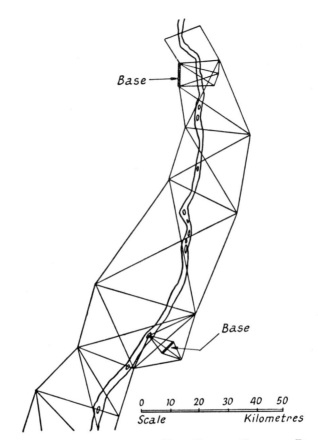

Base

Base

0 10 20 30 40 50
Scale   Kilometres

FIG. 9.14. TRIANGULATION OF NILE VALLEY (CAIRO TO BEBA)
(*Cambridge University Press*)

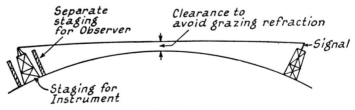

Separate
staging
for Observer

Clearance to
avoid grazing refraction

Signal

Staging for
Instrument

FIG. 9.15

considerations should be kept in mind which may be summarized as follows—

(*a*) Every station should be visible from the adjacent stations and the triangles formed thereby should be well-conditioned, that is to say, as nearly equilateral as possible. No angles should be less than 30°, if at all possible.

(*b*) The size of the triangles will depend on the configuration of the country, but they should normally be as large as possible compatible with the distinct bisection of signals, having regard to the type of theodolite used.

Short sights, in which the signal cannot be accurately bisected, should be avoided. In hilly terrain, longer sights may be obtained by using the higher points as stations; in flat country consideration must be given to the elevation of the instrument and possibly the signal by means of staging (*see* Figs. 9.13 and 9.15).

The cost of such staging adds considerably to the cost of a survey. The height must be such that the line of sight is at least 1·5 m clear of all obstructions.

(*c*) Where choice of station sites exists, the ones most suitable for connexion to subsequent traverse and detail surveys should be used.

The beacons which serve as signals are of two kinds, luminous and opaque. In many countries, luminous beacons become necessary when sighting distant signals, owing to haze, and these are either acetylene or electric lamps, or heliographs which use the sun. Fig. 9.16 shows a typical beacon lamp, by Hilger and Watts, Ltd. A standard type of opaque signal consists of a tripod of timbers lashed together and covered with calico. The stations themselves should be permanently marked, either by a mark in a metal plug set in the upper surface of a masonry pillar (which also serves as a mounting for the theodolite and beacon lamp), or by a mark in a metal plug set in a stone slab about one metre below ground. In the latter case, a second stone is placed at ground level with a mark set vertically above the station mark; thus the buried mark is referred to only if it is feared that the one on the surface has been disturbed.

Reading of Angles. Two methods are available for improving the accuracy of angle measurements, (i) repetition, and (ii) reiteration.

(i) *Repetition* reading consists in measuring single angles a number of times, say n, obtaining the total angle x for n readings by sighting back on the first station with the top plate locked and the bottom plate free. The angle is given by $x \div n$. A common number of repeats is six. The method is for use on vernier instruments, which are not nowadays used in triangulation surveys. It is fully described in Chapter 4 (Theodolite).

(ii) *Reiteration* normally consists in measuring all the angles subtended at a point successively in several rounds of angles. It is sometimes referred to as the *method of rounds*, and is the method to use where possible, i.e. where equal and good all-round visibility exists from the theodolite station. Of the other alternatives, Schreiber's method may be mentioned in which, as well as measuring the angles

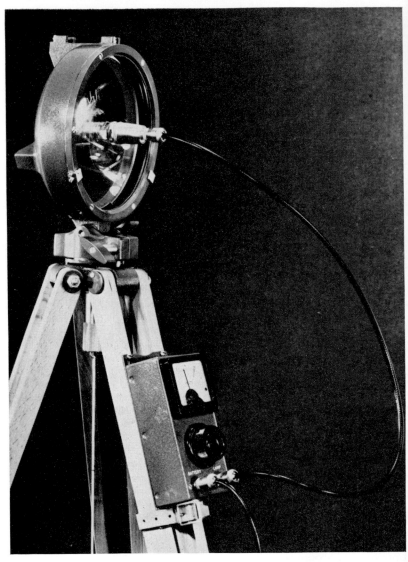

Fig. 9.16. Watts Survey Beacon Lamp
(*Courtesy of Hilger & Watts, Ltd.*)

subtended at the theodolite station by successive stations, the angles between each station and all the other stations are also measured independently. It will be seen that this means a considerable increase in the amount of work at each theodolite station, but has the advantage

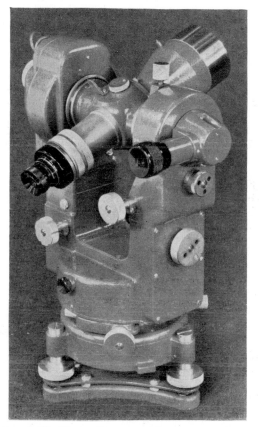

FIG. 9.17. WATTS MICROPTIC THEODOLITE NO. 3
(*Courtesy of Hilger & Watts, Ltd.*)

that even when all-round visibility is not available, work may be possible in the direction in which conditions are satisfactory. In modern geodetic work a glass arc theodolite such as the Microptic Theodolite No. 3 (Fig. 9.17) manufactured by Hilger and Watts, Ltd. will be used. This instrument reads directly to 0·2 second of arc and its telescope has ×40 magnification.

Assuming the method of rounds to be used, a typical programme for measuring the angles at a point could be, referring to Fig. 9.18—

Set up and level the instrument at station *O*. The optical micrometer is set to read approximately zero in the face left position and a pointing is made on the reference station *A*, which will usually be one of the triangulation stations. Then, swinging right, the instrument is pointed on *B* and the reading on the horizontal circle noted; stations *C, D, E* and finally, *A*, are then bisected, ensuring in all cases that the direction of swing to the right is preserved. The face of the instrument may then be changed and pointings made on to stations *E, D, C* and *B*, always swinging to the left. The final reading on *A* should now differ from the first by 180°; this completes one set of readings. The zero setting is now altered to use other parts of the scale and a similar procedure followed, the number of zeros required depending on the degree of precision required. Twelve to sixteen zeros may be taken for primary triangulations and up to eight zeros for secondary triangulations.

Individual angles are deduced from the readings in the usual way and if the difference between any one measurement and the mean for that particular angle exceeds, say, 4 seconds, it should be repeated before leaving the station.

If one of the older micrometer instruments has to be used, the suitable procedure would be exactly as outlined above, except that, as the circle readings are not meaned optically in such instruments, each horizontal circle micrometer would have to be read at each pointing. Some micrometer theodolites have two micrometers and others three.

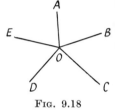

Fig. 9.18

For a fuller treatment of this subject, and indeed the whole study of geodetic surveying, the student is referred to *Geodesy*, by Brigadier Bomford (O.U.P.).

Accuracy of Triangulation

The primary measure of the precision of triangulation is the *average triangular error*, i.e. the average deviation of the sum of the measured angles in the triangles from 180° after correction for curvature.

In a small triangle, whose sides are of the order of one or two miles, the curvature of the earth may be considered negligible, and the three measured angles should sum to 180°. In practice, there will be a difference of a small number of seconds known as the *triangular error*.

Spherical Excess. In larger triangles an error arises from the fact that, though the angles are measured in the horizontal plane, the curvature of the earth throws these planes out of parallel with one another. The three angles of the triangle should now add up to more than 180°, and the excess, known as the *spherical excess*, is easily calculated from the area of the triangle. Then

$$\Sigma(\text{measured angles}) - E + \varepsilon = 180°$$

where *E* is the spherical excess, and *ε* is the triangular error whose value and sign are to be estimated.

The spherical excess is calculated from

$$E = \frac{A}{R^2 \sin 1''}$$

where A = area of triangle, and R = mean radius of the earth.

If a and b are two sides of a triangle, and C is the included angle, we have that $\qquad A = \frac{1}{2}ab \sin C$

For all but the most accurate work, however, the area of the triangle can be estimated as if the area were plane, so that

$$E = \frac{\text{area of triangle in sq km}}{1,000} \times 5 \cdot 09''$$

In geodetic work of the highest accuracy, the average value of ε should be less than one second of arc.

Adjustment. In simple triangles, assuming that all measurements have equal weight (i.e. that all angles have been measured with equal accuracy) the error ε is distributed equally among the three angles.

In the more complex figures, several triangles are obtained whose angles are interdependent upon one another, and these interdependent angles can be corrected simultaneously by the method of least squares to give the most probable values for the angles. Typical solutions by this approach are given in *Advanced Surveying* by A. H. Jameson (Pitman, 1948), whilst an alternative method is given later in Chapter 13.

When these values have been established the lengths of unknown sides can be estimated by spherical trigonometry, or Legendre's method could be adopted. In this method one-third of the spherical excess is deducted from each angle of the triangle and the sides then calculated from the known length by plane trigonometry.

The Satellite Station Problem. It is often convenient to use a church spire or similar tall feature as a signal for sights from other stations, even though it would be extremely difficult to set up the instrument over this particular signal. In such cases the instrument is set up near the signal (which is, say, a church spire, at A in Fig. 9.19) at a satellite station S, and the angle measured at S is adjusted to give the equivalent reading as though the instrument were in fact set up at A. AS might be about 12 m.

BC is known, having been calculated from previous data, and angles ABC and ACB will have been included in the programmes of angular measurement at stations B and C. A value of angle BAC is then estimated: $B\hat{A}C \simeq 180° - (A\hat{B}C + A\hat{C}B)$. From this, approximate lengths of AB and AC are calculated using the sine rule. The distance AS and angles ASB and ASC are now measured to the required accuracy; the problem is then to obtain the best value for angle BAC.

In triangle ABS,

$$\frac{AS}{\sin p} = \frac{AB}{\sin A\hat{S}B}$$

$$\therefore \qquad \sin p = \frac{AS}{AB} \cdot \sin A\hat{S}B$$

Also in triangle ASC,

$$\frac{AS}{\sin q} = \frac{AC}{\sin A\hat{S}C}$$

$$\therefore \qquad \sin q = \frac{AS}{AC} \cdot \sin A\hat{S}C$$

For small angles, $\qquad p'' = \frac{AS}{AB} \frac{\sin A\hat{S}B}{\sin 1''}$ and $q'' = \frac{AS}{AC} \frac{\sin A\hat{S}C}{\sin 1''}$

Now

$$O\hat{A}B = A\hat{S}B + p$$
$$O\hat{A}C = A\hat{S}C + q$$
$$\therefore \qquad B\hat{A}C = O\hat{A}C - O\hat{A}B = (A\hat{S}C + q) - (A\hat{S}B + p).$$

Hence the value reduced to true centre A is obtained. The angles in triangle ABC may now be summed up, and the triangular error determined and distributed.

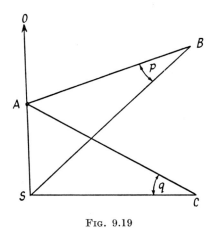

Fig. 9.19

EXAMPLE. The following clockwise angles were observed on three stations A B and C by a theodolite set up at a satellite station S distant 10·40 m from and south of A, as in Fig. 9.19: $A\hat{S}B = 56° 14' 10''$; $A\hat{S}C = 132° 52' 50''$. The approximate lengths of AB and AC were estimated as 3,507·2 m and 3,100·6 m respectively. Calculate the value of $B\hat{A}C$. Take log sin $1'' = \bar{6}\cdot6855749$.

Referring to Fig. 9.19,

	No.	Log
	10·40	1·0170333
	3,507·2	3·5449605
		3̄·4720728
sin 56° 14′ 10″		1̄·9197760
		3̄·3918488
sin 1″		6̄·6855749
	p''	2·7062739

$$\frac{10·40}{\sin p} = \frac{3,507·2}{\sin 56° 14' 10''}$$

$$p = \frac{10·40}{3,507·2} \cdot \frac{\sin 56° 14' 10''}{\sin 1''}$$

$$= 508''$$

Also $\dfrac{10·40}{\sin q} = \dfrac{3,110·6}{\sin 132° 52' 50''}$, so that $q = 505''$

Therefore

$$B\hat{A}C = (A\hat{S}C + q) - (A\hat{S}B + p)$$
$$= 132° 52' 50'' - 56° 14' 10'' + 505'' - 508''$$
$$= 76° 38' 37''$$

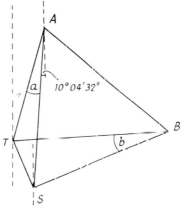

FIG. 9.20

EXAMPLE. It is required to find the bearings of two lines TA and TB from an inaccessible station T. A satellite station S was set up 4·70 m from T in an approximately S–E direction, and from it the theodolite angles were measured—

Telescope pointing on A 360° 00′ 00″
,, ,, ,, B 72° 40′ 00″
,, ,, ,, T 305° 00′ 00″

If point A lies N 10° 04′ 32″ E from station S, what are the true bearings of A and B from T?

AT is 15 km; BT is 16·7 km; log sin 1″ = 6̄·68557. (*Inst. C.E.*, 1953.)

$$T\hat{S}A = 360°\ 00'\ 00'' - 305°\ 00'\ 00'' = 55°\ 00'\ 00''$$

$$T\hat{S}B = 55°\ 00'\ 00'' + 72°\ 40'\ 00'' = 127°\ 40'\ 00''$$

Using the notation of Fig. 9.20 in the formula already given,

$$\therefore \qquad a = \frac{ST}{AT} \cdot \frac{\sin T\hat{S}A}{\sin 1''}$$

$$= \frac{4 \cdot 70}{15,000} \cdot \frac{\sin 55°\ 00'\ 00''}{\sin 1''}$$

$$= 53'' \text{ say}$$

$$b = \frac{ST}{BT} \cdot \frac{\sin T\hat{S}B}{\sin 1''}$$

$$= \frac{4 \cdot 70}{16,700} \cdot \frac{\sin 127°\ 40'\ 00''}{\sin 1''}$$

$$= 46'' \text{ say}$$

Therefore $\quad A\hat{T}B = (T\hat{S}B + b) - (T\hat{S}A + a)$

$$= 127°\ 40'\ 00'' - 55°\ 00'\ 00'' + 46'' - 53''$$

$$= 72°\ 39'\ 53''$$

Bearing TA = bearing $SA + a$

$$= 10°\ 04'\ 32'' + 53''$$

$$= \text{N } 10°\ 05'\ 25'' \text{ E.}$$

Bearing TB = bearing $TA + A\hat{T}B$

$$= 10°\ 05'\ 25'' + 72°\ 39'\ 53''$$

$$= \text{N } 82°\ 45'\ 18'' \text{ E.}$$

Satellite Stations: Accuracy of Measurement. It will be appreciated that the examples selected are to illustrate the problem. In practice, accuracy of measurement must be of a high order when using a satellite station if serious errors and distortion of the framework are not to ensue. Bomford (*op. cit.*) states that if TS/TA exceeds $1/1,000$ significant error is likely to occur in the satellite correction.

The length TS would normally be of the order of, say, 10–15 m, and to retain an accuracy of 0·1 sec, that length must be correct to, say, 5 mm.

EXERCISES 9

1. Show that the pull correction must be modified by a factor of $\pm\ \delta P/P$ and the sag correction by a factor of $\mp\ 2\delta P/P$ if there is an error of $\pm\ \delta P$ in the field tension P.

The pull correction for one bay of a base line was estimated to be 0·0053 m. If an error of $-$ 0·05 mm^2 had occurred during the measurement of cross-sectional area, which was nominally 3·25 mm^2, determine the actual pull correction to be applied.

Answer: 0·0061 m.

2. The following particulars relate to a straight measured base line.

Measured length of suspended span (m)	Rise (m)
23·9986	+ 0·160
24·0012	+ 0·274
24·0014	− 0·132
23·9996	+ 0·338

The length of the tape, when supported on a flat surface was 24 m at 20°C, and under a pull of 89 N. The tape has a cross-sectional area of 3 mm² and a mass of 0·554 kg. Determine the corrected length of base line if the mean temperature during measurement was 16°C and the tension in the suspended tape was 120 N

$\alpha = 0.0000009$ per °C $E = 15.5 \times 10^4 \text{ MN/m}^2$

Answer: 95·9939 m.

3. A steel tape is 30 m long at a temperature of 12°C when laid horizontally on the ground and no pull applied at the ends. The tape is arranged in catenary over two pegs, and marks made at its ends. It is also supported at mid-span. Calculate the actual distance between the marks given that field conditions are: temperature 24°C, pull 120 N, $E = 207,000 \text{ MN/m}^2$, $\alpha = 0.000012/\text{°C}$. The tape has a mass of 1·59 kg and has a cross-sectional area of 6·72 mm². It may be assumed that the pegs are at the same level.

Answer: 30·006 m.

4. A steel tape 50 m long on the flat at 20°C under a pull of 89 N is used to measure a base line which is apparently 1,000 m long. Find the corrected length, given that the field pull was 178 N, the average temperature was 25°C and the tape was supported at mid-span. Take the cross-sectional area of the tape as 5·00 mm², the mass of the tape as 2·32 kg, Young's modulus as 20·7 × 10⁴ MN/m² and $\alpha = 0.000011/\text{°C}$. Allow the total slope correction to be 0·045 m.

Answer: 999·952 m.

5. A copper transmission line, 12·7 mm in diameter, is stretched between two points 300 m apart at the same level, with a tension of 5 kN, when the temperature is 35°C. It is necessary to define its limiting positions when the temperature varies. Making use of the corrections for sag, temperature and elasticity normally applied to base line measurements in catenary, find the tension at a temperature of − 15°C, and the sag in the two cases. Young's Modulus for copper is 68,950 MN/m², its density is 8,890 kg/m³, and its coefficient of linear expansion, 5 × 10⁻⁶/°C. (*L.U.*)

Answer: 5·09 kN, 24·80 m, 25·30 m.

6. Explain the use of a satellite station. The following clockwise angles were observed to three stations A, B and C with a theodolite set up at a satellite station S, distant 4·550 m from A. A, zero. B, 57° 10′ 36″. C, 131° 27′ 40″. The approximate lengths of AB and AC were estimated as 2,460·0 m and 3,090·0 m respectively. Calculate the angle BAC. The value of log sin 1″ is 6̄·6855749. (*L.U.*)

Answer: 74° 15′ 31″.

7. A theodolite was set up at a satellite station O and, with the horizontal angle reading zero, was sighted to the true station C. The horizontal angle to an observed station at A was then read and found to be 121° 56′ 00″, and that to a second observed station B was found to be 144° 54′ 00″. The line OC was 3·733 m in length. Given log CA = 4·2393, and log CB = 4·2316, calculate the angle ACB. log sin 1″ = 6·68557. (*I.C.E.*)

Answer: 22° 57′ 48″.

8. Write an essay outlining the methods used for the accurate measurement of base lines during the past century, paying some attention to modern developments.

CHAPTER 10

HYDROGRAPHIC SURVEYING

HYDROGRAPHIC SURVEYING, so far as the civil engineer is concerned, covers the survey work for projects in, or adjoining, bays, harbours, lakes, or rivers. Generally speaking, the types and purpose of the various branches of hydrographic surveying may be summarized as follows—

1. Measurement of tides for sea coast work, e.g. construction of sea defence works, jetties, harbours, etc., for the establishment of a levelling datum, and for reducing soundings.

2. Determination of bed depths, by soundings:

(i) for navigation, including the location of rocks, sand bars, navigation lights, buoys, etc.,

(ii) for the location of under-water works, volumes of under-water excavation, etc.,

(iii) in connexion with irrigation and land-drainage schemes.

3. Determination of direction of current in connexion with:

(i) the location of sewer outfalls and similar works,

(ii) determination of areas subject to scour and silt,

(iii) for navigational purposes.

4. Measurement of quantity of water, and flow of water—in connexion with water schemes, power schemes, flood control, etc.

Normally the civil engineer is not concerned with navigation, work in connexion with this being carried out by such bodies as the Admiralty, U.S. Coast and Geodesy Survey, etc.; the remaining aspects of hydrographic surveying are, however, of vital concern to the engineer, and will be dealt with here. It will be apparent from this brief outline that though some of the work is fundamental, much of it is of a specialized nature in practice, and is carried out by specialists in these fields.

If we exclude item (4) from the above list of the various branches of hydrographic surveying (since it is concerned mainly with hydraulics), the fundamental task is the preparation of a plan or chart showing physical features above and below water, and involves—

1. Vertical control. A chain of bench marks must be established near the shore line, and these serve for setting and checking tide gauges, etc., to which the soundings are referred.

2. Horizontal control. When making soundings of the depth of a river bed or a sea bed the location of the sounding vessel is made by reference to fixed control points on shore, and the accurate establishment of this shore framework is of the utmost importance.

3. Determination of the bed profile by soundings and use of the fine wire sweep.

4. Location of all irregularities in shore-line islands, rocks, etc. by normal surveying methods.

Vertical Control—Tide Measurement—Datum Lines

According to the requirements of the particular survey, soundings will be reduced to one of two datum lines—

(*a*) The land-levelling datum, which in Great Britain is the *Ordnance datum*, this being the one generally used in civil engineering construction, since it enables the levels to be directly related to those of the adjoining shore installations; (though for the work itself an arbitrary datum below the level of the lowest work may be adopted so that all levels are positive).

(*b*) The tidal datum, which is generally used for navigation purposes. The usual level adopted is that level of the water surface below which the tide rarely falls—mean low water ordinary spring tides (M.L.W. O.S.T.), and has the name *Chart datum*.

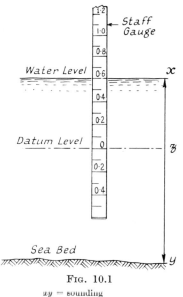

FIG. 10.1

xy = sounding
zy = $xy - xz$
= reduced sounding

Tide Gauges. Since we are not concerned with the needs of navigation in this chapter, it is necessary to deal only with vertical control using the land-levelling datum. This involves the location of tide gauges at intervals over the area to be surveyed, the number employed depending on the accuracy required. For example,[*] if the tide levels at each end of the area under survey differ by a maximum of 0·5 m at any given time, and an accuracy of 0·1 m is required in the sounding, then 5 gauges should be used, and any particular set of soundings reduced by reference to the gauge nearest the sounding area.

This difference in tide level, or *tidal gradient*, may be determined by initially setting up gauges at each end of the survey area and comparing simultaneous readings over a complete tidal cycle.

The gauges themselves may be either non-self-recording or self-recording. The simplest of the former type is the *staff gauge*, which is merely a vertical staff about 150 mm broad, with painted graduations covering sufficient length to deal with the highest and lowest known tides. It is not necessary for the zero to be set exactly at Ordnance datum. To calibrate the staff, a line of levels is taken, from the nearest bench mark, on to a graduation on the staff as fixed. In this way the actual R.L. of the staff zero may be determined and thence the correction which, when applied to the staff gauge readings, will reduce them to O.D. Fig. 10.1 shows a typical staff gauge.

Such gauges are often difficult to read owing to wave action at the

---

* *Hydrographic Surveying*, by Lieut.-Cmdr. A. D. Margrett, R.D., R.N.R., F.R.G.S. (Kelvin Hughes (Marine) Ltd., London).

water surface, and *float gauges* can be used in which the staff is attached to a float which is enclosed inside a box open to the water at its lower end. The box acts as a stilling chamber. Otherwise, the float may be attached to a counterweight by means of a steel tape, suitably graduated and passing over a pulley.

The *weight gauge* removes the need for the float, and in the type developed by the U.S. Geological Survey* a graduated stainless steel tape is mounted on a reel and has a weight at its free end. The reel is attached to an insulated bracket. One terminal of a $4\frac{1}{2}$-V battery connects to the reel via a sensitive voltmeter, and the other terminal to ground, so that as the weight touches the water surface the voltmeter needle shows a deflection.

These instruments naturally have to be read at fixed and frequent intervals so that when fairly lengthy records are required a *self-recording gauge* must be used, of which the hydrographic type manufactured by R. W. Munro & Co. Ltd., London, shown in Fig. 10.2, is typical. This instrument is of course float operated, and a permanent record is obtained on the chart, time being recorded along the drum, and height by its rotation.

For a full description of the theory of tides see, for example, the Admiralty *Manual of Tides*.

Horizontal Control—The Shore Framework

The precise nature of shore control depends on the method used to locate the soundings, which in turn depends to some extent on whether sea coast, river, or estuary, is being surveyed. Taking the typical case in which soundings are taken along a stretch of coast, and sextant readings are made, as in Fig. 10.3, to locate them, a series of beacons must be located. Where available, of course, salient features such as lighthouses or churches can be used, in conjunction with beacons. Location of the control points is by closed traverse or by triangulation. Should the area have been surveyed already, it may be possible to obtain co-ordinates of salient features and trig. stations, etc., but this previous survey must, of course, be of at least the same accuracy as the hydrographic survey being carried out. When the sextant is being used, the control points should be placed so that there are always three stations visible from the sounding points which will not subtend angles of less than 30° or more than 115° (*see* Chapter 5—notes on the sextant).

For estuaries, rivers and inlets, networks of the type shown in Fig. 10.4 are employed. In addition to providing controls for the soundings, the beacons (or other control points) form the framework for the remainder of the shore survey. This may be done by any of the usual methods; i.e. closed traverses run between the control points, plane tabling, or tacheometry. In picking up the shore line, high and low water lines are located: the former may be obtained from the position of deposited material and the latter from tidal observations.

---

* "Stream-Gauging Procedure." Geological Survey Water-supply Paper 888 (U.S. Dept. of the Interior)

FIG. 10.2. PORTABLE WATER LEVEL RECORDER

(h. W. Munro & Co., Ltd.)

Sounding—Determination of Bed Profile

This is one of the main operations in hydrographic surveying and corresponds to levelling in land surveying. The measurements are made from a boat and corrections for the constantly varying water level must be made by reference to the tide gauges when, for example, tidal waters are being sounded. Hence water level measurement

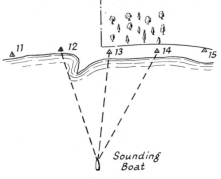

FIG. 10.3

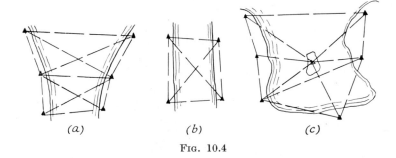

FIG. 10.4

and sounding are two inseparable steps in the one operation, in all cases where the water level is changing.

Sounding may be carried out by the following methods—

1. Direct. (*a*) Sounding rods; (*b*) Sounding leads on graduated lines.
2. Indirect. Echo sounders.

The sounding is usually carried out from a flat-bottomed boat of low draft, those for sea work being larger and equipped with power.

(1) Direct Methods

(*a*) **Sounding Rods.** Where the currents are not strong, graduated wooden poles may be used to measure the bed depth. This method is limited to depths of about 5 m. In strong currents it is difficult to maintain verticality of long sounding rods.

(*b*) **Sounding Lines.** For greater depths than 5 m a lead line—a leaden weight attached to either a stretched and graduated hemp line or to a metal chain—is used. Such a line may be incorporated in a *sounding machine*, in which a flexible wire is used, the amount paid out being measured by a friction-driven roller and shown on dials.

Soundings are required when discharges and currents are being measured by means of a current meter (*see* later), and in this case the current meter acts as the sounding weight and the heavy-duty electric cable, which connects it to the surface, as the sounding line. The

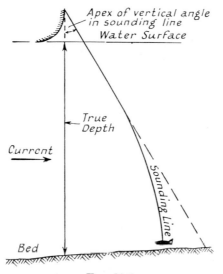

FIG. 10.5

amount paid out may be indicated on dials, or in shallow waters by graduated rods to which the meter is attached.

If line sounding is used in deep, swift-flowing water, the length of line paid out may be greater than the true depth due to drag, as shown in Fig. 10.5. The correction to be made is in two parts, i.e. part to the straight unwetted length of line, and part to the wetted length. Tables giving such corrections (which are always negative) are given in U.S. Geological Survey Water-supply Paper 888 (*op. cit.*).

(2) Indirect Method

Echo Sounding. This method is somewhat analogous to the method of measuring base lines by means of the measured time interval which a light impulse of known velocity takes to traverse the line and back again, which was described in Chapter 9. In the echo sounder, a sonic or supersonic impulse is transmitted by an oscillator fitted in the bottom of the sounding boat (or slung over the side). The return (echo) pulse is picked up by the receiver which, by an electronic arrangement, records the time interval between transmission and reception as a depth. The method has certain limitations which are mentioned later, but, excluding capital cost of the equipment, these limitations are probably less than in the direct methods, while the accuracy is much higher. The Kelvin Hughes MS 36M Hydrographic Echo Sounder sends out supersonic impulses and has an operating range up to 100 m. It normally operates with basic depth scales of 0 to 20 m (shallow) and 0 to 40 m (deep), the pulse repetition frequency being greater in the

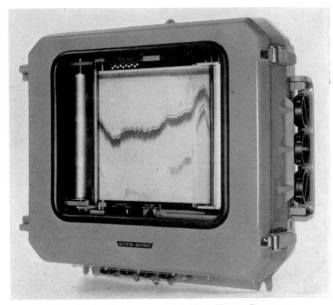

FIG. 10.6. RECORDER OF MS36 ECHO SOUNDER
(Courtesy of Kelvin Hughes)

former case. With phasing of scales, effected by electrical switching, a range of 30 to 50 m is possible in the first instance and a range of 60 to 100 m in the second. It thus covers the range of all engineering hydrographic surveying requirements so far as sounding as such is concerned, i.e. sounding pure and simple; direct sounding in stream gauging and current measurement where the current meter acts as its own sounding line will, of course, continue to be used.

The record of depth is made by a stylus on a moving band of dry paper and it can be viewed through a glass window as shown in Fig. 10.6; the draught of the vessel is compensated for, so that the transmission is effectively from water level. Either inboard or outboard transducers can be connected to the power/transmission unit, and in the latter case transmitter and receiver transducers are laid on opposite sides of the keel.

Calibration. The drive motor of the MS36M Sounder has a controlled speed of 3,000 r.p.m. for a speed of sound of 1,500 m/s. The speed of sound is not constant, however, but varies with water temperature and degree of salinity, and since a change in the speed of sound will produce an error in the recorded depth, the echo sounder is adjusted to meet local conditions by altering the operating rev/min. There are three methods of calibration—

(i) By calculation. The surface temperature of the water in

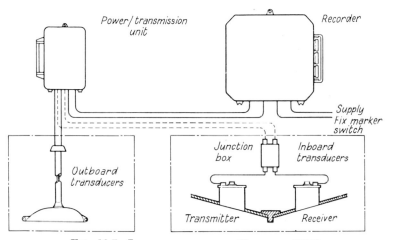

FIG. 10.7. INTERCONNEXION OF UNITS OF MS36
(*Courtesy of Kelvin Hughes*)

degrees Celsius (T), and the salinity in parts of NaCl per 1,000 (S), are determined, whence the velocity of sound in m/s (V) is given by

$$V = 1,410 + 4\cdot21\ T - 0\cdot037\ T^2 + 1\cdot14S$$

This local velocity is used to calculate the operational revolutions (R) from the aforementioned relationship between speed and sound velocity.

(ii) By direct calibration. A target is lowered into the water to a series of known depths and the echo sounder is adjusted to read those depths.

(iii) When the bottom is smooth, the recorder may be calibrated by direct sounding with the lead line, but such a method is subject to error.

Special Points in Echo Sounding. These limitations are not generally serious, but a knowledge of them assists in interpreting the records. One of the main points to note is that the impulses do not form a true beam, but have a conical shape, the main strength of the impulse being in an acute-angled cone at the centre. On steep slopes, therefore, echoes are received from the area of slope covered by this central high intensity zone, the strongest signals being from vertically below, with signals of decreasing strength towards the edges of the beam. These stray readings can upset the record, and to cut them out the instrument is used at its lowest sensitivity when over a sloping bed. The cone radius increases with depth so that potholes can be missed. The instrument will always record the minimum depth, therefore projections above the mean bed level falling into the beam will be recorded.

Care must be taken when operating close to jetties and quay walls, since the return echo from the wall may blank out the bottom echo. Air bubbles passing under the oscillators will cause a reflection, so that the instrument must be carefully positioned with regard to pockets of air bubbles which may occur under the bottom of the moving boat.

When sounding in shallow water with an inboard-mounted oscillator (the transmitter being on one side of the keel and the receiver on the other) the magnitude of the errors will depend on the distance between the two instruments. A correction table is easily drawn up by solving the right-angled triangle linking separation and measured depth.

Sampling and Sweeping. These are two operations often associated with sounding, and of them the former is probably the more likely to be of interest to the civil engineer.

If samples of the bed bottom are required, a sounding lead is used which has a hollowed-out base filled with tallow. For larger samples, special grabs, e.g. the Dutch grab or the Binkley Silt Sampler, are used.

Sweeping is used to check that no underwater obstructions have been missed, and the usual method employed in open water is the *fine wire drag*. It is held between two boats, one of which is equipped with a friction brake drum. The boats steam slowly in parallel paths about 100 m apart with the drag—a length of piano wire—suspended at a constant depth. When the wire touches an obstruction, the man on the friction brake feels the tension in the supporting cable increase. He signals the boats to stop, and the angles of the supporting cables to the horizontal are measured. The depth of the drag is known and hence a fix can be made of the obstruction.

Location of Soundings

Many methods are employed depending on the extent of the water under survey, and some of them will be considered here. As a broad classification, the methods may be divided into (1) those in which the sounding vessel is secured, used in channels or narrow waters, and (2) those in which the vessel is free, used in open and wide waters.

1. Soundings in Channels

With rivers and channels which are narrow enough for a line or steel wire to be slung across, soundings taken with the sounding vessel secured probably constitute the most accurate method. The cable supports on each bank must be accurately located; the cable is graduated, so that the boat can be pulled or manoeuvred to the appropriate station and the sounding made. This technique is usually adopted when stream flows are being measured with the current meter, which is dealt with later in the chapter. The boat used should be heavy enough to damp out wave action; maximum width of water for this method may be taken at 300 m. (By running between an anchored boat and the shore the method can be used for off-shore work, but it is not so convenient as other methods.)

2. Soundings in Open Water

In the general case, where the survey vessel is not secured, there are two problems: (i) to keep the boat on a known course, so as to obtain systematic sounding, known as *conning* the vessel, and (ii) to locate the soundings so that they may be charted.

(i) **Conning the Survey Vessel.** This task is of course mainly one of seamanship. One of the most common methods is to fix markers (poles, beacons, etc.) on shore, as shown in Fig. 10.8, a method suitable for work in rivers and open sea areas up to 5 km off shore where working on the larger scales, e.g. 1/2,500. The vessel is run down each track,

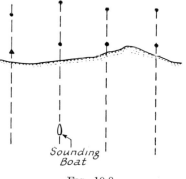

Sounding
Boat

FIG. 10.8

the steersman lining himself in with the successive pairs of shore stations which have been fixed in advance and which must be easily identifiable.

The other method in common use is to steer on the compass, but this is suitable only for smaller scales of working, say 1:10,000 or less.

(ii) **Fixing the Soundings.** The above methods serve to guide the steersman. The survey vessel is accurately and continuously fixed by other methods, some of which are given below.

Note that as well as reading angles for the horizontal fix, when working in tidal waters, the times at which the soundings are taken are also noted so that by reference to the appropriate tide gauges, the observed depths may be linked to the datum (*see* above under Vertical Control).

(a) THE STATION POINTER. Referring to the previous section on horizontal control, which describes the accurate fixing of beacons and prominent points on shore, fixes are taken by sextant by observing two simultaneous horizontal angles between three stations on the shore. These fixes are then plotted by the station pointer, as shown in Fig. 10.9, the principle being self-explanatory. The angles are set off on the pointer to approximately the same accuracy as that with which they are measured, and the required point obtained by pricking through at the intersection of the three arms of the pointer.

The three-point problem has already been dealt with in Chapter 6 (notes on the Plane Table), and the student is referred back to the principles governing the relation between the station positions and the observer's position insofar as they affect the strength of the fix. A

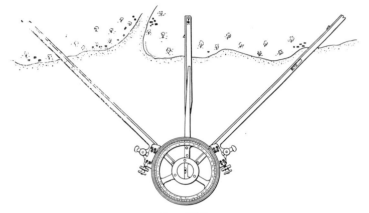

Fig. 10.9

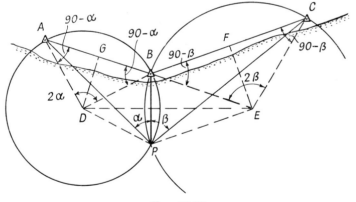

Fig. 10.10

well-conditioned fix by sextant can be taken to have an accuracy of the order of 100 m insofar as location is concerned for sightings up to 25 km.

(*b*) RESECTION. This method is identical with the last except that it is required to plot the fix to greater accuracy than is possible with a station pointer.

The problem is, therefore, referring to Fig. 10.10, to locate *P*, given the co-ordinates of positions A, B, and C, and angles $A\hat{P}B$, $C\hat{P}B$, measured by sextant as α and β.

D and E are the centres of the circumscribing circles through points ABP and CBP respectively; AB and BC are known or can be calculated from the co-ordinates.

D and E are located by either setting off angles of $(90 - \alpha)$ from A and B and $(90 - \beta)$ from B and C respectively, or by calculation.

In the latter case

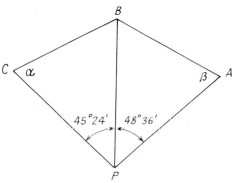

FIG. 10.11

$$A\hat{D}B = 180 - 2(90 - \alpha) = 2\alpha = 2A\hat{P}B$$

If DG be the perpendicular bisector of AB

then $\quad\quad\quad A\hat{D}G - G\hat{D}B - \alpha$ since triangles AGD and BGD are congruent.

Thus $\quad\quad\quad AD = AG$ cosec $\alpha = \frac{1}{2}AB$ cosec α

and $\quad\quad\quad GD = AG$ cot $\alpha = \frac{1}{2}AB$ cot α

Thus D (and similarly E) can be located. The intersection of the two circles gives P since α and β are simultaneously subtended by AB and BC respectively.

In the $\triangle$'s ABD, CBE, the lengths of AB, BC and the angles are known, so that it is possible to calculate the co-ordinates of D and E. Then, in $\triangle DPE$, the length of side DE and the angles are known, so that the co-ordinates of P may be calculated. The following alternative method is however often preferable.

The angle $A\hat{B}C$ being known or computed, P may be located by determining all angles in triangles ABP and BCP and then calculating sides AP, BP and CP.

EXAMPLE. A, B and C are three shore stations on a coastline (Fig. 10.11), and P is a sounding point at sea. $AB = 400$ m, $BC = 381$ m, $A\hat{B}C = 122°$ 30′, $A\hat{P}B = 48°$ 36′, and $B\hat{P}C = 45°$ 24′. A and C are respectively east and west of BP. B and P are respectively north and south of AC. Calculate the distances AP, BP, and CP.

Let $\qquad$ $B\hat{C}P = \alpha$ and $B\hat{A}P = \beta$

$$\alpha + \beta = 360° - 48° \, 36' - 45° \, 24' - 122° \, 30'$$
$$= 143° \, 30'$$

Therefore $\qquad$ $\alpha = 143° \, 30' - \beta$

Now $\qquad$ $\dfrac{BP}{\sin \alpha} = \dfrac{381}{\sin 45° \, 24'}$ and $\dfrac{BP}{\sin \beta} = \dfrac{400}{\sin 48° \, 36'}$

$\therefore \qquad BP = 381 \, \dfrac{\sin \alpha}{\sin 45° \, 24'} = 400 \, \dfrac{\sin \beta}{\sin 48° \, 36'}$

$\therefore \qquad \sin \alpha = \dfrac{400}{381} \, \dfrac{\sin 45° \, 24'}{\sin 48° \, 36'} \, \sin \beta$

$$= \sin (143° \, 30' - \beta)$$

whence $\qquad \beta = 72° \, 02'$

and $\qquad \alpha = 71° \, 28'$

Therefore $\qquad C\hat{B}P = 180° - 45° \, 24' - 71° \, 28' = 63° \, 08'$

$$A\hat{B}P = 180° - 48° \, 36' - 72° \, 02' = 59° \, 22'$$

$\dfrac{AP}{\sin 59° \, 22'} = \dfrac{400}{\sin 48° \, 36'} \quad \therefore \ AP = 459 \text{ m}$

$\dfrac{BP}{\sin 72° \, 02'} = \dfrac{400}{\sin 48° \, 36'} \quad \therefore \ BP = 507 \text{ m}$

$\dfrac{CP}{\sin 63° \, 08'} = \dfrac{381}{\sin 45° \, 24'} \quad \therefore \ CP = 477 \text{ m}$

P may now be located by striking arcs equal in length to these values.

(*c*) USING CIRCLE PLOTTING SHEETS. This method eliminates the use of the station pointer and enables positions to be plotted immediately as on a graph. It is a graphical development of the previous method. Selecting the shore control stations, we require (*see* Fig. 10.10) the centres and radii of the circles passing through the pairs of points used for sighting. Thus, three points, as in Fig. 10.10, may be used, i.e. the two angles observed have a common centre station; or four control stations may be used.

Considering one pair of control stations, A and B (Fig. 10.12) we require to know AD and the location of D for various values of α. Join A and B, and construct the perpendicular through the mid-point E. The length AB is known, and as previously

$\qquad\qquad ED = AE \cos \alpha$—which gives the position of D

and $\qquad\qquad AD = AE \operatorname{cosec} \alpha$

$\qquad\qquad\qquad = $ radius of the circle on whose circumference AB subtends α

A range of values is chosen for α, and the appropriate curves plotted. The procedure is then repeated for values of β subtended by the other control points, and a graph is obtained, as in Fig. 10.13.

Referring to Fig. 10.13, three control points, the middle one of which is a lighthouse, have been chosen and a series of circular arcs passing through pairs of points for different values of α and β are given. Three runs have been indicated and individual soundings numbered; for example the angles subtended at 22/23 were 45° 24′ and 31° 25′ respectively and the sounding position is obtained interpolating between the curves given by whole degrees.

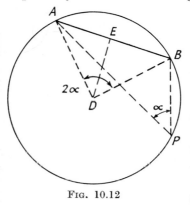

FIG. 10.12

(*d*) THE SUBTENSE BOARD. When soundings are being made within 60 m of a jetty or river bank, instead of using the taut wire a special subtense board may be erected on the bank and the boat brought in at right angles to the shore with the sextant, set at a fixed angle of 3°, sighted on the board. The general principle is shown in Fig. 10.14. The subtense board is set up so that the lower sighting mark, which is viewed directly through the horizon glass of the sextant, is at the observer's eye-level as he stands in the boat. The height of the graduations above the sighting mark are such that they subtend 3° at known distances from the board, so that, with the sextant set at 3°, as each graduation (viewed through the index glass) becomes coincident with the sighting mark (viewed directly) the observer knows his distance from the board. This fixing is rapid enough for use with the echo sounder; when in tidal waters the run must be made rapidly so that no correction for change in water level is necessary.

The heights of the graduations above zero for a 3° subtense bar are—

Height from bottom (m)	Number	Giving fix at (m)
3·144	1	60
2·882	2	55
2·620	3	50
2·358	4	45
2·096	5	40
1·834	6	35
1·572	7	30
1·310	8	25
1·048	9	20
0·786	10	15
0·524	11	10
0·262	12	5

The bar may be hinged at line No. 6.

(e) FIXING BY THEODOLITE. In this method two shore-based theodolites, conveniently placed, take simultaneous sights on to the sounding vessel. The principles involved are straightforward, but the method has the disadvantage of requiring two skilled surveyors, who take the readings on receiving signals from the boat.

The use of (a) theodolites and (b) the Tellurometer and Hydrodist in the location of boreholes for the Channel Tunnel study has been

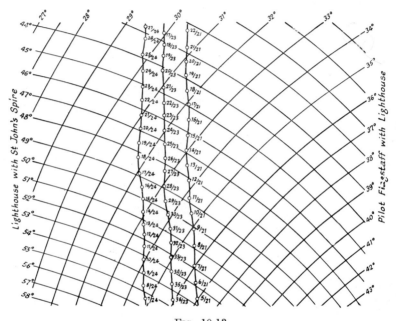

FIG. 10.13

(Courtesy of Kelvin Hughes, Ltd.)

discussed by C. A. Murray in Paper G3 given to the Conference of Commonwealth Survey Officers in 1967. This paper contains much interesting information on observational procedures.

The Hydrodist operation is based on that of the Tellurometer, a master station being located at sea on a vessel, or on a drilling rig, and a remote station established on shore; one remote station only can be related to one master.

(f) FIXING BY RADAR. For fixing deep soundings, out of sight of land, the Shoran (short-range) radar system is used. Radio waves are sent out from the sounding vessel and returned from two accurately-located shore base stations. By measurement of the time interval elapsing in each case and knowing the velocity of the radio wave, the distance can be found from the ship to the ground stations. The fixes lie within an area of uncertainty of 10 m over a distance of 80–100 km. Fixes up to 800 km can be made using Loran (long-range) radar.

Current Measurement

Currents are measured by (a) floats, or (b) current meter.

(a) Floats. Many types of float are available, ranging from small surface floats with flags on them to the double float which has a perforated cylinder or a canvas vane suspended at a known depth below the

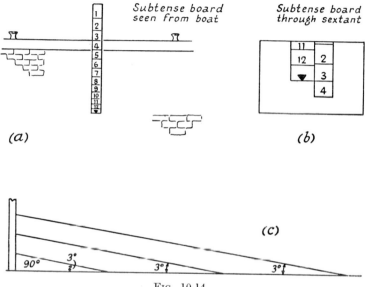

(a)   (b)

(c)

Fig. 10.14

surface from a small floating buoy. An ordinary 2 m ranging pole can also be used, as it floats upright with about one-third showing. Flags of different colours are attached (small enough to avoid acting as sails) and the floats released at intervals.

The fixing of the positions of the floats may be done by simultaneous theodolite observations from the shore, or, better, by frequent visits to each float in turn by a surveying boat, whose position is fixed by sextant observations on the shore control points. The time of observation is noted in each case, so that the rate of drift can be calculated.

(b) By current meter. This instrument may be slung from a buoy (the readings being either recorded on tape or transmitted continuously by radio) or from a boat. In either case it is obvious that it must be attached to something static, since the speed of the current and its direction can only be measured by water flowing past the measuring device (either a cupped wheel, horizontally mounted, or a propeller mounted with its axis along the line of flow). A further description of the current meter is given in the next section.

Measurement of Flow—Stream Gauging

Commonly employed direct methods used for the gauging of rivers and streams are—

(a) Current meter, (b) Floats, (c) Chemical methods, (d) Weirs and notches.

(a) **The Current Meter.** One of the most widely used types is the Price current meter, devised in 1882 by W. G. Price, a civil engineer

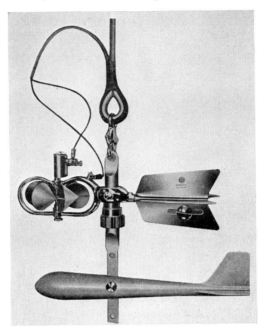

Fig. 10.15. Water Current Meter Mk IV
(Courtesy of Hilger & Watts, Ltd.)

employed by the Mississippi River Commission; its main feature is a bucket wheel mounted on a vertical axis, the pivot bearing of which has an air pocket to prevent silt accumulation. Fig. 10.15 shows the model made by Hilger & Watts, Ltd. The tail fin holds the head of the meter upstream, and an electric cable, which may be used to suspend the meter, connects a contact breaker, operated by the rotating bucket wheel, to the surface. The number of revolutions of the bucket wheel in a given time may thus be measured, this rate of rotation being proportional to water velocity. The calibration of the instrument is carried out either by the manufacturer or alternatively at the Hydraulic Research Station say. The one illustrated can be calibrated to measure from 0·03 m/s to 4·50 m/s to within ± 1 per cent.

To use the meter for discharge measurement, a station is chosen on the stream or river, the suitability of the station being governed by the following requirements—

(1) Channel should be regular in shape and straight up- and down-stream of the section;

(2) Channel bed should be free of obstructions;

(3) Flow should be as streamlined as possible, since eddy currents affect the meter.

A graduated rope or wire is pulled taut across the river at the station and divided into equal intercepts from 1 m to 5 m depending on the width of the river (note that bridges may be used as stations). At each

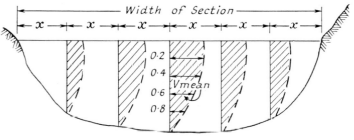

Fig. 10.16

of these points the current meter is lowered to the water and the velocity measured at, say 0·2, 0·4, 0·6, and 0·8 of the depth. In deep water, or with heavy sinkers, special gauging reels can be attached to the meter. It will be noticed that the velocity, which at the river bed tends towards zero due to skin friction, rises to a maximum at 0·2 depth of about 1·2 times the mean velocity. By plotting these velocities for each vertical section, as shown in Fig. 10.16, the mean velocity, V_{mean}, in each case may be found—

$$V_{mean} = \frac{\text{area enclosed by curve (hatched)}}{\text{depth at that position}}$$

Note, by plotting this value of V_{mean} on the velocity/depth diagram for the vertical section concerned, that V_{mean} occurs in each case at *about* 0·6 × depth. Thus, if time allows only one velocity measurement at each vertical section, this measurement should be made at 0·6 × depth.

The value V_{mean} having been obtained for each vertical section, it is possible to obtain mean velocities for the water passing through the trapezoids into which the cross section is divided, and hence by integration, using planimeter, Simpson's rule, or trapezoidal rule, the discharge is determined from—

$$Q = A \cdot V$$
$$= \Sigma a \cdot V_{mean}$$

The areas are in m², and the velocities in m/s, so that Q is in m³/s. This is known as the area-velocity method.

Alternatively, "contours" may be drawn on the cross section, joining up points of equal velocity. The areas of these curves are measured by planimeter, and these are treated as equidistant cross-sectional areas of a solid, the volume of which gives the discharge. Yet another way is to use the "spot height" analogy assuming the cross section to be subdivided into a series of areas, the velocities of flow through these being the mean of the velocities at the corners.

(*b*) **Floats.** Although the current meter is by far the most widely used of the area-velocity methods for calculating discharges (and normally it is the most accurate method), there are times when floats are used instead—notably when excessive velocities, depth, and floating drift prohibit the use of the current meter.

Surface floats give the velocity of the surface water only, and apart from the sensitivity of such light floats to wind, the choice of coefficient to convert $V_{surface}$ to V_{mean} ranges from about 0·7 to 0·95. The results are thus of doubtful accuracy.

The double float already mentioned can be used to give velocities at different depths (though allowance must be made for the effect of the surface float) the calculation of discharge thus being exactly as described for the current meter.

Float velocities are measured by releasing the floats at the appropriate point upstream and then timing them along the measured distance to a second station downstream.

(*c*) **Chemical Methods.** These methods involve the introduction of a chemical into the stream, and out of many variations, the two most frequently used are the *salt-velocity* and the *salt-dilution* methods.

SALT-VELOCITY METHOD.* Salt in solution increases the electrical conductivity of water. Two sets of electrodes a known distance apart in a stream of constant cross-section are connected to a recording galvanometer which records the changes in electrical conductivity of the stream with respect to time. Under normal conditions the graph given is a more or less horizontal straight line, but when a volume of salt solution is injected at the upstream electrodes (which are in the form of pipes), the graph shows a rectangular jump. Later on, when the salt gets down to the second pair of electrodes, a second jump occurs. The time of transit is taken as the time between the centres of area of the two jumps, and dividing this time into the volume of water between the two stations gives the discharge.

SALT-DILUTION METHOD.† A salt solution of known concentration is added at a constant rate to the stream to be gauged, and, by analysis, the subsequent dilution of the solution is determined. The samples are taken far enough below the entry point for complete mixing and uniform distribution to have taken place. No measurement of area or

* "The Salt-velocity Method of Water Measurement," by C. M. Allen and E. A. Taylor, *American Soc. Mech. Eng. Trans.*, Vol. 45, 1923.
† "Gauging Water Flow by the Salt-dilution Method," S. Hutton and E. Spencer, *Proc. Instn. Civil Engrs.*, Aug. 1960.

distance are necessary, and the method is pre-eminently suitable for use in turbulent mountain streams. The weight of salt that passes in each second at the point where samples are taken must equal the combined weights of salt normally present, and the salt added in solution, i.e.

$$WX + W'X' = (W + W')X''$$

where W is the wt. of water discharged per sec, W^1 is the wt. of salt solution added per sec, X is the percentage (by weight) of natural salt in the stream, X' is the percentage (by weight) of salt in concentrated solution and X'' is the percentage (by weight) of salt in the sample after mixing.

Therefore

$$W = W' \frac{(X' - X'')}{(X'' - X)}$$

X'' must be uniform at all points in the cross-section and $(X'' - X)$ must be found accurately: the salt used should be detectable in small quantities and should be stable in the water. When obtaining samples, bottles can be immersed in the stream at the relevant cross-section or a hand pump can be used for drawing off. Sodium dichromate is a suitable chemical according to Hutton and Spencer, although at concentrations of 30 parts per million it is toxic to fish. Five parts per million have been used to achieve an accuracy of $\pm$ 1–2 per cent in measurement of flow. Larger quantities of sodium chloride would be needed since it is present in natural waters and in this case X should not exceed $0.15 X''$.

(d) **Weirs and Notches.** These may be used to measure the discharge of liquids flowing under gravity, i.e. rivers or similar channel flows. A notch, which can be taken as an orifice with its upper edge free, is usually used to measure flows from reservoirs or tanks, and of streams of modest discharge; flows in sewers are often established by V-notches.

A weir, which may be considered as a large notch, extends across the stream at right angles to the flow (though side weirs set parallel to the flow are often used in storm-water overflows in sewerage practice).

The free water surface is drawn down as it passes over the weir, and the water level before drawdown occurs, referred to the top of the weir, is the *head* of water over the weir. Weirs may be *suppressed*—when they extend across the full width of the approach channel; or *contracted*—when they do not extend across the full width. In this latter condition end contractions are induced.

Weirs may be sharp-crested or broad-crested and this is defined by the nature of the face (crest) over which the water flows. In the latter case, often formed by masonry or concrete, the sheet of water is in contact with the crest over much of its area and the discharge is greater than for a corresponding sharp-crested type, which could have been made of thin stainless steel plate.

The water level is drawn down as it passes over a notch or weir, and the effective head of water, upon which the quantity of flow depends, must be referred to the level before draw-down occurs. Head is best measured by a float and on connexion to a drum type recorder a permanent and continuous record of head, or rate of flow, may be obtained. It is advisable, if possible, to check the discharges as computed by the weir formula by current-meter observations or a similar method; this procedure is frequently adopted to calibrate large weirs.

The following formulae are available and the student is referred to the various text books on hydraulics* for their derivation.

RECTANGULAR NOTCH. For a simple small rectangular notch, of breadth B, the discharge Q for a head H is given by—

$$Q = \tfrac{2}{3}Cd \cdot B\sqrt{2g} \cdot H^{\frac{3}{2}} \qquad . \qquad . \qquad (10.1)$$

The coefficient of discharge, Cd, may be found by experiment. It varies slightly with head but usually for civil engineering purposes a mean value of $\tfrac{2}{3}Cd\sqrt{2g}$ is determined from experimental results.

V- OR TRIANGULAR NOTCH. The wetted length of such a notch depends directly on the head. If the apex angle is θ, it can be shown that—

$$Q = \frac{8}{15}Cd\sqrt{2g} \cdot \tan\frac{\theta}{2} \cdot H^{\frac{5}{2}} \qquad . \qquad . \qquad (10.2)$$

An average value for Cd is 0·6.

RECTANGULAR WEIR. The formula (10.1) does not apply for large notches or weirs and an empirical formula given by Francis is usually adopted. Neglecting the velocity of approach the discharge can be calculated from

$$Q = 1·83(B - 0·1\,nH)H^{\frac{3}{2}} \qquad . \qquad . \qquad (10.3)$$

where $\qquad\qquad B = $ breadth of weir

and $\qquad\qquad n = $ number of end contractions

n is zero for a suppressed weir. B should be greater than $3H$ and there are certain other limits imposed with regard to the height of the crest above the channel bottom and the position of the sides of the weir.

The velocity of approach (V) of the water to the weir may be taken into account by allowing an additional head $h = \dfrac{V^2}{2g}$ for the kinetic energy of the water. The still water head (H_1) is thus

$$H_1 = H + \frac{V^2}{2g}$$

and this figure can now be substituted in the Francis formula (10.3) to give

$$Q = 1·83(B - 0·1\,n\,H_1)(H_1^{\frac{3}{2}} - h^{\frac{3}{2}}) \qquad . \qquad . \qquad (10.4)$$

EXERCISES 10

1. Two stations A and B are 846 m apart. From theodolite stations P and Q on opposite sides of AB the following angles were observed—

* e.g. *Hydraulics*, by E. Lewitt (Pitman).

$$A\hat{P}Q = 61° 12';\qquad B\hat{Q}P = 53° 28';$$

$$Q\hat{P}B = 44° 11';\qquad P\hat{Q}A = 41° 29'.$$

Calculate the distance between stations P and Q. (*L.U., B.Sc.*)
Answer: 713 m.

2. *A*, *B*, and *C*, are three stations on a coastline used to fix the position of a buoy, *P*, at sea, which lies on the opposite side of *AC* to *B*. $AB = 482$ m, and

$BC = 344$ m. The seaward angle $A\hat{B}C = 143° 30'$, and angles APB and BPC are found to be 45° 36' and 40° 48' respectively. Find dimensions AP, BP, and CP.
Answer: 671 m, 518 m and 453 m respectively.

3. A survey has to be carried out as a preliminary to the construction of a small harbour off a rocky coast. Describe how you would execute this work, the equipment you would need, and the data you would collect.

4. Describe a method of finding approximately the discharge of a river, stating the requirements of a site for the relevant measurements.
Calculate the discharge of a river, given the following measurements made with a flow meter—

Distance across river from one bank (m)	0	10	20	30	40	50	60	70
Depth of bed (m)	0	0·7	1·2	1·5	1·8	1·5	0·9	0
Rate of flow at 0·6 depth (m/s)	0	0·15	0·24	0·30	0·36	0·33	0·24	0

Answer: 21 m³/s approx.

5. A 90° V-notch is used to gauge the flow of a sluggish stream. Determine the flow corresponding to a head over the weir of 0·6 m.
Describe briefly another method suitable for estimating the flow of mountain streams.
Answer: 0·4 m³/s approx.

6. Describe with the aid of sketches

(*a*) how the profile of the bed of a tidal river, approximately 100 m wide and having a minimum depth of 5 m, may be determined;
(*b*) three methods of locating soundings off shore.

7. *A*, *B* and *C* are three shore stations used to fix the positions of a float *O* at sea, the float lying on the opposite side of *AC* to *B*. *AB* is 562 m, *BC* is 481 m and the seaward angle $A\hat{B}C$ is 133° 43'.
If angles $A\hat{O}B$ and $B\hat{O}C$ are found to be 38° 44' and 43° 22' respectively, determine dimensions by which *O* could be located on a chart.
Answer: $AO = 898$ m; $BO = 700$ m; $CO = 484$ m.

CHAPTER 11

PHOTOGRAMMETRY

PHOTOGRAMMETRY IS THE NAME GIVEN to that branch of surveying in which plans and maps are based on, and prepared from, photographs taken from ground or air positions. Some basic principles are given herein, together with descriptions of certain techniques and instruments.

Ground Photogrammetry

The initial developments were in ground photography. The Frenchman Laussedat prepared a survey in 1861, and during the nineteenth century the method was used in Canada and Switzerland, among other countries, particularly for the plotting of mountainous terrain: plotting machines were devised early in the twentieth century.

One interesting value of terrestrial photographs is that they can be used as a check on the condition of buildings nearby proposed work to be undertaken, should there be a possibility of damage occurring, with resultant claims. They give a record of the previous condition for comparison with the condition after completion.

Ground photography is carried out using a phototheodolite which consists basically of a camera of known focal length, together with a theodolite. One type of instrument which may be encountered is shown in Fig. 11.1. Its theodolite, which rotates independently of the camera, allows (a) the determination of the orientation of the camera axis, (b) the angular measurements needed to fix the camera stations. A frame carrying the principal horizontal and vertical wires or marks is pressed against the plates placed in the camera so that these principal wires (or marks) will appear on the negative, and can then serve as axes for co-ordinate measurement, the intersection of the axes giving the principal point.

The collimation line of the camera joins the centre of the object glass to the intersection of the wires or lines joining the marks, and the vertical axis of the theodolite lies in the vertical plane of collimation of the camera. Thus when the horizontal scale reading is zero, the collimation line of the theodolite and the vertical plane of collimation of the camera are parallel. To fix one point in plan will require overlapping photographs from at least two camera stations, which may be, say, at the ends of a traverse line. The distance between the stations and the bearing of the line joining them will thus be known. After setting up, usually with the plate vertical, the instrument is so arranged at each station in turn that the photographs to be obtained will contain the required point C, as in Fig. 11.2a. A very important point should be located by three photographs from three different stations. The plates are now developed and the point C located on each, its co-ordinates

364

(x, y) being measured from the horizontal and vertical axes. Sufficient photographs are taken from each station so as to collect the detail required and they usually overlap photographs taken from other

Fig. 11.1. P 30 Photo Theodolite
(Courtesy of Wild Heerbrugg, Switzerland)

stations. Angles A_1AB and B_1BA, which denote the directions of the camera axes, must be measured at the relevant station for each photograph and carefully noted in the field book. The directions of the axes to obtain these overlaps may be convergent, as shown in Figs. 11.2 and 11.3, or may be parallel to each other, as indicated later.

In plotting point C on the plan to be completed from the photographs, line AB is plotted (ab) to the scale required, and lines aa_1 and bb_1 indicating the directions of the camera axes, then drawn in at the measured angles (Fig. 11.3).

The positions of the plates are then plotted at a convenient distance,

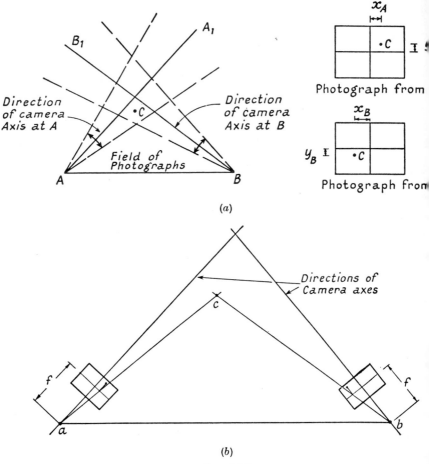

(a)

(b)

FIG. 11.2

representing the focal length of the camera objective, and at right angles to aa_1 and bb_1. These positions need not be to the same scale as ab represents AB. Co-ordinates x_A and x_B are plotted to the scale by which aO_A and bO_B represent the focal length, giving c_1 and c_2 respectively. Rays are drawn in through a and b and then the point of intersection gives c, the plan position of C.

It is more convenient to plot C from the print than the negative but some errors may occur due to shrinkage or distortion of that print. Enlargements of the original may be made in order that errors in measurement of the co-ordinates may be reduced, and important features may have enlargements on glass.

If the print be used, then its position is plotted as indicated by the dotted lines in Fig. 11.3, and C is plotted on each as $c'_1 o'_A = x_a$ and $c'_2 o'_B = x_b$.

The photographs themselves may be used for plotting the plan directly. Each photograph is positioned as in Fig. 11.2b such that its principal vertical line lies in the direction of the camera axis at the station and such that the intersection of the axes, the principal point, is at a distance f from the plotted position of the station. A line may now be drawn, or a straight-edge laid, parallel to the vertical line through a detail point C, and the position at which it meets the horizontal line is marked. A line may now be drawn through the plotted

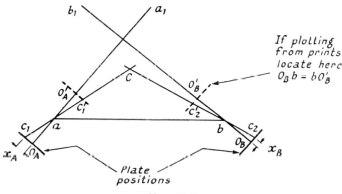

Fig. 11.3

position of the station and the point on the principal horizontal line just obtained, and it will intersect a similar line obtained from the second photograph at the plan position of point C.

Levelling or contouring may also be carried out from the photographs; if the distance from the camera to the point under consideration is great, the correction for curvature and refraction as given in Chapter 3 will be required.

Having found the plan position C (Fig. 11.4) of a certain point, it is required to determine the elevation, H, of that point with respect to the horizontal plane containing the line of collimation of the camera. If C_1 is the position of the point in elevation, then, from similar triangles,

$$\frac{H}{y_a} = \frac{AC''}{OA} = \frac{AC}{AC_1} = \frac{AC}{\sqrt{x_a^2 + f^2}}$$

$$\therefore \qquad H = \frac{y_a}{\sqrt{x_a^2 + f^2}} \cdot AC$$

EXAMPLE. In a survey carried out by phototheodolite, a base line AB running west to east was found to be 240 m long, and a station C was observed. On a *photograph* taken from A with angle $BAC = 52°$, a point S was observed to be 12·45 mm to the left of the principal vertical line and 6·20 mm above the horizontal line. On a photograph taken from B with angle $ABC = 48°$, point S was observed to be 6·80 mm to the left of the vertical line. If the focal length of the camera was

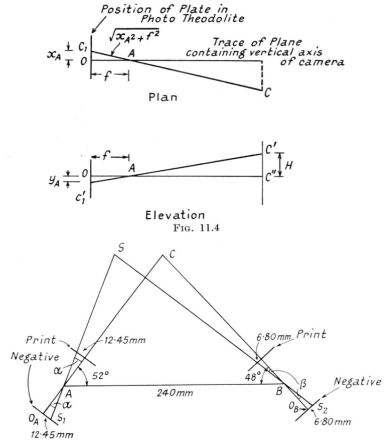

Fig. 11.4

Fig. 11.5

165 mm and the reduced level of the camera axis at A 89·80 m above datum, determine the co-ordinates of S relative to A and the reduced level of S.

Points will be located on the plan as shown in Fig. 11.5.

$$\tan \alpha = \frac{12·45}{165} = 0·0754 \qquad \tan \beta = \frac{6·80}{165} = 0·0412$$

$$\alpha = 4° \, 18' \qquad\qquad \beta = 2° \, 21'$$

$$S\hat{A}B = 4° \, 18' + 52° \qquad S\hat{B}A = 48° - 2° \, 21'$$

$$= 56° \, 18' \qquad\qquad = 45° \, 39'$$

By the sine rule,

$$\frac{SA}{\sin 45° \, 39'} = \frac{240}{\sin (180° - 45° \, 39' - 56° \, 18')}$$

$$SA = 175 \cdot 4 \text{ m}$$

Then the co-ordinates of S will be

175·4 sin 56° 18′ N and 175·4 cos 56° 18′ E

or 145·9 m N and 97·3 m E

Difference in level (S higher than A)

$$= \frac{y_A}{\sqrt{x_A{}^2 + f^2}} \cdot AS$$

$$= \frac{6 \cdot 20}{\sqrt{12 \cdot 45^2 + 165^2}} \times 175 \cdot 4$$

$$= 6 \cdot 6 \text{ m}$$

∴   Level of $S = 89 \cdot 8 + 6 \cdot 6$

$$= 96 \cdot 4 \text{ m}$$

EXAMPLE. The horizontal angle between two points A and B (Fig. 11.6) was measured directly at a station C and found to be 23° 42′. A phototheodolite was set up at C and, on a photograph, A was found to be 30·48 mm to the left of the vertical hair and 12·70 mm above the horizontal hair, while B was found to be 38·80 mm to the right of the vertical hair and 19·05 mm below the horizontal hair. Determine the focal length of the camera lens and the difference in level between A and B if $AC = 86$ m and $BC = 68$ m.

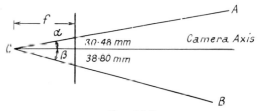

FIG. 11.6

From direct measurement $\alpha + \beta = 23° \, 42'$.

$$\tan (\alpha + \beta) = \frac{\tan \alpha + \tan \beta}{1 - \tan \alpha \tan \beta}$$

∴ $$\tan 23° \, 42' = \frac{\dfrac{30 \cdot 48}{f} + \dfrac{38 \cdot 80}{f}}{1 - \dfrac{30 \cdot 48}{f} \times \dfrac{38 \cdot 80}{f}}$$

$$f^2 - 157 \cdot 81 f - 1{,}182 \cdot 6 = 0$$

$$\therefore \qquad f = 165 \text{ mm}$$

$$H_A = \frac{y_A}{\sqrt{x_a{}^2 + f^2}} \, CA = \frac{12\cdot70}{\sqrt{165^2 + 30\cdot48^2}} \times 86 = 6\cdot5 \text{ m}$$

$$H_B = \frac{y_B}{\sqrt{x_b{}^2 + f^2}} \, CB = \frac{19\cdot05}{\sqrt{165^2 + 38\cdot80^2}} \times 68 = 7\cdot6 \text{ m}$$

A is higher than B by 14·1 m

This example illustrates a method by which the focal length of the camera may be roughly determined or checked.

From a station, a photograph is taken of two points which subtend a known measured angle at that station. The co-ordinates of the points are obtained, and then inserted in the formula which has been derived above. If the relative levels of A, B, and the camera axis at C, are known, then a check will also be given on the accuracy of the horizontal hair.

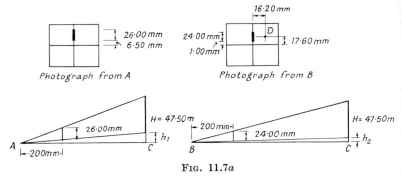

Fig. 11.7a

EXAMPLE. A phototheodolite having a focal length of 200 mm (Figs. 11.7a and b) was used at two stations A and B having co-ordinates (0, 0) and (0, 152·0) m station B being 9·1 m higher than A. In both cases station C, a tower 47·5 m high to the north, is on the vertical centre line of the photograph. In that from A it measures 26·00 mm and its base is 6·50 mm above centre, while in that from B it measures 24·00 mm and its base is 1·00 mm above the centre.

Determine the co-ordinates of C, the level of the tower base relative to A, and the directions of the sights from A and B. There is some doubt as to the horizontality of the sight from B. Check for any error: and then determine the co-ordinates and level above A of a point D located 16·20 mm right and 17·60 mm up in the photograph from B, if it is in line with the tower as seen from A. Would it be visible from A? (B.Sc. (Ext.), London, 1953.)

By similar triangles,

$$\frac{200}{32\cdot50} = \frac{AC}{H + h_1} \quad \text{and} \quad \frac{200}{25\cdot00} = \frac{BC}{H + h_2}$$

also $\qquad \dfrac{200}{6\cdot50} = \dfrac{AC}{h_1} \quad \text{and} \quad \dfrac{200}{1\cdot00} = \dfrac{BC}{h_2}$

whence $AC = 365 \cdot 4$ m $BC = 395 \cdot 8$ m

                     $h_1 = 11 \cdot 88$ m $h_2 = 1 \cdot 98$ m

now $CB^2 = CA^2 + AB^2 - 2CA \cdot AB \cos C\hat{A}B$

         $395 \cdot 8^2 = 365 \cdot 4^2 + 152 \cdot 0^2 - 2 \times 365 \cdot 4 \times 152 \cdot 0 \cos C\hat{A}B$

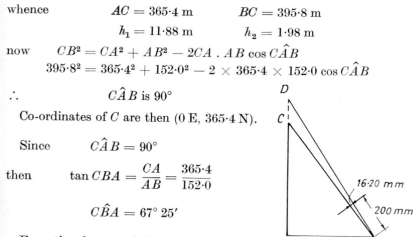

∴ $C\hat{A}B$ is 90°

Co-ordinates of C are then (0 E, 365·4 N).

Since $C\hat{A}B = 90°$

then $\tan CBA = \dfrac{CA}{AB} = \dfrac{365 \cdot 4}{152 \cdot 0}$

             $C\hat{B}A = 67° \, 25'$

Fɪɢ. 11.7*b*

From the photograph from B the tower base is apparently 1·98 m above B, whereas it should be (11·88 − 9·10) or 2·78 m higher. Thus the sight of B was inclined upwards, so that an error of 0·80 m in level is given in a distance of 395·8 m, i.e. 1 in 495, or 0·40 mm difference on the photograph.

Now $\tan CBD = \dfrac{16 \cdot 20}{200}$

∴ $C\hat{B}D = 4° \, 35'$

and $AD - AB \tan (4° \, 35' + 67° \, 25') = 152 \cdot 0 \tan 72° \, 00'$

                  $= 467 \cdot 8$ m

The co-ordinates of D are (0, 467·8). Apparent difference in level of D with respect to the horizontal axis of the instrument at B

$$= BD \times \frac{y_D}{\sqrt{f^2 + x_D{}^2}}$$

where $BD = \sqrt{467 \cdot 8^2 + 152 \cdot 0^2} = 491 \cdot 9$ m

now $y_D = 17 \cdot 60$ mm (measured) or 18·00 mm (corrected) and since

           $x_D = 16 \cdot 20$ mm and $f = 200$ mm

therefore $BD \dfrac{y_D}{\sqrt{f^2 + x_D{}^2}} = 44 \cdot 1$ m, when $y_D = 18 \cdot 00$ mm

Then the level of D is 44·1 m above B, i.e. 53·2 m above A, and so D will not be seen from A, since the tower is 47·5 m high and its base is 11·9 m above the horizontal axis at A.

Wild P32 Terrestrial Camera. The P30 Phototheodolite is being replaced by the P32 Terrestrial Camera in the Wild manufacturing programme. This camera is normally mounted above a T2 theodolite, fitted with an adaptor for the purpose, and it can therefore rotate with the telescope either about the trunnion axis or about the vertical axis (*see* Fig. 11.8). It can, however, also rotate about its own optical axis through 360°, with fixed stops at 90° intervals, so that photographs can be produced with either their longer sides horizontal or their shorter sides horizontal, therefore giving further utilization of the picture format. Both cameras use ground-glass screens for view finding, but whereas the P30 is restricted to the use of glass plates for photography, the P32 can also accept cut or roll film.

General Procedure

1. A reconnaissance must be undertaken, particularly for surveys for plans to scales greater than 1 in 20,000, in order that (*a*) the points selected for phototheodolite stations give satisfactory intersections, (*b*) points selected as stations on the control survey give suitably-conditioned geometrical figures. During this period the surveyor will check that the area will be completely covered by the photography.

2. The control survey will then be conducted, and this depends upon the area of the survey. A major triangulation scheme (with an accurately measured base line) may be required, and a minor triangulation network may be based on this. These stations may then serve as the phototheodolite stations if required and extra phototheodolite stations may be obtained by resection from three triangulation stations, a photograph being taken from the chosen station to include the three triangulation stations. Their co-ordinates are measured on the plate and the station located by means of a solution to the three-point problem. If the area be small in extent, one network may suffice, and the base line could be measured by means of the subtense bar, or a band used in conjunction with pegs driven flush with the ground, not in catenary. Levelling of the stations may be arranged by trigonometrical levelling, tacheometry, or from the photographs.

3. For plotting purposes a prominent point should appear on the common overlap of two photographs. A good intersection of the camera axes is essential for plotting points in the manner previously described and very important points should appear on three photographs from three separate stations. Dead ground from the chosen instrument stations will require subsidiary instrument stations for complete coverage. Where the ground is flat and tending to fall away from the camera the method of ground photogrammetry tends to be uneconomical owing to the large amount of dead ground and as a general principle the ground should slope towards the phototheodolite. The number of stations required will be determined by the general character of the area being surveyed, and difficulties due to low cloud may preclude certain stations in mountainous country. The plates may be oriented for plotting at a station using the following method on the survey—

Set the horizontal circle to zero, sight a camera station or triangulation station as reference mark and then register bearings of the camera axis from this reference mark as the separate photographs are taken.

FIG. 11.8. WILD P32 TERRESTRIAL CAMERA MOUNTED ON T2
THEODOLITE

(Courtesy of Wild Heerbrugg, Switzerland)

A sketch may be made for each exposure at a station showing the view taken and the angle read may be booked with this.

Stereo Photogrammetry

In this method, pairs of photographs are taken from stations at each end of a line with the instrument set up normally so that the plates are in the same vertical plane and parallel to the line joining the stations. Thus the camera axes when set up at each station in turn must be made parallel by setting them at right angles to the line, which is achieved by sighting the theodolite along the line, whereupon the horizontal circle of the instrument shown in Fig. 11.1 reads 90° or 270°. (Short base

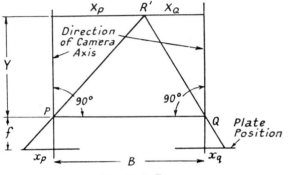

FIG. 11.9. PLAN

photogrammetry, for short ranges, may be carried out by setting two similar cameras at each end of, and normal to, a bar mounted on a tripod.) PQ should be reasonably parallel to the terrain being surveyed.

The pairs of photographs are then placed in a stereo-comparator, a stereoscopic relief obtained, and co-ordinates of required points measured, or are placed in a plotting machine, and the points themselves plotted. A typical plotting machine is the Zeiss Terragraph.

In the field, the length of the line joining the two stations may be measured tacheometrically (preferably by subtense methods), as may be the relative levels between the stations. The bearings of some prominent features which will appear on the photographs are determined and serve as a check on the photography. The general procedure is as previously indicated, the major difference to note being that intersections of camera axes are not considered in this method.

A series of relationships may be obtained for plotting as shown in Fig. 11.9.

By similar triangles,

$$\frac{X_P}{x_p} = \frac{Y}{f}$$

and

$$\frac{X_Q}{x_q} = \frac{Y}{f}$$

then
$$\frac{X_P}{X_Q} = \frac{x_p}{x_q}$$

But $X_P + X_Q = B$, which is measured independently,

$$\therefore \qquad X_P + X_P \cdot \frac{x_q}{x_p} = B$$

and so
$$X_P = B \frac{x_p}{x_p + x_q} \text{ and } X_Q = B \frac{x_q}{x_p + x_q}$$

Write $x_p + x_q = p$ the parallax) *see* p. 371), then

$$X_P = B \frac{x_p}{p} \text{ and } Y = X_P \frac{f}{x_p} = \frac{Bf}{p}$$

thus the co-ordinates of R with respect to P may be determined.

Note that errors in Y due to errors dB, df and dp can be assessed as follows—

$$dY = \frac{f}{p} dB = Y \frac{dB}{B}$$

$$dY = \frac{B}{p} df = Y \frac{df}{f}$$

$$dY = \frac{-Bf}{p^2} dp = \frac{-Y^2}{Bf} dp$$

Thus if $\frac{dY}{Y}$ is not to exceed $1/1,000$, say, and if dp is not less than 0.01 mm, then Y should not be greater than $16.5\ B$ if $f = 165$ mm. Usually this maximum value for Y is restricted to about $10\ B$. The value given to dp can be taken as the accuracy of stereoscopic measurement; a study of Fig. 11.9 will show that p increases as R approaches PQ and accordingly, if the maximum value of p be taken as 40 mm we have the "minimum range" condition. This can be evaluated by the reader as $4.2\ B$, say, when $f = 165$ mm, and so we can take it that, in general, R should lie between $5\ B$ and $10\ B$ from PQ.

$\frac{dB}{B}$ and $\frac{df}{f}$ can be neglected providing good standards of measurement and calibration obtain; similar relationships can be established for X_P.

Again by similar triangles (Fig. 11.10),

$$\frac{Z_P}{z_p} = \frac{PR'}{\sqrt{x_p^2 + f^2}} = \frac{Y}{f}$$

Thus
$$Z_P = Y \frac{z_p}{f} \text{ and } Z_Q = Y \frac{z_q}{f}$$

and $(Z_P - Z_Q)$ indicates the difference in level between the instrument horizontal axes at P and Q.

Note that the whole of the terrain between P and Q will not be photographed. The extreme rays through P and Q intersect between these points and the area enclosed by this intersection and PQ will not be covered. It is preferable that P and Q be at similar altitudes.

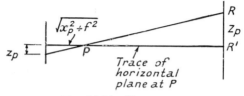

FIG. 11.10. ELEVATION

EXAMPLE. Two photographs are taken with a phototheodolite from stations P and Q, 100 m apart, the lines of collimation being 90° to PQ in each case. A point R appears on the photograph from P as 12·78 mm to the right of the vertical hair, and 10·48 mm above the horizontal hair, while on the photograph from Q it appears as 24·88 mm to the left of the vertical hair, and 9·05 mm above the horizontal hair. Q is east of P.

Calculate the co-ordinates of R from P, and the difference in level of the two collimation planes if the focal length of the camera is 165 mm.

Refer to Figs. 11.9 and 11.10.

$$x_p = 12\text{·}78 \text{ mm} \quad z_p = 10\text{·}48 \text{ mm}$$
$$x_q = 24\text{·}88 \text{ mm} \quad z_q = 9\text{·}05 \text{ mm}$$
$$\frac{X_P}{12\text{·}78} = \frac{Y}{f} \text{ and } \frac{X_Q}{24\text{·}88} = \frac{Y}{f}$$

$$p = x_p + x_q = 12\text{·}78 + 24\text{·}88 = 37\text{·}66 \text{ mm}$$

$$\therefore \qquad X_P = 100 \times \frac{12\text{·}78}{37\text{·}66} = 33\text{·}93 \text{ m}$$

$$Y_P = \frac{165}{37\text{·}66} \times 100 = 438\text{·}1 \text{ m}$$

$\therefore$ Co-ordinates of R are (33·9, 438·1) m w.r.t. P

$$Z_P = \frac{z_p}{f} = 438\text{·}1 \times \frac{10\text{·}48}{165} = 27\text{·}8 \text{ m}$$

$$Z_Q = Y\frac{z}{f} = 438\text{·}1 \times \frac{9\text{·}05}{165} = 24\text{·}0 \text{ m}$$

and so, since these values refer to R, which is of course fixed in position, P must be lower than Q by 3·8 m.

USES OF STEREO PHOTOGRAMMETRY. Amongst the uses may be listed the following—

(i) survey of reservoir sites
(ii) contouring building sites

   (iii) geological work on vertical faces

   (iv) recording of architectural features and conditions

   (v) recording of traffic accidents or construction accidents, etc.

For areas up to 25 km² it is a most economical method and for areas between 25 km² and 100 km² it may be as economical as aerial photogrammetry. For larger areas the latter method is probably more economical.

Aerial Photogrammetry

A survey plotted from photographs taken from above the ground was made by Laussedat in 1858, and by the end of the century other work on these lines had been carried out. A great deal of photographic surveying and reconnaissance was carried out during the World War 1914–1918, and from that time onwards progressive steps have occurred in the techniques of flying, photography, and plotting.

   Photographs from the air may be used for both surveying and compilation of topographical conditions, but for the photographs to give a true plan certain conditions must be fulfilled, namely: (i) the ground on the photograph should be horizontal, (ii) the camera must not be tilted from the vertical when the exposure is made, (iii) the camera lens and photographic material should be as perfect as possible and there should be no atmospheric refraction. In addition, when flying at high altitude, the curvature of the earth is of some account.

   In the following sections vertical photography only will be considered.

Principles

   Fig. 11.11 *a* represents a photograph taken over undulating ground, the camera being directed vertically downwards. Rays from points on the ground pass through the perspective centre O of the lens, and the images of those points appear on the negative, i.e. A on the ground appears at a, p is the photograph principal point, P is the ground principal point, and is the plumb point for a vertical photograph. The principal point p lies at the base of the perpendicular dropped from O on to the plate and is located by the intersection of lines joining the collimation or *fiducial* marks which are usually found at the corners or on each of the boundaries of the photograph (*see* Fig. 11.17). If the ground were level, as shown by the dotted line through P, the scale for that photograph would be $f{:}H$ $\bigg($since triangles Oda and $OD'A'$ are similar,

then $\dfrac{da}{D'A'} = \dfrac{f}{H}\bigg)$. But if a horizontal plane were drawn through A, a further pair of similar triangles could be obtained from which the scale would be f/H_1 and so on for all points not at datum. Thus the scale at a point on the photographs depends upon the height of that point above the chosen datum.

   Consider the side of the high building BC and its consequent image cb on the negative. B is vertically above C, and in plan the two coincide, but on the photograph the side of the building *cb* would be observed as well as the roof, and this building would appear to be leaning outwards from the centre of the photograph. It may be shown that the

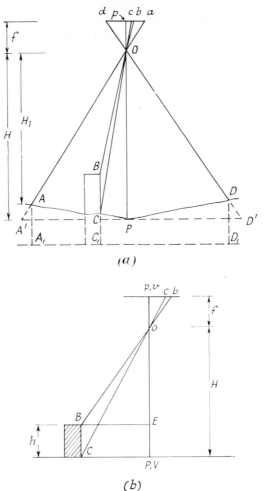

(a)

(b)

(a) The aerial photograph is a central perspective; the map is an orthogonal projection of the terrain, ground point *A* being positioned thereon at A_1

distortion of the image is proportional to the distance from the photograph plumb point which in this particular case coincides with the principal point.

Fig. 11.11*b* shows the building *BC*, shown for the sake of clarity with its base on the datum plane, and with *B* at a height *h* above that plane.

So long as pOP is truly vertical, then v, V (or p, P), B, C, c and b are all contained in one vertical plane, since B is vertically above C. This vertical plane intersects the plane of the negative, and since two planes intersect in one line only, v, c and b must be collinear and the displacement of b from c is radial from the plumb point v.

Consider the similar triangles vbO, EBO,

$$\frac{vb}{vO} = \frac{EB}{EO}, \text{ i.e. } \frac{vb}{f} = \frac{EB}{H-h}$$

Also from similar triangles vcO, VCO,

$$\frac{vc}{vO} = \frac{VC}{VO}, \text{ i.e. } \frac{vc}{f} = \frac{VC}{H} = \frac{EB}{H}$$

$$\therefore \quad \frac{vb}{vc} = \frac{H}{H-h} = \frac{vb}{vb-bc}$$

$$\therefore \quad \frac{vb}{bc} = \frac{H}{h}$$

Thus, the distortion due to height BC

$$= bc$$
$$= \frac{h}{H} \cdot vb$$

This expression only holds for a truly vertical photograph. Height distortion however is radial from the plumb point whether the photograph be vertical or tilted from the vertical.

If the height H varies during flying, then the scales of adjoining photographs will vary, and the line between camera stations at exposure will be inclined to the horizontal. Similarly, if H be incorrectly measured, then an error in the scale would occur. Statoscopes assist in a more accurate relative determination of H, one type recording variations of height from a datum starting height.

It will be more economical to photograph at a smaller scale and then enlarge from two to six times, depending upon the photographic material used. If, say, a plan at a scale of 1 in 5,000 is required on a 230 mm × 230 mm format, the flying height H would be obtained from the expression,

$$f/H = 1/5,000$$

and   if $f = 152$ mm., then $H = 760$ m

Refer to Fig. 11.11a, $ad = 0.230$ m

$$\frac{0.230}{D'A'} = \frac{f}{H} = \frac{1}{5,000}$$

$$\therefore \qquad D'A' = 1,150 \text{ m}$$

Therefore area covered by that *single* photograph

$$= 1,150 \times 1,150 \text{ m}^2$$

$$= 132 \cdot 25 \text{ hectares}$$

Now if the flying height be 3,800 m, with the same lens, the scale will be $1:25,000$, and $D'A' = 5,750$ m with an area covered by the single photograph of 3,306·25 hectares. Thus there will be an economy in time and materials by flying at the greater height. If a lens of focal length 210 mm and a picture size 180 mm square were used, to give a coverage of 132·25 hectares, the flying height would be about 1,340 m.

EXAMPLE. Vertical photographs at a scale of $1:20,000$ are to be taken of an area whose mean ground level is 500 m above mean sea level. If the camera has a focal length of (*a*) 210 mm, (*b*) 152 mm, find the flying height above mean sea level.

$$(a) \; f = 210 \text{ mm} \qquad \frac{1}{20,000} = \frac{0 \cdot 210}{H - 500}$$

$$\therefore \qquad\qquad H = 4,700 \text{ m}$$

$$(b) \; f = 152 \text{ mm} \qquad \frac{1}{20,000} = \frac{0 \cdot 152}{(H - 500)}$$

$$\therefore \qquad\qquad H = 3,540 \text{ m}$$

Aircraft Tilt. Small tilts from the vertical in unknown directions are unavoidable, and the average angle of tilt (from the vertical) generally will not be below 2° without gyroscopic aid, but it may be reduced to say $\frac{1}{2}$° when using an automatic pilot. Tilt errors cause distortion, and they must be eliminated before or during plotting.

Fig. 11.12*a* shows the negative tilted at an angle θ to the horizontal. O is the perspective centre of the camera lens. V and v are the ground and photograph plumb points, and are in the same vertical line with O. P and p are the ground and photograph principal points. iOI is the bisector of angles pOv and POV; I and i are termed the ground and photograph *isocentres*. Points such as i and I, v and V, which are on any ray through O, are known as *homologous points*. Triangles iOp and IOV are similar, since $i\hat{p}O$ and $I\hat{V}O$ are both right angles, whilst $i\hat{O}p$ equals $I\hat{O}V$, and the remaining angles $p\hat{i}O$ and VIO must then be equal. Hence iOI makes the same angles with both negative and ground planes. Fig. 11.12*b* shows perspective conditions when the negative is inclined at θ to the horizontal. The negative or plate principal line through p when produced meets the ground plane at M. Mp is perpendicular to the intersection LMN of the ground and the negative plane, and its projection MP on the ground is also perpendicular to LMN, whilst both lines (together with point O and the plumb points v and V) are contained in a vertical plane termed the *principal plane*. The horizontal plane containing O gives the horizon trace JK on the negative, and any horizontal line parallel to this on the negative is termed a *plate parallel*.

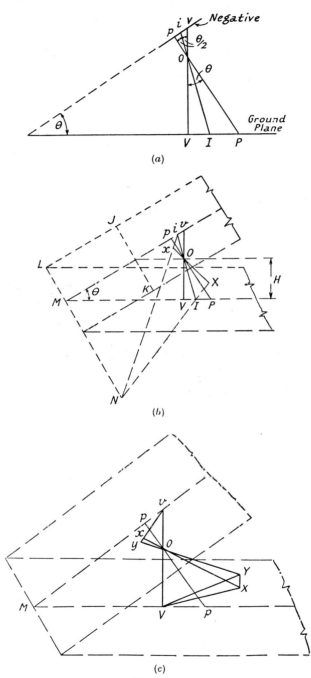

(a)

(b)

(c)

Fig. 11.12

Now $$p\hat{O}i = \frac{\theta}{2}$$

and $$Op = f$$

$\therefore$ $$Oi = f \sec \frac{\theta}{2}$$

$$Ov = f \sec \theta$$

The scale given at the isocentre, i, and at any point on the plate parallel through i is equal to Oi/OI

But $$\frac{Oi}{OI} = \frac{f \sec \dfrac{\theta}{2}}{H \sec \dfrac{\theta}{2}} \text{ since } p\hat{O}i = V\hat{O}I = \frac{\theta}{2}$$

$$= \frac{f}{H}$$

which is the scale given if the photograph were vertical. This is not true at any other plate parallel as may be readily verified by checking the scale at say p and v.

The image of a point X on the ground plane is given at x, so that the points I, X, x and i all lie in one plane. This plane, the ground plane and the plane of the negative intersect at one point only,* which is denoted as N.

It has been shown earlier that the line iOI joining the isocentres makes the same angle with the ground plane as with the negative, with a result that the triangle iMI is isosceles.

Hence $$iM = IM$$

and $$M\hat{i}I = i\hat{I}M$$

$$= 90° - \frac{\theta}{2}$$

Now $$i\hat{M}N = I\hat{M}N = 90°$$

Therefore triangles iMN and IMN are congruent.

$\therefore$ $$M\hat{i}N = M\hat{I}N$$

Thus in Fig. 11.12*b* the angle subtended at I by X and P on the ground equals the angle subtended at i by x and p on the negative, and by extension of the analysis it will be seen that in general angles subtended at the ground isocentre by points on the ground are equal to angles subtended at the isocentre by the corresponding images of these points.

In Fig. 11.12*c* the displacement of the image of Y with respect to the image of X is shown to be radial from the plumb point on the negative since Y is vertically above X, so that Y, X, V, O, v, x and y are all

---

* See *Projective and Analytical Geometry*, by Todd (Pitman).

contained in a vertical plane which intersects the negative to give the straight containing v, x and y.

Thus in the tilted photograph,

(i) distortions due to height are radial from the plumb point,
(ii) distortions due to tilt alone are radial from the isocentre.

An assumption adopted is that both distortions are radial from the principal point provided that the tilt angle is small—say less than 2° or 3°—and the ground height variations are small compared to the flying height. This assumption is used in radial line plotting and makes for considerable simplification, since the principal point can be located by means of the fiducial marks whereas the plumb point cannot be located without a knowledge of the plan positions and heights of at least three points.

LENSES. Camera lenses may be classified as superwide-angle, wide-angle, normal-angle and narrow-angle. Superwide-angle lenses are those types having angular fields of the order of 120° such as the Wild Super-Aviogon, which has a focal length of 88·5 mm. According to Messrs. Wild it gives a maximum distortion of $\pm$ 0·05 mm but this can be eliminated during measurement, together with the effects of earth curvature and refraction, by the use of correction plates in the photographic system. Such lenses are used for mapping at scales of say 1:50,000 to 1:100,000 and they result in coverage of an area three times as large as that given by a wide-angle lens of focal length 152 mm, assuming the same effective flying height. Wide-angle lenses, which are used to a great extent, have angular fields of about 90° and produce distortions less than $\pm$ 0·01 mm, with a very even intensity of illumination. Normal-angle lenses refer to those with fields of say 60°, focal lengths of the order of 210 mm, and formats of the order of 180 mm square. Narrow-angle lenses, which are of value in urban surveys, but do not give a good impression of relief, have angular fields of 40° or less.

CAMERAS. Cameras may be obtained using either film or plates, the former being in more general use. At present most film used in this country is black and white panchromatic on a polyester base, but colour film is also available and is said to be of advantage in so far as interpretation is concerned. Panchromatic emulsions are sensitive to all colours of the spectrum, particularly in the blue to violet range. As light passes from ground to film it may be dispersed by airborne dust particles and water particles, resulting in diffused light being presented at the camera. This tends to be in the above range of the spectrum and, in order that sharp images of the ground detail be produced, this diffused radiation must be removed. Filters may be used to eliminate the light of short wavelength or, alternatively, infra-red emulsions in, conjunction with suitable filters, can be adopted. In hazy conditions therefore infra-red photography is of advantage as it also is, compared to panchromatic, for applications such as forestry studies.

The Wild RC10 Universal Camera takes vertical photographs, 230 mm square, on panchromatic, infra-red or colour film. Amongst its components are the mount, lens cone, drive unit, view-finder telescope and control unit.

The mount is formed by a two-part base plate and box-type housing, cardanically mounted. Levelling corrections against tip and tilt up to $\pm 5°$ can be introduced, together with a drift correction up to $\pm 30°$.

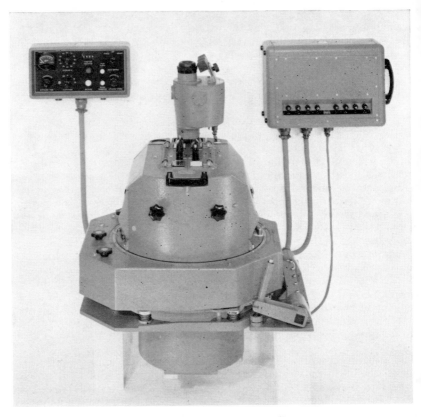

FIG. 11.13*a*. WILD RC10 FILM CAMERA
The Control unit and circuitry unit are to the left and right respectively.
(*Courtesy of Wild Heerbrugg, Switzerland*)

These may be applied manually or by remote control, through servo motors fitted to the mount, from the NF2 Navigation Sight. Either the Universal-Aviogon Wide-angle lens or the Super-Aviogon II (120°) can be incorporated into the camera and they can be exchanged during flight. The lens cones are equipped with shutters giving exposure times of 1/100 to 1/1,000 second. In this, as in all aerial cameras, it is essential that the inner orientation remains constant: four corner fiducial marks are standard, but four mid-side marks can also be

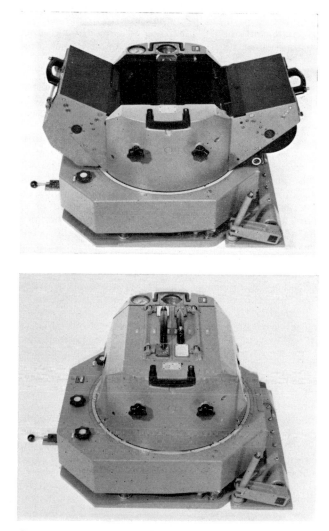

FIG. 11.13*b*. WILD RC10: INSERTION OF CASSETTE INTO
DRIVE UNIT

The cassettes are first hinged to the drive unit in a tilted position to feed the film from one to the
other.

(*Courtesy of Wild Heerbrugg, Switzerland*)

provided. These allow the photograph centre to be established and this should coincide with the principal point p in Fig. 11.11a.

The drive unit contains the means to advance the film and to flatten it against a pressure platten during exposure. Also shown on the photograph, with the fiducial marks on the focal plane frame, are a counter, altimeter and a watch, amongst other instruments. Take-up and feed spools of film are accommodated in separate interchangeable half-cassettes, which are initially hinged on to the drive unit to feed the film into the take-up cassette and then swung into the operating position as shown in Fig. 11.13b.

A view-finder telescope is inserted into the drive unit and two reticules can be seen: one shows the ground coverage and the other moving reference lines which are synchronized with the speed of the image of the ground and which allow overlap control. All elements essential for operation of the camera are contained in the Control Unit, i.e. main switch, diaphragm selection, shutter speed, overlap settings. This unit must be placed near the operator, although he may be at the NF2 Sight and be operating the camera by remote control: this sighting device contains the relevant reticules for overlap control and navigation.

In the United Kingdom, Messrs. Williamson manufacture the F49 Mk IV Air Survey Camera (*see* Fig. 11.14) which gives 230 mm square photographs, like the Wild RC10: the camera has three major components—magazine, body, and optical unit. Two mountings are available, each with anti-vibration isolators, and one of these has been designed to allow remote control. The magazine capacity is such that some 300 exposures can be obtained on film, panchromatic, infra-red or colour.

A multi-element lens of reverse telephoto type, and of focal length 152 mm, had been produced by Messrs. Wray for this camera, and Messrs. Williamson state that no image point along the two diagonals of the frame will be displaced by more than $\pm$ 10 microns (0·01 mm): full calibration data are supplied. The shutter is of the rotating type, having speeds from 1/250 to 1/1,000 second and, like the diaphragm, is remotely controlled; exposure is controlled by Intervalometer, the minimum interval being 5 seconds. The use of the reseau or register glass has been pioneered in the U.K. for analytical (triangulation) applications of photogrammetry, and such a glass is incorporated into this camera, the film being pressed against it at exposure. There are 529 crosses at 10 mm intervals on the register glass over the 230 mm $\times$ 230 mm format, and the centre cross coincides with the centre of best symmetry to within 0·005 mm. Each glass is calibrated so that the position of each cross is known to within $\pm$ 0·001 mm.

Survey Flying

The photographs are usually taken vertically in a series of strips: there is normally a minimum overlap of 60 per cent in the direction of flight between adjoining photographs, and an overlap of 25 per cent may be made laterally between strips so that all may be connected together. Some fore and aft tilt may reduce the 60 per cent overlap. To prevent the aircraft drifting off course under the action of cross winds, the aircraft must be directed so that it will be kept along the correct line, but in this case the photographs will not form a continuous strip unless the camera is turned so that the photographs are taken squarely on course. Sighting devices, previously mentioned, will allow

Fig. 11.14. Williamson F49 Mk IV Survey Camera
(Courtesy of Williamson Manufacturing Co. Ltd.)

for the necessary corrections. Errors in direction mean a wastage of effort and material, and rigid control is necessary. Course may be set by compass or on to a distant landmark, but it is difficult to hold on to a course in lengths over, say, 20 to 25 km without intermediate marks.

Tracking or controlling the flying of straight and parallel lines may be improved by using the Decca Navigator. The aircraft flies along pre-determined lines planned at known spacing so that adjoining strips of photographs will have the desired lateral overlap. In the Decca system, stable stationary wave patterns, hyperbolic in form, are generated by pairs of radio transmitting stations, the patterns depending upon frequency and distance apart of the stations. A radio receiver carried on the aircraft gives information from which the aircraft's plan position can be fixed when used in conjunction with a chart on which hyperbolic patterns are marked, the positions of the transmitters being known. In this fix, two sets of patterns are required, and the second set is provided by a further transmitting station, the chart showing both sets of patterns. It will be noticed that only three transmitting stations have been mentioned; one acts as master and the others as slave with the master and one slave acting as a pair.

The basis of one method of control is to mark the flight lines on the chart and the aircraft is then directed along them, being controlled or checked by an observer in the aircraft making continuous "fixes" which may be plotted on that chart. Should there be any deviation off course, the observer may then give the correction to the pilot. The accuracy of control depends upon a rapid interpolation of the receiver's information and consequent plot. Actually the short delay involved in plotting by interpolation on a hyperbolic grid would not allow sufficiently accurate tracking, and a simplified chart, a track graph, is used. Essentially this is a Decca grid converted into a rectilinear grid, which is much more convenient for rapid interpolation and plotting the receiver's information. The flight line is marked on the track graph as before, and lines showing the maximum permissible deviation off course may also be plotted either side of these lines.

Stereoscopic Pairs. Longitudinal overlap of 60 per cent at least is needed, 50 per cent being required for stereoscopic examination, and the extra 10 per cent allowing some margin against tilt errors.

Fig. 11.15 illustrates three successive equivalent positives taken from air stations 1, 2 and 3, respectively. X, Y, and Z, are ground principal points, and X_1, Y_1, and Z_1 are photograph principal points. The distances between the air stations are termed air bases, B. It will be seen that there is a common overlap of 20 per cent between the three photographs, and that the ground principal point Y, which is vertically below the photograph principal point Y_1, given by the collimating marks, will appear on the three photographs, appearing at Y' and Y'' on photographs 1 and 3 respectively. Thus by similar triangles,

$$\frac{f}{H} = \frac{X_1 Y'}{XY} = \frac{X_1 Y'}{B}$$

$$\therefore \qquad X_1 Y' = \frac{B}{H} f$$

The portion of photograph not overlapped by its immediate neighbour, i.e. b in Fig. 11.15, may also be related to the other quantities, since its corresponding ground dimension is B, and then by similar triangles,

$$\frac{b}{B} = \frac{f}{H}$$

$$\therefore \qquad b = \frac{B}{H} f = X_1 Y'$$

A similar situation applies to the lateral overlap of strips; this overlap is of the order of 25 per cent. If w be the lateral portion of the photograph *not* overlapped, and W is the corresponding dimension on the ground,

$$\frac{w}{W} = \frac{f}{H}$$

The net area of ground covered by one photograph is $W \cdot B$.

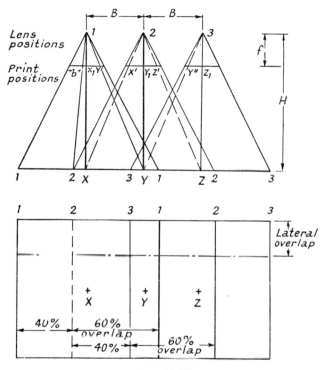

Fig. 11.15

It would then seem that the number of photographs to cover an area A would be $A/W \cdot B$, but this would give only a preliminary estimate, since the number required will depend upon the arrangement by which the strips of photographs cover the area.

EXAMPLE. An area of 150 km² at datum level is to be photographed at a scale of 1 in 10,000 from the air, using a camera of focal length 152 mm, the photos being 230 mm square. A longitudinal overlap of 60 per cent and lateral overlap of 25 per cent must be provided. If the operating speed of the aircraft is 250 km/h, find,

(i) flying height of the aircraft, and interval between exposures,
(ii) number of prints required if the flying strips are 15 km in length.

Let H be the flying height.

$$\frac{f}{H} = \frac{1}{10,000}$$

$$H = 10,000\,f = 1,520 \text{ m, since } f = 0\cdot152 \text{ m}$$

If B be the air base,

$$\frac{b}{B} = \frac{f}{H}$$

$\therefore$
$$B = (0\cdot40 \times 0\cdot230) \times 10,000 = 920 \text{ m}$$

Therefore time between exposures $= \dfrac{920}{250,000/3,600} = 12\cdot9$ sec

If W be lateral unlapped portion on ground,

$$W = w \times \frac{H}{f} = 0\cdot75 \times 0\cdot230 \times 10,000$$

$$= 1,725 \text{ m}$$

The width of the area $= 150/15 = 10$ km, and the spacing of flight lines will be 1,725 m.

Then number of spaces $= \dfrac{10,000}{1,725} = 6$ say, and allowing for a flight along each side for complete coverage, there will be 7 strips.

Number of photographs required in a strip

$$= \frac{15,000}{920} = 17, \text{ say}$$

Add 1 in order that the ends are covered.

Then the number of photographs $= 19 \times 7 = 133$

If an estimate of the number of photographs were required, formula $A/W \cdot B$ gives

$$\frac{150,000,000}{920 \times 1,725} = 95, \text{ say}$$

In stereoscopic observation, pairs of photographs are placed in an instrument arranged so that the left eye observes the left-hand picture, and the right eye observes the right-hand picture only. The two images combine in the brain to give a single image of the common overlap, which gives an impression of relief in a manner similar to an observer himself looking over the terrain; i.e. the flat pictorial effect from the photograph is transferred into an appearance of depth in space and relative heights are then obtainable from the stereoscopic pairs.

In Fig. 11.16a O and O_1 are successive air stations B apart, from which vertical photographs have been taken. Consider the top and bottom of the pylon XY with images (Fig. 11.16b) at b, c and a, d respectively. These images are projected in Fig. 11.16c to B, C and A, D on the line PQ' $P'Q$, which contains the principal points P and Q

and their conjugate images Q' and P'. In Fig. 11.16a OC' has been drawn parallel to O_1C, and OD' is parallel to O_1D. The points have

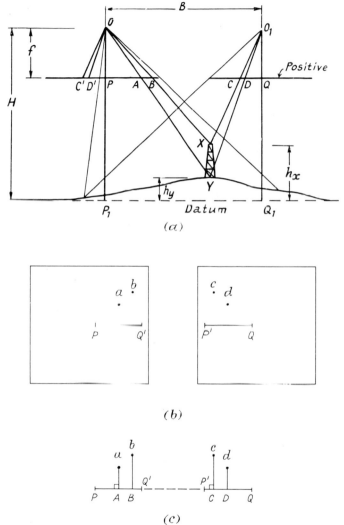

(a)

(b)

(c)

Fig. 11.16

been moved with respect to the principal points P and Q on the photographs, $D'A$ and $C'B$ giving the movements of the images of Y and X respectively. These movements in the direction parallel to the air base give the *absolute parallax* of Y and X, denoted by p_y and p_x. It will be seen that $D'A = D'P + PA = DQ + PA$, and $C'B = CQ + PB$.

Triangles $OD'A$ and YO_1O are similar, and so

$$\frac{OO_1}{D'A} = \frac{H - h_y}{f}$$

and since triangles $OC'B$ and XO_1O are similar,

$$\frac{OO_1}{C'B} = \frac{H - h_x}{f}$$

$$D'A = p_y = f\frac{B}{H - h_y} \quad [\text{parallax of } Y]$$

$$C'B = p_x = f\frac{B}{H - h_x} \quad [\text{parallax of } X]$$

In each case the parallax depends upon $(H - h_y)$ or $(H - h_x)$ and in general all points at a height of h_x and h_y above datum must have the same absolute parallax on this pair of photographs. An expression for the difference in level between X and Y may be determined as follows—
Let the difference in parallax between X and Y be dp,

$$dp = p_x - p_y = fB\left[\frac{1}{H - h_x} - \frac{1}{H - h_y}\right]$$

$$= fB\left[\frac{h_x - h_y}{(H - h_x)(H - h_y)}\right]$$

Since $\quad\quad \dfrac{fB}{H - h_x} = p_x$

then $\quad\quad dp = p_x\dfrac{h_x - h_y}{H - h_y} = P_z\dfrac{dh}{H - hy}$

or $\quad\quad dh = \dfrac{dp}{p_x}(H - h_y) = $ difference in level between X

and Y

If h_y be taken as small compared to H, then

$$h_x - h_y = dh = dp\frac{H}{p} = \frac{H^2}{fB}dp \text{ nearly.}$$

B is the average distance between the two principal points as located in both photographs, and the difference in level between two points may thus be determined from the difference of parallax of the pairs of images measured in the direction of the air base. The lines joining the principal point and image of the other principal point on each photograph must be in prolongation as in Fig. 11.16c.

The equations derived are based on the assumption that there is no tilt and no variation in flying height above datum. If one or both of these assumptions be not complied with, errors will be introduced by use of the equations, since apart from parallax in the direction of the air base—the x-direction—there will be parallax in a direction at right angles—the y-direction and known as y-parallax or *want of correspondence*. This can be expressed in the form

$$dp_y = y_1 - y_2$$

where y_1 and y_2 are the distances from the x-axis or air base to a point on each of two adjoining photographs.

When parallax differences are being found the points under consideration should be near to each other on the common overlap; otherwise the level difference derived may be quite incorrect. If points of known height are located in the overlap, an estimation of the errors occurring may be obtained.

EXAMPLE. In a pair of overlapping vertical air photographs the mean distance between the two principal points, both of which lie at datum level, is 89·60 mm. At the time of photography the aircraft was 3,200 m above datum, and the camera had a focal length of 150 mm. In the common overlap, a tall chimney 120 m high with its base at datum level is observed. Determine the difference in absolute parallax measurements for top and bottom of the chimney.

Refer to Fig. 11.15,

$$\frac{X_1 Y^1}{X Y} = \frac{f}{H}$$

$$\therefore \quad \frac{89 \cdot 60}{B} = \frac{150}{3,200}$$

$$B = 1,911 \text{ m}$$

Refer to Fig. 11.16 and take $h_y = 0$,

$$p_y = \frac{fB}{H} = 150 \times \frac{1,911}{3,200} = 89 \cdot 60 \text{ mm}$$

Note. Points at same level have same absolute parallax.

$$p_x = f \frac{B}{H - h_x} = 150 \times \frac{1,911}{3,200 - 120}$$

$$= 93 \cdot 07 \text{ mm}$$

$$\therefore \qquad \text{Difference in absolute parallax} = 3 \cdot 47 \text{ mm}$$

EXAMPLE. A photographic survey is carried out to a scale of 1 : 20.000, a camera with a wide angle lens of focal length 152 mm being used with a 230 mm × 230 mm format, and 60 per cent overlap. Find the difference in height given by an error of 0·10 mm in measuring the parallax of a point.

$$\frac{b}{B} = \frac{f}{H}$$

$$\therefore \quad \frac{0 \cdot 4 \times 0 \cdot 230}{B} = \frac{1}{20,000} \qquad \therefore \quad B = 1,840 \text{ m}$$

Fig. 11.17(a). Aerial Photograph of a Railway Station and Surrounding Land

(Courtesy of the British Transport Commission)

Now
$$\frac{dh}{dp} = \frac{H^2}{fB} = \frac{fB}{b^2}$$

and if $dp = 0.10$ mm, then

$$dh = \frac{0.152 \times 1,840}{(0.4 \times 0.230)^2} \times 0.1 \text{ mm} = 3.3 \text{ m}$$

The values derived assume that there is no tilt and no variation in flying height above datum. A false parallax is given by tilt and for accurate work the measurements should be corrected for that tilt*. If the photographs are set up in the same relative positions they occupied in the air at exposure, then a point on the ground, its image on each photograph, and the two air stations, are in the same plane (the basal plane). Each ground point can be fixed by a basal plane which must include the air base, and which can be located by a rotation about that base. If the relative positions are not the same, the five points will not be coplanar, giving a "want of correspondence."

Plotting

Fig. 11.17a shows an aerial photograph of a railway station and surrounding land, the plot derived from it is shown in Fig. 11.17b.

Various devices and methods are available for plotting and heighting. Amongst the simple instruments for heighting is the Folding Mirror Stereoscope manufactured by Casella (Fig. 11.18), which is portable and about 460 mm by 230 mm in size, and which deals with photographs 230 mm square. A large inclined mirror faces downwards on to one photograph, while a second inclined smaller mirror facing upwards passes rays reflected from the larger mirror to one eye. A second arrangement of mirrors over the other photograph passes rays to the other eye. Thus each eye sees a separate photograph and the common overlap gives a relief model.

The Folding Mirror Stereoscope is provided with a stereometer or parallax bar for measurement of relative heights. This consists of two glass rectangles, each with an identical mark on the centre, carried in frames attached to the bar. The left-hand rectangle or graticule may be clamped and unclamped, while the right-hand graticule may be moved 12.5 mm by a micrometer reading to 0.01 mm. The principal point of each photograph is found, and the principal point of the right-hand photograph identified and marked on the left-hand photograph, and the principal point of the left-hand photograph marked on the right-hand photograph. The photographs are separated to give satisfactory fusion by a distance which may be specified by the manufacturer. They can then be rotated about pins through their respective principal points until the two transferred principal points are collinear with the principal points as in Fig. 11.16c. Viewing should always be parallel to this base.

---

* *See* papers by E. H. Thompson in *Photogrammetric Record*, October 1954 and October 1968, and by B. D. F. Methley in that journal dated April 1970.

The photographs should be evenly illuminated, and should be secured to prevent movement, the pins then being removed.

The right-hand mark in the centre of its run is laid over a point of detail on the right-hand photograph, and the left-hand mark adjusted over that point on the left photograph. When this graticule is clamped, both glasses should be flat so that the marks do not cast shadows. The marks should appear fused together when viewed through the stereoscope, and rotation of the micrometer, which alters the separation of the marks, causes the fused mark to appear to move vertically. It is made to touch the point selected, at ground level. The micrometer is then read, and the mean of several readings then obtained. Another

Fig. 11.18. Folding Mirror Stereoscope
(*C. F. Casella & Co., Ltd.*)

point is now treated, and the difference of parallax of the two determined. The parallax bar must always be used parallel to the base line, When several pairs are to be examined, the repeated movement of the stereoscope and its consequent alignment with the base line becomes tedious and an instrument may now be obtained which has an arm capable of moving in two directions only, i.e. a parallel guidance mechanism. The stereoscope is now mounted on the arm and is thus always set with the base line. A stereometer can also be readily attached to the arm.

Messrs. Hilger and Watts, Ltd. manufacture a folding mirror stereoscope which can observe a 75 per cent overlap and which can accept a parallel guidance unit (Fig. 11.19). This unit is attached to a supporting base of the stereoscope, the two being mounted on a board, and the simple pantograph linkage moves the right-hand picture with respect to the left-hand one. It will be noted that the stereometer has been replaced by a parallax measuring system which allows an unobstructed view of the stereomodel.

Light spots are injected into the optical system of the stereoscope, the right-hand one being moved in the X direction, i.e. direction of

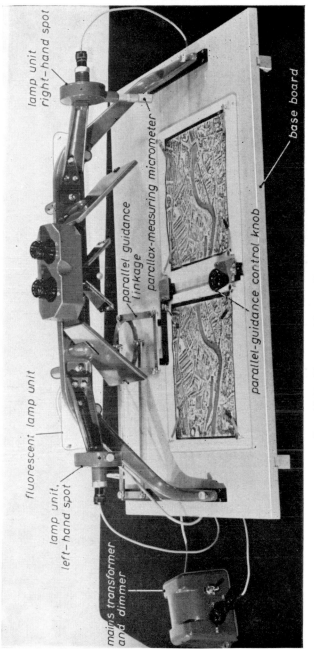

lamp unit
right-hand spot

parallel guidance
linkage

parallax-measuring micrometer

base board

parallel-guidance control knob

fluorescent lamp unit

lamp unit,
left-hand spot

mains transformer
and dimmer

Fig. 11.19. Folding Mirror Stereoscope
(Courtesy of Hilger & Watts, Ltd.)

PQ in Fig. 11.16c, under the control of a micrometer. As with the stereometer, the two spots are fused stereoscopically to touch the points of detail at ground level and differences of micrometer readings allow the reduction of heights.

EXAMPLE. In a pair of overlapping photographs (mean base length 89·84 mm) mean ground level is 70 m above datum. Two nearby points are observed and the following information obtained—

Point	Height above Datum	Parallax Bar Reading
X	55 m	7·34 mm
Y		9·46 mm

If the flying height was 2,200 m above datum and the focal length of the camera was 150 mm find the height of *Y* above datum.

The air base *B* can be found from the expression $\dfrac{b}{f} \simeq \dfrac{B}{H_m}$ where H_m is the flying height with respect to mean ground level and *b* is the mean photographic base.

Thus
$$\frac{89\cdot84}{105} \simeq \frac{B}{2,130}$$

∴
$$B \simeq 1,276 \text{ m}$$

Now
$$p_x = \frac{fB}{H - h_x} = 150 \times \frac{1,276}{2,145}$$
$$= 89\cdot22 \text{ mm}$$
$$p_y = p_x + dp = 89\cdot22 + (9\cdot46 - 7\cdot34)$$
$$= 91\cdot34 \text{ mm}$$

Also
$$dp = p_y \frac{dh_{xy}}{H - h_x}$$

∴
$$dh_{xy} = \frac{2\cdot12}{91\cdot34} \times 2,145$$
$$= 50 \text{ m, say.}$$

Thus *Y* is 105 m above datum.

If that height was known to be so, we could directly write that a height change of 50 m was equivalent to a parallax change of 2·12 mm, i.e. 23·6 m/mm of parallax, and so if the reading for an intermediate point was 8·18 mm a heighting difference of $0\cdot84 \times 23\cdot6 = 20$ m above *X* would be implied.

Stereoscopic examination of photographs at the reconnaissance survey stage of extensive engineering projects, such as, say, highway construction, allows the evaluation of various controlling factors, such as river crossings and swampy land, and of relative heights by stereo-meter. Possible routes can be located and marked on the photographs where they appear in stereoscopic correspondence with the topography, and comparison can then be made in the field.

Plotting Machines. There are many types of plotting machine available at the present time, and these can be categorized in various ways. For instance a machine may be referred to as of "first-order," which implies that it is of wide applicability, being suitable for large-scale surveys, and of maximum precision. Such a machine should produce a three-dimensional model of the terrain from two successive photographs and thence ensure the accurate plotting and heighting of features in the model without calculations, or the determination of co-ordinates of points in the model. Two central projections must be arranged and the corresponding rays caused to intersect to form the model, the plot then being produced orthogonally. Projection can be effected either mechanically or optically or by optical-mechanical means as in the Thompson-Watts Plotter described shortly. In the first the path of the rays is characterized by "space rods" or similar, whilst in the second case light rays represent the paths. The two photographs can be positioned by relative orientation (with reference to each other) and then by absolute orientation (with reference to ground points).

The purpose of the first operation is to produce a model geometrically similar to the terrain, whilst the second operation transforms the model scale to the required scale and orientates the model into the given co-ordinate system. Here, theoretically, the positions and height of two ground points must be known together with the height of a third point, but as a rule four points, in the model corners, are determined and perhaps one in the middle. For each new model fitted on, three more ground points should be provided. All such points can be obtained (and surveyed) after the photography, but if there is little firm detail in the terrain they will need to be pre-marked. It will be apparent that the accuracy of the photogrammetric work will depend upon that of the ground survey, and that this can be an expensive part of the whole. "Second-order" instruments provide exact solutions of the relative and absolute orientation problems but are not capable of plotting at such large scales as the "first-order" machines.

To minimize the amount of ground control the models can be connected by aero-triangulation or bridging. If the first model of a strip is set up as described above using the first two photographs, the second model can be formed using the third photograph which is oriented with respect to the second, since these two form a pair as well. The scale of the first model is thus carried along the strip, but since errors in scale and altitude can occur, three ground control points can be considered at say the sixth and the eleventh model. This technique is quite suitable for road construction surveys, but since heighting accuracy is very important in this context, bench marks as additional control marks could be established at frequent intervals near the proposed alignment.

In lieu of the analogue type of approach previously described there is also an analytical approach in which the co-ordinates of points imaged in the photographs (or diapositives) are measured and the corresponding ground co-ordinates are ultimately derived mathematically via the digital computer. This method is used by the Ordnance Survey to provide control points from the existing second-order triangulation network, these points then being used to control stereoplotting.

The Hilger and Watts SB50 Stereocomparator can be used to measure image co-ordinates on a stereopair, and it will accommodate reseau photography. Register plates are required in the instrument, the diapositives being held in close contact with their undersides. Alternative register plates, gridded at 10 mm intervals, are required if a reseau is not incorporated in the flight camera.

Measurements are based on the 10 mm sided squares with subdivision to 0·001 mm, and the co-ordinates, and parallax measurements, are automatically recorded in a manner suitable for the computer.

THOMPSON–WATTS PLOTTER MODEL 2. This first-order plotter (Fig. 11.20a) has been designed for small- to large-scale mapping using the principle that relative and absolute orientation should be independent functions; it is also suitable for aerial triangulation. Diapositives are placed in the projectors which are dimensional replicas of the air camera. These projectors are carried by beams which are so connected that they rotate through exactly the same angle about a common axis; one beam can be rotated relatively to the other thereby providing differential rotation ($\Delta\omega$) between projectors for relative orientation. Other possible rotations include those of the diapositives in their own planes (κ) and of the projectors about their trunnion axes (ϕ), which are at right-angles to the beam axis.

Linkages are attached to two telescopes, which can rotate about parallel axes at right-angles to the beam axis. Their directions are parallel to the lines of sight of the telescopes and in effect they extend the respective lines of sight. The lower ends of these links are attached to carriages which can move along the D bridge; this bridge is parallel to the beam axis and it can move in a direction perpendicular to that axis. Base settings are arranged at this level with the aid of space rods which separate the carriages. The space rods are carried on the Z carriage which is moved along the Z slide by means of a screw controlled by the foot wheel. This particular slide is carried by, and is at right angles to, the X carriage, which moves on the X slide.

Thus if the carriages are positioned on the D bridge for a given separation whilst the D bridge is at such a distance from the beam axis that the telescopes point at homologous images, the two links will lie in the basal plane. They are in effect analogues of the rays from the air stations to the ground point, providing that the beam has been suitably rotated, since the links are in a vertical plane and the basal planes containing the ground points must be brought into this plane. This is effected via the Y carriage. A roller attached to the Z carriage works against the edge of a second bridge known as the Z bridge, which is again parallel to the beam axis and which is connected to the Y slide. The Y carriage can be controlled by the operator to move along the Y slide and, by a linkage the beam is caused to rotate about its axis. All image points can thus be set stereoscopically, X and Y co-ordinates obtained from drum readings and heights Z measured from an optically projected scale.

Fig. 11.20c. Thompson–Watts Plotter
(*Courtesy of Hilger & Watts, Ltd.*)

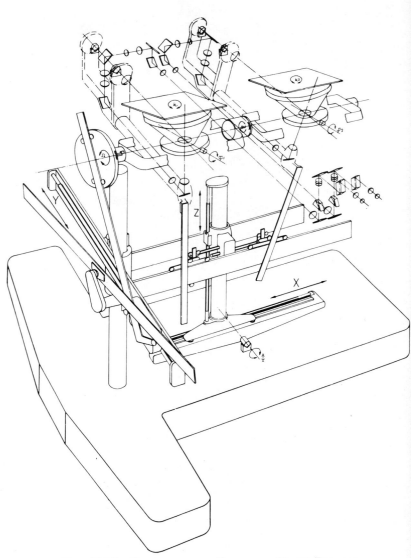

FIG. 11.20b. DESIGN OF THE THOMPSON–WATTS PLOTTER

(Courtesy of Hilger & Watts, Ltd.)

Radial Line Plotting. Topographical maps may be prepared graphically from the assumption that rays drawn to two points, say 2 and 5 in Fig. 11.21a, from the photograph principal point subtend the same angle at that principal point as they actually do on the ground at the ground principal point.

Refer to Fig. 11.21a, principal points P, Q, R, have been located on the photographs by small crosses, and the positions of the images of other principal points are now marked on adjacent photographs, i.e.

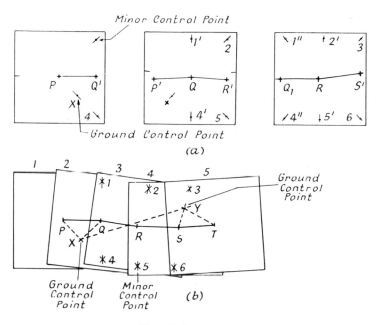

FIG. 11.21 (a and b)

Q is represented by Q' and Q_1 on the photographs at each side. By this means the base lines may be drawn on each photograph, connecting PQ', $P'Q$, and QR', etc. Q' could be located by inspection of the two photographs, and when PQ' and $P'Q$ are drawn they should pass through the same detail on each photograph. If the photographs be positioned such that $P'Q$ lies collinearly over PQ' then the two will be in accordance, and if Q_1R is laid over QR', the third photograph will be oriented with respect to the other two.

Minor control points on the overlap of three photographs are chosen. These points (1, 4, etc.), marked by short rays, should be spaced away from the base lines approximately opposite principal point positions and need to be easily and accurately identifiable. Note that these minor control points are chosen on the photographs and not on the ground. It will be seen from the strip of photograph in Fig. 11.21*b* that rays drawn to the minor control points will intersect as the strip is built up. Actually the graphical minor control plot is made on a transparent sheet, placing it over the photographs in turn, the final effect being similar to Fig. 11.21*b* insofar as the triangulation is concerned.

A line is drawn on the sheet over base line PQ' produced, P and 1 are marked, and ray P 4 drawn on the sheet, which is then placed over photograph 2 with the base line drawn on the sheet over QP'. Photograph 2 is moved, keeping the two base lines collinear, until the ray Q 1' drawn on the photograph passes through the plotted position of 1 on the sheet. This now fixes Q, and the intersection of rays P 4 and Q 4' on the photographs fixes 4 on the sheet. A line may now be drawn over QR' and short rays drawn to 2 and 5. The procedure is now repeated using photograph 3 with the relevant base line on the sheet collinear with RQ_1, and R is fixed by ensuring the intersections of rays R 1'' and R 4'' on photograph 3 pass through the plotted positions of 1 and 4 on the sheet. Triangles of error could occur if tilts were large, and would have to be adjusted, but if tilts remain small then these triangles should not be of any account. This procedure is now carried on for the strip. The scale is not yet known and so ground control points X and Y appearing on at least two photographs are required. Short rays PX and QX', etc., will plot the positions of X and Y on the minor control plot. It is then possible either to determine the scale for this plot knowing true distance XY and enlarge or reduce to the required scale, or to convert the plot to the required scale providing X and Y have been plotted at the required scale on another sheet. The various lateral strips must be brought to the common scale, and this entails the use of tie points common to both strips.

Detail points are now chosen on the photographs. These points must be identifiable on at least two photographs so that rays drawn from each principal point give an intersection on the plot by the method described, i.e. arrange base line PQ over PQ' in photograph 1 with P on the plot over P in the photograph, and rays drawn to the detail points. Next QP is laid over QP' in photograph 2 with the two points Q coincident, rays being drawn from Q to the detail points on photograph 2 to give an intersection with the rays drawn previously. The detail points are shown by small crosses. Topographical detail may now be filled in by tracing from the photographs using the detail points as control, making small adjustments in the scale where necessary when plot and photograph scales are not exactly the same.

It will be apparent that the minor control points form a rhomboid chain when connected up to their three relevant principal points. Accordingly angles such as $2\hat{Q}R, 4\hat{Q}R, 2\hat{R}Q, 4\hat{R}Q$ can be measured by a suitable device, i.e. the Wild RT1 Triangulator, and relationships between QR, RS, etc. can then be obtained. Reference may be made to *Outline of Photogrammetry* by K. Schwidefsky (Pitman, 1959) for this aspect of analytical radial triangulation.

Ideally if the ground were flat and the camera axis were vertical

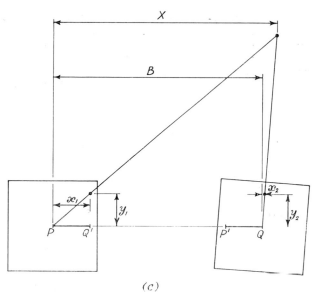

(c)

Fig. 11.21c

co-ordinates of ground points could be estimated as follows, PQ' and $P'Q$ being collinear as in Fig. 11.21c, and Y being the northing w.r.t.p.

$$\frac{X}{x_1} = \frac{Y}{y_1} \quad \text{and} \quad \frac{(X - B)}{x_2} = \frac{Y}{y_2}$$

$$\therefore \quad \frac{Xy_1}{x_1} = \frac{(X - B)y_2}{x_2}$$

$$X = \frac{Bx_1y_2}{x_1y_2 - y_1x_2}$$

$$Y = \frac{By_1y_2}{x_1y_2 - y_1x_2}$$

If b is the mean value of PQ' and $P'Q$ we can write

$$x = \frac{bx_1y_2}{x_1y_2 - y_1x_2}$$

and

$$y = \frac{by_1y_2}{x_1y_2 - y_1x_2}$$

These latter two expressions give preliminary photographic co-ordinates which can be used directly to derive ground co-ordinates. Since these may not be related to the photography in so far as directions of axes are concerned, the well-known transformation equations

$$X = ax + by + X_0 \quad \text{and} \quad Y = -bx + ay + Y_0$$

can be adopted, where x and y relate to the photographic measurements and X and Y to the ground measurements: X_0 and Y_0 allow for displacement of the origins of the two systems.

EXAMPLE. The image co-ordinates of three points A, B, and C and of the principal points P and Q on two overlapping photographs were determined as follows—

	x (mm)	y (mm)	x (mm)	y (mm)
P	0·0	0·0	− 76·2	0·0
Q	+ 76·0	0·0	0·0	0·0
A	+ 10·6	+ 60·5	− 66·0	+ 59·0
B	+ 11·2	− 6·3	− 64·5	− 6·7
C	+ 14·5	+ 34·3	− 61·5	+ 33·7

If the ground co-ordinates of A and B are 79,000 m E, 92,940 m N and 78,910 m E, 92,760 m N respectively, estimate those of C. (*U. of Salford.*)

From the data it will be seen that b = photo base = 76·1 mm and so for A,

$$x = \frac{76\cdot1 \times 59\cdot0 \times 10\cdot6}{10\cdot6 \times 59\cdot0 + 66\cdot0 \times 60\cdot5} = 10\cdot3 \text{ mm}$$

$$y = \frac{76\cdot1 \times 59\cdot0 \times 60\cdot5}{10\cdot6 \times 59\cdot0 + 60\cdot5 \times 66\cdot0} = 58\cdot8 \text{ mm}$$

In the same way the preliminary co-ordinates for B and C can be derived as $x = 11\cdot9$ mm, $y = -6\cdot7$ mm for B, and $x = 14\cdot3$ mm, $y = 33\cdot9$ mm for C.

Conversion of the co-ordinates now gives—

$$79,000 = \quad 10\cdot3a + 58\cdot8b + X_0 \text{ for } A$$
$$92,940 = -10\cdot3b + 58\cdot8a + Y_0$$
$$78,910 = \quad 11\cdot9a - 6\cdot7b + X_0 \text{ for } B$$
$$92,760 = -11\cdot9b - 6\cdot7a + Y_0$$

$$\therefore \quad a = 2\cdot71, \ b = 1\cdot56, \ X_0 = 78,881 \text{ and } Y_0 = 92,797$$

Thus we can establish the co-ordinates for C as

$$X = 2\cdot71 \times \quad 14\cdot3 + 1\cdot56 \times 33\cdot9 + 78,881$$
$$= 78,973 \text{ m E}$$
$$Y = -1\cdot56 \times 14\cdot3 + 2\cdot71 \times 33\cdot9 + 92,797$$
$$= 92,867 \text{ m N}$$

Slotted Templates. In the method previously described rays to control points are drawn on the sheets, but in the slotted template method slots are cut in templates of celluloid or other suitable material, and take the place of these rays. A template approximately 0·5 mm thick is cut for each photograph and is made slightly larger than that photograph.

On each template the photograph principal point, minor control points, ground control points, if any, and the points corresponding to the principal points of adjoining photographs are pricked through and numbered.

A hole is punched in each template at the photograph principal point (Fig. 11.22), together with slots radial to that point and centred on those points mentioned in the previous paragraph. The slots are made the same width as the centre hole.

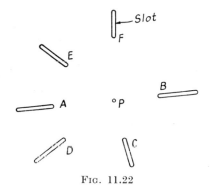

F<small>IG</small>. 11.22

The circular punch forming the centre hole has a retractable needle with a fine point which is arranged in the prick mark. The cutting edge of the punch is made to touch the template surface, the punch is given a blow with a hide hammer, and this results in a small hole being driven as required. Through this hole, and from underneath, a stud is passed and a straight-edge is then threaded over it so that the straight-edge is in line with the hole centre. Using a needle held in a ruling block, fine lines about 10 mm long are marked on the template to pass through the other marked points. A knife punch then cuts the slots on the radial lines as follows. The principal point hole is passed over a pin attached to a sliding carriage which moves in the longitudinal axis of the punch. The template is arranged so that one of the marked radial lines is beneath the punch, which can then form the slot. A retractable fine needle also assists the location of the punch over the radial lines. The template surface is illuminated and a small telescope is arranged so that its field of view shows the needle pointing downwards with its image reflected from the surface. The needle tip can easily be placed over the line which contrasts with the illuminated surface.

The templates overlapping longitudinally and laterally are set out on a special floor or dais in the order in which the photographs are taken; the slots act as the rays in the graphical method, and the intersection of slots forms a hole through which a stud may be passed. A fine hole in the centre of the stud marks the exact intersection of the radial lines.

Studs in the floor, which carries a grid, give the ground control points, and the template assembly is adjusted until those slots relating to the ground control points slip over their relevant studs without deformation of the assembly. Then those studs representing principal and minor control points will "fall into line" as far as scale is concerned. Headless pins are now inserted through the fine holes in the stud centres, and the intersections are now fixed on the floor after which the assembly may be removed.

The co-ordinates of the minor control points and principal points can be scaled from the grid lines or pricked through on to transparent sheets and detail plotting may be carried out as in the radial line method.

A great saving in time and ground control is gained by this method, and it may be used for topographical scales up to say 1 in 10,000.

FIG. 11.23. RADIAL LINE PLOTTER
(*Hilger & Watts, Ltd.*)

Radial Line Plotter. This instrument (Fig. 11.23) functioning on the radial line method, is manufactured by Hilger & Watts Ltd., and can be used to plot detail from pairs of aerial photographs.

Basically the plotter consists of two circular tables, taking photographs 230 mm square, each with a glass cursor and radial arm. Each radial arm engages with a pin on a plotting bar which carries the plotting pencil, and although the bar may move in both x and y directions it is so controlled that it is always parallel to the air base.

A photograph is placed on each table with the cursor above and the radial arm beneath, and with its principal point, located by a pin passing through the cursor glass, over the centre of the radial arm. The photographs are rotated to make the air bases collinear, and then viewed stereoscopically, when the cursor lines are seen to be apparently intersecting at a point which will be plotted by the pencil. Thus, as the plotting bar is moved, the radial arms swing round, the point of intersection moves over the detail in the overlap of the two photographs, and this detail is then plotted on the sheet. No intersection is possible in the area round the air base and the tables are moved in the y direction so that the radial arms can swing from a false centre, the detail plot thus being completed.

The instrument is very useful for plotting detail when the control points, etc., have been established by slotted templates or survey. An alignment jig is available for positioning the plotter, using the principal points if these have been located on the plan. This jig consists of a plate, carrying two vertical needle holders, about 125 mm apart, which is attached to the plotting bar with a needle each side of the pencil, and with needles and pencil in a line parallel to that bar. The plotter can now be positioned so that its axis on which the photograph principal points lie is parallel to the flight lines given by joining successive plotted principal points.

The photographs are illuminated by adjustable lamps and the bar linkages make use of ball races and low friction material for smooth action. By means of a sleeve on the plotting bar the plotting scale can be varied from about half photograph scale to over full photo scale, and allowing for some loss of cover at the photograph edges the scale may be about twice photo scale.

The Stereotope

Messrs. Zeiss have produced this instrument primarily for the production of topographic maps at scales between 1 : 25,000 and 1 : 100,000 from normal angle and wide angle photography; it can also be used for plotting from photography which has been based on a horizontal axis (i.e. from phototheodolite surveys). The main components are—

    (*a*) a mirror stereoscope
    (*b*) a photo carriage with parallel guides and parallax sleeves
    (*c*) two built-in computers
    (*d*) a pantograph
    (*e*) an aluminium base plate on which the stereotope is mounted.

If so required the mirror stereoscope can be used as a separate instrument, when the photographs are placed on a metal plate and secured at each corner by permanent magnets. As will be seen in the illustration it has obliquely-positioned eyepieces and monoculars which can be swung into position by means of a hinged pivot.

When used in the stereotope, the stereoscope is fastened to the anchor block of the parallel guides of the photograph carriage, and also, with the latter, to the base plate. As mentioned previously in the descriptions of parallax measurement with a stereometer and a stereoscope, the stereometer is moved over the photographs which remain in a fixed position. In the stereotope, however, the measuring marks (engraved on circular glass plates) remain fixed whilst the photographs are moved below them on the parallel guided photograph carrier. At the left- and right-hand sides of this carriage are found the y-parallax and x-parallax sleeves respectively. Using the latter, the right-hand photograph can be displaced with respect to the left-hand one which remains fixed relative to the carriage. These parallax sleeves also serve as handles for moving the picture carriage, and whilst the y-parallax sleeve eliminates y-parallax, the x-parallax sleeve is graduated and serves as a measuring device.

One computer provides corrections for distortions due to tilt. Four correction screws on the front face of the carriage act in such a way that each lifts or lowers one corner of the observed stereo-model independently of the others. The x-parallax sleeve is pivoted on the take-off of this computer and receives the correction values. The corners are adjusted in height by means of the correction screws until

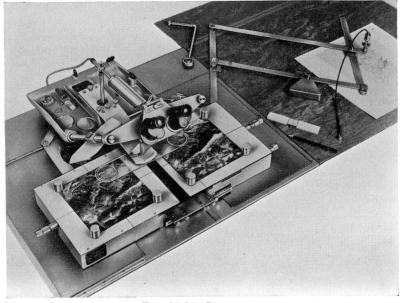

FIG. 11.24. STEREOTOPE
(*Courtesy of Zeiss-Aerotopograph*)

the corresponding parallax values are obtained when the spatial measuring mark is brought into position.

Below the four screws is a slide which can be locked and this is used to set the picture base on the second or perspective computer, which corrects horizontal position errors (since ground relief, as previously discussed, influences the scale of an aerial photograph which is based on central perspective). A map, on the other hand, requires orthogonal projection and the perspective computer transmits the necessary corrections to the pantograph take-off. A supplementary gear coupled to this second computer arranges for rectification of the planimetry which will be affected by picture tilt. Conversion of photoscale to the required map scale is carried out by the pantograph.

Four control points in the model corners are required for orientation and a fifth point in the centre will ensure increased accuracy. These points may be determined by either ground measurement or aerial triangulation. Having determined the principal points, and knowing the flying height, the following steps are required for orientation—

(*a*) computation of the nominal parallax setting for the four control points;

(*b*) photograph positioning in the instrument;

(*c*) setting of nominal parallaxes on the first computer and setting the corresponding correction screws;

(*d*) setting the picture base distance on the second computer;

(*e*) setting the pantograph to the two left-hand control points;

(*f*) setting the gear for tilt correction to the two right-hand or upper and lower control points.

It is claimed that for the production of maps, one first-order plotting machine, determining control points by aerial triangulation, in combination with about ten stereotopes for mapping, gives a favourable arrangement. For plotting to the picture scale the accuracy of the planimetry is $\pm$ 0·2 mm whilst the heighting accuracy is $\pm$ 0·05 per cent of the flying height.

Rectification

Rectified photographs may be used for the preparation of controlled mosaics and map revisions. If the photograph is tilted, the scale is greater at one side than the other, and distortion occurs. This may be eliminated by rectification, depending upon the configuration of the ground, by printing a new picture on a tilted surface.

For clarity in rectification the following conditions are to be satisfied—

(*a*) the planes of the negative and the rectified print, and the plane through the perspective centre of, and at right angles to, the axis of the lens, must intersect in a straight line;

(*b*) the general condition of object and image distances from the lens and its focal length must hold, i.e. $\dfrac{1}{f_1} + \dfrac{1}{f_2} = \dfrac{1}{f}$.

Condition (*a*), illustrated in Fig. 11.25, is known as the *Scheimpflug condition*, and sufficient movements must be provided in the rectifier to satisfy this, and to allow a change of scale.

The rectifier can be used with large-scale photographs in which the area covered by each is small, since areas of no slope or uniform slope throughout are not usually of great extent. Four points are fixed in each photograph and plotted to scale. The enlarger is then adjusted so that those points projected from the negative coincide with the control points plotted, and a rectified photograph is then made. When the photographs cover ground which is not flat or of uniform slope throughout but rougher, then the distortion on one print cannot be removed by this method of rectification, although a possible solution might be to break down the whole into many separately rectified areas.

Revision of maps up to a scale of 1 in 2,500 can be effected using the existing map as control. Each photograph is enlarged and rectified so that photographic detail coincides as nearly as possible to existing detail on the map. Nearby common points are then used as local control, and revision of detail made by direct tracing.

A plan can be plotted from rectified photographs by a plumb point radial triangulation using the plumb point as centre, instead of the principal point, which is correct for height distortions. Heights may be found by parallax; contouring will require some height control in the overlap with interpolation by inspection under a stereoscope.

A graphical method can be adopted to locate the map positions A, B, C and D of four points a, b, c, and d suitably positioned over that photograph (Fig. 11.26). The map position F of a point f on the photograph can be easily located using the paper strip method.

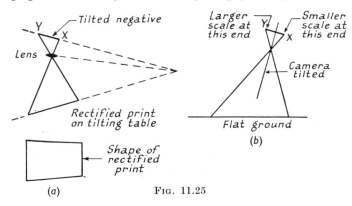

Fig. 11.25

The intersection of rays emanating from one image, say a to b, c, d and f, is marked on a strip of paper and identified. A similar procedure is adopted at d. These strips can now be positioned over the corresponding lines drawn on the map, so that the lines previously marked lie over the respective lines to B, C and D or A, B and C. The position of F can be fixed using rays Af_1 and Df'.

If many points are to be located, then grids can be drawn on both map and photograph by producing corresponding lines on them. Detail can now be transferred from photograph to map by reference to the network.

Mosaics. If a set of overlapping vertical photographs is joined together on a backing board, a rough map, or aerial view, of the area is obtained which has no uniformity of scale. This is called an uncontrolled mosaic, which allows preliminary inspection of conditions in the area and ensures full photographic coverage. If the rectified photographs be brought to the same scale or thereabouts, and fitted to ground control points, the mosaic is said to be controlled, and is superior to the uncontrolled mosaic. The ground control points may be obtained by the normal methods of surveying or by slotted template. The former method could supply ground control points around the boundary of the area with perhaps some intermediate points, the remainder could be supplied by the latter method.

The mosaic can be of great assistance at the reconnaissance and planning stage as it contains all that which could be seen directly from the aircraft at the time of photography. The photographs show such points as the type and density of any development, land use, existing access and that required for a highway scheme, alterations in topography, etc. It reveals information in far greater detail than can be plotted on a map.

Photographs by normal-angle or wide-angle cameras can be used, but distortions due to relief in hilly country can cause displacement in the mosaic, and the normal-angle lens would then be adopted. In addition, only the centre portions of the photographs should be used.

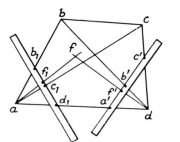

Photograph

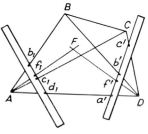

Map

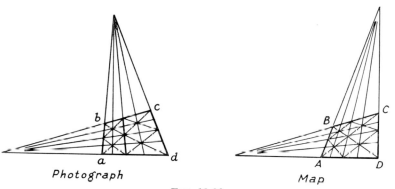

Photograph   Map

FIG. 11.26

Deker in his article, "The Aerial Mosaic," suggests that for the normal-angle camera, giving a picture size of 18 cm × 18 cm, an image size of 12 cm × 12 cm be adopted and that the basic scale for a topographic mosaic should not be less than 1 in 25,000. Depending upon the aircraft flying height, some enlargement may be required to achieve this scale. When the controlled mosaic has been assembled, a grid system can be supplied, in the margin, say, and the names of towns can be printed on and sheet titles and scales given in this margin.

If required, features such as rivers, roads, woods, etc., can also be emphasized by the use of symbols, and the mosaic can then be termed a photomap.

STEREOSKETCH. This instrument produced by Hilger & Watts is for the purpose of plotting detail directly from the aerial photograph on to existing maps, without correcting height distortions. As shown in Fig. 11.27, it is somewhat reminiscent of the Radial Line Plotter (p. 408). The two photographs (of size 230 mm) are positioned for stereoscopic fusion when viewed through the binocular system, and the height of the table, which can be slightly tilted, is adjusted by hand wheel until the model has the size of the map base. A scale on the left

FIG. 11.27. STEREOSKETCH

KEY

1. On-off switch	8. Drawing board illuminator
2. Dimming rheostats	9. Drawing table
3. Photograph illuminators	10. Height control (drawing table)
4. Wing mirrors	11. Foot rest
5. Binocular head	12. Indicator scale (supplementary lens powe
6. Mounting bar	13. Photo tables
7. Supplementary lenses	14. Carriers for supplementary lens and/or mask

(Courtesy of Hilger & Watts, Ltd.)

of the instrument indicates that supplementary lens which has to be placed in one of the carriers under the binocular head, so that the drawing table is focused with the photographs: a mask is fitted in the other carrier, so that the map is viewed by one eye, and detail from the photographs is traced directly on to the map.

Urban developments can be plotted in map revision, but the same distortions mentioned in the previous sections really call for the areas to be reasonably flat.

Differential Rectification

As mentioned, even when tilt effects are eliminated, the influence of relief will still induce distance errors in rectified photographs. However, the process of differential rectification resulting in the ortho-photo* allows correct planimetry. In this process small surface elements are successively rectified, the scale being continuously modified. Generally the terrain, i.e. model viewed, is divided into parallel strips which are subdivided into these small elements by means of a slit which is kept in contact with the model. A sheet of film lies near the slit, and the model image, free from relief displacement and at the correct scale, can be exposed thereon. Since the operator adjusts the height of the slit to conform to the model profile, heighting information can be obtained. Often, however, the points at which required contour values are attained on each profile are registered by a dropped line plot: this can be given by a series of lines and gaps or lines of different thicknesses. The ends of these lines can then be connected to produce contour lines.

The Zeiss Jena Differential Rectifier is taken in this section as an example. This unit, when complete (*see* Fig. 11.28*a*), consists of the Orthophot B differential rectification device, and the Orograph dropped line drawing equipment, linked to the Topocart B stereocarto-graphic instrument, which is basically for plotting at medium to small scales.

The Orthophot B is fitted to the rear of the Topocart and during rectification the right hand beam of the Topocart is split, being used for both visual observation and rectification. The slit (Fig. 11.28*b*) exposes on to sheet film carried by a rotating cylinder and it travels in the Y direction whilst the cylinder movements give steps in the X direction. Slits of different width are available to cater for terrain shapes; the slit will be set at the mean height of the terrain it covers and full coverage is only possible on flat ground. Depending upon the direction of the slopes, points may be double-imaged or omitted altogether in consecutive passes. This effect is minimized by narrowing the slit or strip, but this then consequently demands more profiles scanned over the same model.

The Orograph consists of a special drawing head and control unit; the control equipment is connected by synchro-transmitter to the Z spindle of the Topocart and the movements are divided into preselected height differences corresponding to the required contour interval. The drawing head is attached to the Topocart's drawing table and plots plans of the contour lines as shown in Fig. 11.29.

Relative and absolute orientation of the stereopair is initially arranged in the Topocart, the Orthophot and Orograph being switched off. Then the Orthonegative can be prepared as indicated and the dropped line plot can be made by layer engraving on foil or glass plate.

Orthophoto prints and maps can be made from the Orthonegative

---

* *See*, for example, *International Symposium on Photo Maps and Orthophoto Maps*, Canadian Institute of Surveying and International Society for Photogrammetry, Working Group IV, 3 September, 1967.

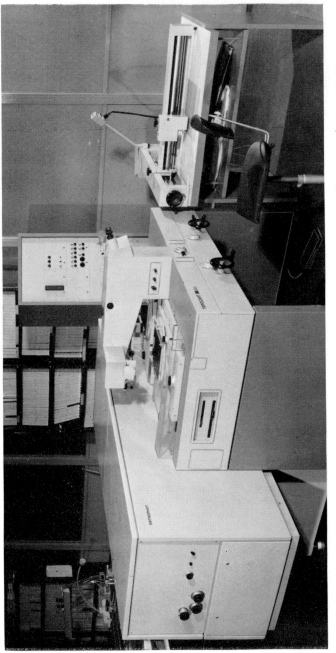

Fig. 11.28a. Topocart B Topographical Plotter with Orthophot and Orograph Attachments

(Courtesy of C. Z. Scientific Instruments Ltd. and Fairey Surveys, Ltd.)

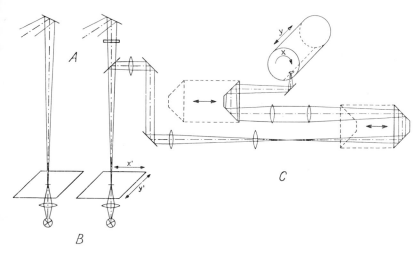

FIG. 11.28*b*. OPTICAL PATH OF ORTHOPHOT B
A: Eye-pieces of Topocart. B: Topocart *B*. C: Orthophot *B*.
(*Courtesy of C. Z. Scientific Instruments Ltd.*)

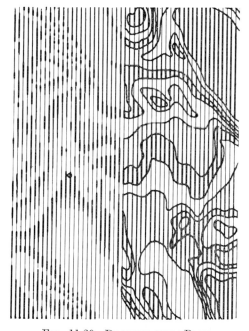

FIG. 11.29. DROPPED-LINE PLAN
(*Courtesy of C. Z. Scientific Instruments, Ltd.*)

afterwards, and it is advantageous when rectifying in this manner if the photo scale can be converted directly to the map scale. Zeiss provides a magnification range of $\times 0.7$ to $\times 5$ in this context.

It will be realized that the Orthophoto map is a topographic map on which the various features of the terrain are shown by photographic images. Lines and symbols can be overprinted for clarification but the map does not appear the same as the conventional line map. When these latter are produced by photogrammetric methods a selection of details given in the stereomodel has to be made by the operator, whereas all details are presented by the orthonegative.

Some Applications of Aerial Photogrammetry

The civil engineer is often concerned with mapping at scales between 1 in 500 and 1 in 2,500, and photogrammetry can be readily applied to such tasks when considering highway and railway projects, pipe line locations, industrial layouts and catchment problems. Plans at a scale of 1 in 500 can be produced with contours at intervals of 0·5 m, and in fact mean square errors in plan and height positions of some $\pm\, 0.1$ m are obtainable at that scale with modern cameras, film and first-order plotters.

Some mention has been made of the use of the stereoscope and stereometer in the preliminary reconnaissance survey of a highway project, when photographs at scales between, say, 1 in 20,000 and 1 in 5,000, depending upon the topography, can be scrutinized together with any existing maps of the terrain.

The location stage deals with the selection of an alignment and in the U.K., the existing 1/2,500 sheets could be revised and contours superimposed at, say, 2 m to 5 m intervals for studies in respect of motorways or ring roads round large cities. Of course it is also possible to carry out an entirely new survey to produce contoured plans, to that scale or to, say, 1/5,000 to 1/10,000 in less developed areas. A Digital Terrain Model (D.T.M.) could also be produced: this can be defined as a statistical sampling of the X, Y and Z co-ordinates of the terrain, i.e. ground surface, so that by interpolation the Z co-ordinate of any point can be accurately estimated from its X, Y co-ordinates. The morphological shapes must therefore be properly represented by the co-ordinated points chosen, and not only should those Z co-ordinates be established to the required accuracy, but the points must be sufficiently numerous and adequately arranged. Spatial co-ordinates of these characteristic points may be stored in a computer to allow the automatic interpolation of other points. The simplest arrangement takes the form of a square grid of spot heights, and this is generally taken to be appropriate for plane or uniformly sloping areas. It can be prepared by scanning the stereomodel, with automatic recording of the data on to tape. In rolling terrain the contour lines may be scanned, so that X, Y co-ordinates change for constant Z values, but in difficult conditions where sudden changes or breaks occur it might well be that single points at a higher density will have to be determined along the line of the break itself. According to K. M. Keir in a paper *"Photogrammetry and Current Practice in Road Design"* presented to the 11th Congress of the International Society for Photogrammetry, half-metre heighting accuracy will suffice at this stage.

Using the available information horizontal and vertical alignments can be studied and approximate earthworks computed so that the most favourable route line can be established and the design stage then put in hand. In this phase, construction plans are prepared at scales of, say, 1/1,250 and 1/500 from photography at scales of the order of 1/3,000 to 1/4,000.

A relatively narrow band of interest may be studied out of the strip photographed. Ground control can be put down to a fairly dense network or alternatively a geodetically supported aerial triangulation may be effected, pre-marked bench marks being established at frequent intervals near the proposed alignment. These provide control over lateral tilt errors which might arise in the bridging.

In addition to the plans, which might be contoured at 0·5 m or 1 m intervals, cross-sections, or a close-net D.T.M., will be derived with levels to an accuracy of + 0·1 m. In urban areas the former are probably the more likely to be produced since the terrain is artificial to an appreciable degree, but in rural areas the latter may be preferred and cross-sections can be interpolated from the D.T.M. This approach in fact allows additional cross-sections to be obtained without returning to photogrammetric methods, should there be slight changes in alignment or should the need for extra cross-sections arise. Keir states that the D.T.M. is also suited to the computation of volume by the columnar method discussed in Chapter 5.

The British Integrated Program System for Highway Design (Dept. of Environment) uses a square grid, the band of interest being divided into blocks of 41 × 41 nodes. A variable grid interval is easily introduced thereby allowing more frequent Z co-ordinates to be obtained in broken ground.

A device which may be used in establishing either the cross-sections directly or the D.T.M. is the Watts Locatorscope, which is mounted in the drawing pencil carriage on the plotting table of the Thompson–Watts plotter. If a plan is placed on that table and is oriented with respect to the model viewed in the plotter, the operator can set the index mark to the plan position of a point by observing in the viewing screen of the Locatorscope and without leaving the plotting machine. He can then arrange for the plotter index to be set on the corresponding point in the stereomodel and the spatial co-ordinates of the point may then be registered manually or by an automatic recording device for computer processing.

F. J. Brunnthaler delivered a paper *Application of Orthophotos to Civil Engineering* to the 11th International Congress for Photogrammetry, and made some reference therein to the role of the orthophoto in highway design. At the design stage he instances the production of orthophotos at 1/2,000 scale prepared from 1/5,600 photography with a heighting accuracy for contours constructed from drop lines of

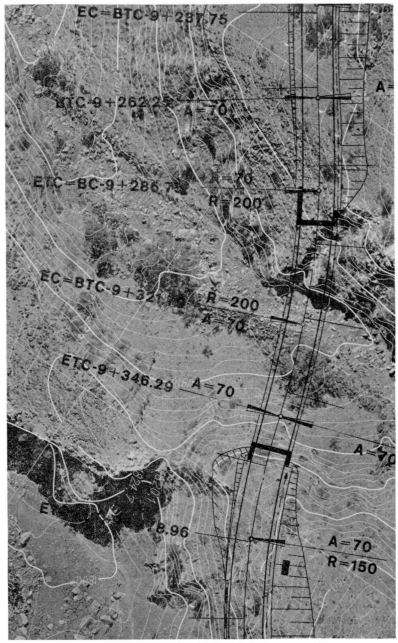

FIG. 11.30. PART OF ORTHOPHOTO COMBINED WITH CONTOUR
LINES, GRID LINES AND ROAD DESIGN DATA

Scale 1/1,000.
(*Courtesy of Hansa Luftbild, Münster*)

± 0.42 m. The road centre line, beginning and end of curves, side-widths, etc., can be superimposed on the orthophoto to produce a design orthophoto map as shown in Fig. 11.30; enlargement and reproduction of orthophotos can be arranged without loss of detail to $1/1,000$ scale. In this study Brunnthaler quotes a cost saving of the order of one-third when compared with the production of a conventional line map.

EXERCISES 11

1. Two photographs are taken with a phototheodolite from stations P and Q 160 m part, the lines of collimation being at right angles to PQ in each case. A point R appears on the photograph from P as $32\cdot38$ mm to the right of the vertical hair and $5\cdot65$ mm below the horizontal hair, and on the photograph from Q as $36\cdot42$ mm to the left of the vertical hair and $10\cdot80$ mm below the horizontal hair. Q is to the right of P and the focal length of the instrument is 165 mm.

Calculate the co-ordinates of R from P as origin, and the difference in levels of the two collimation planes. (*L.U., B.Sc.*)

Answer: $383\cdot7$ m N; $75\cdot3$ m E; $12\cdot0$ m.

2. A phototheodolite is set up and levelled, and using it as a theodolite the angle γ subtended at the instrument by two distant points A and B (selected so that they will show on a photograph taken with the instrument) is measured. Next using the phototheodolite as a camera, a photograph is taken from the same stations to include both points A and B. The images a and b respectively of these points on the plate are found to be at horizontal distances x and y on opposite sides of the mark of the vertical cross-wire.

Show that the focal length of the camera lens is

$$\frac{x+y}{2\tan\gamma} + \sqrt{\frac{(x+y)^2}{4\tan^2\gamma} + xy} \qquad (L.U., B.Sc.)$$

3. What is meant by the radial assumption? Describe the radial line method of plotting from aerial photographs and discuss the slotted template assembly.

4. In a pair of overlapping vertical air photographs the mean distance between the two principal points, both of which are at datum level, is $89\cdot6$ mm. At the time of photography the aircraft was 2,500 m above datum and the camera had a focal length of 150 mm. In the common overlap a tall chimney, whose base is at datum level, is observed, and the difference in absolute parallax measurements of the top and base found to be $2\cdot40$ mm. Estimate the height of the chimney.

Answer: $65\cdot2$ m.

5. An area of 220 km^2 is to be photographed at a scale of 1 in 8,000 from the air using a camera of focal length 150 mm, the photographs being 230 mm square. A longitudinal overlap of 60 per cent and a lateral overlap of 25 per cent must be provided. If the operating speed of the aircraft is 225 km/h find

(*a*) the flying height of the aircraft and the interval between exposures;

(*b*) the number of prints required if the flying strips are 16 km long.

Answer: (*a*) 1,200 m; $11\cdot8$ sec, (*b*) 253.

6. If a stereoscopic measuring accuracy of $0\cdot01$ mm is possible and an accuracy of 1 in 1,000 is to be attained in plotting a point y m from the base line in a terrestrial stereophotometric survey, find the relationship between y and the base length B. A phototheodolite of focal length 165 mm was used in the survey.

Answer: $B = y/8\cdot25$.

CHAPTER 12

FIELD ASTRONOMY

THE EARTH ROTATES about its polar axis in an anti-clockwise direction approximately once every twenty-four hours mean time (i.e. civil clock time), and also traces out an elliptical path around the sun once every year. The polar axis is inclined at an angle of approximately $23\frac{1}{2}°$ to the normal to the plane of orbit round the sun. The velocity of the earth along this path is not constant, and according to Kepler's law an arm connecting the earth to the sun would sweep out equal areas in equal times.

Apart from the earth, other bodies (planets) move in definite paths around the sun with a periodic time of travel round the sun depending upon their distance from the sun—the further away from the sun the longer being the time taken. The sun and its planets and satellites, together with some small planets known as asteroids, comprise the *solar system*.

Fixed stars are those whose relative positions appear to have very little change with time, and which also appear unchanged in position when observed from any point on the earth's surface. They are at a great distance from the earth (well outside the solar system) and are classified in constellations or groups. If a fixed star were continuously observed from a point on the earth's surface, it would appear to describe a circle about a fixed point in the sky, known as a *celestial pole*, once in approximately twenty-four hours clock time. All fixed stars, if they could be followed by the eye throughout their path, would seem to rotate in this way, and due to their great distance from the earth they would seem to be moving over the surface of a sphere—known as the *celestial sphere*. The earth is taken to be stationary at the centre of this sphere, the stars rotating around the celestial pole. Actually, of course, the stars' apparent movement is due to the earth's rotation about its polar axis. All the stars are assumed to lie on the surface of the celestial sphere, which is of such a large radius that it is immaterial which point on the earth is assumed to be the centre. When we consider that the nearest star, "α Centauri," is about 20 billion miles away, this assumption is obviously justified.

Fig. 12.1 shows the celestial hemisphere for an observer in the northern hemisphere, the centre of the earth being taken at the centre of that hemisphere. Z is the zenith of the observer, and is a point on the celestial sphere directly above the observer. P is the celestial north pole, and WEE_1 is the celestial equator. NE_1SW is the *true horizon* of the observer, N being the north point of the horizon. In the northern hemisphere the celestial pole is north of the zenith, while in the southern hemisphere (Fig. 12.2) the pole is south of the zenith.

The *sensible horizon* (Fig. 12.3) may be taken as the plane in which the level tube of a theodolite lies when it has been accurately set up.

When star observations are being carried out, the two horizons are taken to coincide, but when the sun is being observed a correction must be applied to reduce altitudes measured from the sensible horizon to the true horizon which passes through the earth's centre. Both

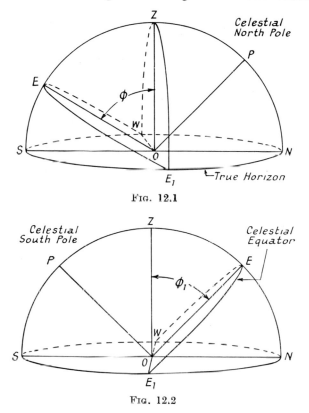

Fig. 12.1

Fig. 12.2

horizons are at right angles to the line connecting the observer to his zenith.

The *latitude* of the station is $E\hat{O}Z$. Fig. 12.4 illustrates the trace of the equatorial plane EE on the earth's surface. It is at right angles to the polar axis, $p_n p_s$, and the latitude of the station Z_1 is $E\hat{O}Z_1 = \phi$. If OZ_1 and Op_n were produced upwards to meet the celestial sphere, the zenith Z, and the celestial pole P, would be obtained (refer to Fig. 12.1). The sensible horizon at Z_1 would be a plane tangential to the earth's surface. The celestial meridian of an observer is the circle passing through $SEZPN$ in Fig. 12.1, and corresponds to the terrestrial meridian $p_s E Z_1 p_n$. The celestial meridian passes through the zenith of the station and the celestial poles, and the great circle through Z at right angles to this meridian is the *prime vertical.*

Longitude is the angle, measured eastwards or westwards, between

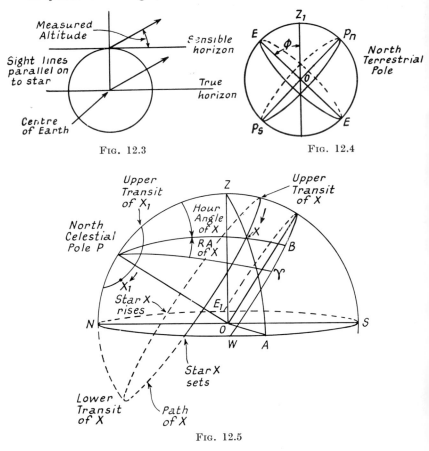

Fig. 12.3

Fig. 12.4

Fig. 12.5

the meridian of the station and the meridian for Greenwich. Longitude is of great account in time measurements.

Altitude, Azimuth, Declination, Right Ascension and Hour Angle

In the northern hemisphere, stars appear to rotate anti-clockwise around the north celestial pole when looking towards that pole, and their altitudes vary. When a star (or the sun) X, sets, it passes down below the horizon, and when it rises, it climbs over the horizon.

In its progress, it will be seen from Fig. 12.5 that the star sets and then crosses the observer's meridian ($NPZS$) at a point termed *lower transit,* recrosses the horizon when it rises, and then recrosses the meridian at *upper transit,* whence it returns towards the horizon. At upper transit the body is at maximum altitude, and this occurs on the same side of the pole as the zenith: at lower transit the star is at a minimum altitude, and may or may not be visible. Star X is not visible at lower transit, but star X_1 will be visible at upper and lower transit, i.e. it will never set. This star is a *circum polar star* and has a co-declination which is smaller in value than the latitude of the observer.

ALTITUDE. The altitude of X is angle $X\widehat{O}A$, being measured from the sensible horizon along the circle ZXA. $X\widehat{O}Z$ is known as the *co-altitude* of the star.

AZIMUTH. The angle between the observer's meridian and the circle ZXA is known as the *azimuth* of the body. It is usually measured

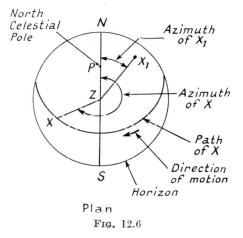

Plan

FIG. 12.6

clockwise from the north from 0° to 360° (Fig. 12.6). Both azimuth and altitude of the star vary, and are not used as co-ordinates for fixing the position of the celestial body as are declination and right ascension.

DECLINATION. The angle from the celestial equator to the star $(X\widehat{O}B)$ is known as the *declination*, and this is measured along the declination circle of the star. The declinations of many stars may be found from the Star Almanac, and show little variation, i.e. the stars maintain paths virtually parallel to the equator. The sun's declination, however, varies throughout the year. Declination is measured from 0° to 90° north or south, depending which side of the celestial equator the star lies. Angle $X\widehat{O}P$ is the *co-declination* of the star.

The *declination circle* of the star is the great circle passing through X and the celestial north and south poles.

RIGHT ASCENSION (R.A.). The right ascension is that angle (measured in the opposite direction to the travel of the stars or sun, i.e. anti-clockwise in the northern hemisphere) between a point on the celestial equator (termed the First Point of Aries, ♈) and the declination circle of the body. This angle is measured *either* from 0° to 360° *or* 0 hours to 24 hours, since right ascension is linked with time measurements—note $15° = 1$ hr, $1° = 4$ min, and $1' = 4$ sec.

The First Point of Aries is the point at which the sun crosses the celestial equator at the spring equinox (about 22nd March). The sun has northerly declination from 22nd March to 22nd September, and a southerly declination from 22nd September to the following 22nd March. Values of right ascension may be found in the Star Almanac.

HOUR ANGLE. The hour angle (H.A.) of a star (Fig. 12.7) is the angle between the declination circle of the star and the meridian of the observer, and is measured in the direction of the star's motion, from 0° to 360°, *or*, more usually, 0 hours to 24 hours. At upper transit the

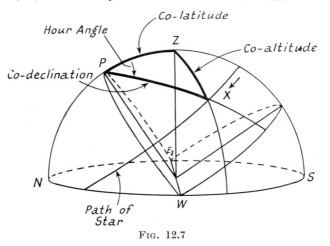

FIG. 12.7

star's hour angle is zero and the hour angle will increase steadily as the star traverses its path, the hour angle at lower transit being 12 hours.

EXAMPLE. Find the true altitude of the star Betelgeuse (declination N 7° 24′ 03″) at upper transit on 1st January, 1954, at a place in latitude 53° 29′ 30″ N.

Draw a semicircle $NPZES$ (Fig. 12.8) with centre O to represent the meridian of the observer at O. Draw a vertical OZ to give zenith Z and mark N and S as the north and south points of the horizon of the observer. Set off $Z\hat{O}E = \phi = 53° 29′ 30″$ and OP at right angles to OE, giving P (north celestial pole) north of zenith. Mark off X such that $X\hat{O}E$ = declination = 7° 24′ 03″ north of equator OE.

Then altitude at upper transit

$$= X\hat{O}E + E\hat{O}S$$
$$= 7° 24′ 03″ + (90° - 53° 29′ 30″)$$
$$= 43° 54′ 33″$$

At this instant the hour angle of Betelgeuse is 0 hr, and its azimuth reckoned clockwise from North is 180°. At lower transit this star cannot be seen since angle XOE is less than $(90° - \phi)$, and would be below horizon NS, if plotted at lower transit.

EXAMPLE. (a) Find the altitude of the sun at upper transit at the same place given that the declination is 2° 20′ 00″ S.

(b) At what latitude in the southern hemisphere would the sun have the same meridian altitude at the same instant?

(a) Set off sun S_1 south of equator, so that $S_1\hat{O}E = 2° 20′ 00″$ (Fig. 12.9).

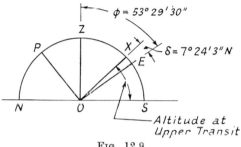

FIG. 12.8

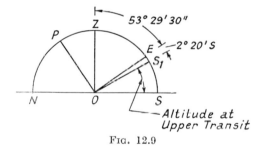

FIG. 12.9

Then altitude of sun = 36° 30′ 30″ − 2° 20′ 00″
= 34° 10′ 30″ in the south.

(b) In the southern hemisphere the sun will be at upper transit in the northern parts of the celestial sphere.

Set off $S_1\hat{O}N = 34° 10′ 30″$ (Fig. 12.10). Mark off $S_1\hat{O}E =$ declination = 2° 20′ 00″ south. Then $E\hat{O}N = 90° - \phi = 31° 50′ 30″$.

∴ $\phi = 58° 09′ 30″$ S

Note that the south celestial pole is south of the zenith.

EXAMPLE. Find the altitudes at upper and lower transit of a star (declination N 63° 15′ 36″) at a place in latitude 53° 30′ 00″ N.

At upper transit (Fig. 12.11),

$$N\hat{O}X_1 = \text{Altitude} = 53° 30' 00'' + 26° 44' 24'' = 80° 14' 24''.$$

At lower transit,

$$N\hat{O}X_2 = \text{Altitude} = 53° 30' 00'' - 26° 44' 24'' = 26° 45' 36''.$$

Note: This star is circumpolar since its co-declination is less than the latitude.

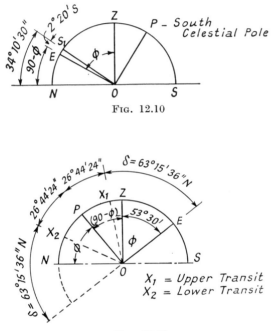

Fig. 12.10

Fig. 12.11

Elongation. If the progress of a star, which has a co-declination smaller than the latitude of the observer, is followed by instrument, it will be found to have maximum bearings both in the east and west. After having its maximum altitude at upper transit, such a star observed in the northern hemisphere will move westwards and be found to have a maximum western bearing from the observer. It is then said to be at *western elongation*; the line of sight from the observer will be tangential to the path of the star. The star then moves round towards the observer's meridian, crossing it at lower transit, and the bearing then steadily increases until it is a maximum in the east at *eastern elongation* (*see* Fig. 12.12).

Astronomical Triangle. Whatever the position of a star or the sun (except when on the obrsever's meridian), a spherical triangle is formed between the celestial pole P, the observer's zenith Z, and the body X itself. Then in Fig. 12.13 the following relationships obtain—

$$\cos Z = \frac{\cos PX - \cos PZ \cos XZ}{\sin PZ \sin XZ}$$

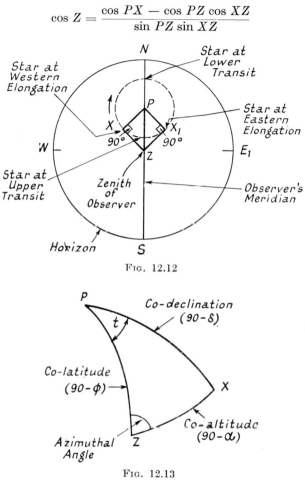

Fig. 12.12

Fig. 12.13

and

$$\tan \frac{Z}{2} = \sqrt{\frac{\sin (S - XZ) \sin (S - PZ)}{\sin S \sin (S - PX)}}$$

where

$$2 S = XZ + PX + PZ$$

Similar relationships may be obtained for the angle P. Now when star X is at elongation, then $\hat{X} = 90°$, and we have the following relationships—

$$\sin Z = \frac{\cos \delta}{\cos \phi}$$

$$\sin \alpha = \frac{\sin \phi}{\sin \delta}$$

$$\cos P = \frac{\tan \phi}{\tan \delta}$$

In field astronomy, most of the problems reduce to the solution of the spherical triangle ZPX.

Time. A sidereal day is the interval between successive upper transits of the First Point of Aries. *Local sidereal time* is the interval

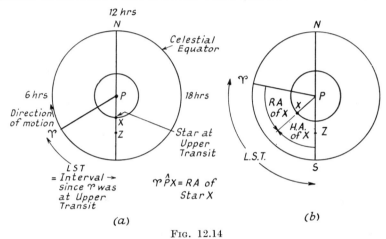

Fig. 12.14

which has elapsed since the First Point of Aries was at upper transit at that place (Fig. 12.14a). At upper transit, the hour angle of First Point of Aries (Υ) is zero. Since the right ascension of a star is measured from First Point of Aries, then it will be seen that the local sidereal time of transit of that star is the interval which has elapsed since First Point of Aries was at upper transit, and will equal the right ascension of the star.

Also from Fig. 12.14b, it can be seen that local sidereal time = R.A. of star + H.A. of star.

A solar day is the interval between successive upper transits of the sun, and the hour angle of the sun leads to *local solar time* or *local apparent time*. *Local apparent noon* occurs when the sun is at upper transit. The intervals between the successive transits of the sun are not equal (due to the elliptic orbit of the earth and certain other factors), whereas those of a star are virtually so (*see* Jameson's *Advanced Surveying*).

For this reason, a *mean solar day* is used in time measurement. The mean sun may be considered as a point moving at uniform speed round the equator, when the earth is considered stationary.

Greenwich Mean Time or *Universal Time* is shown in the Star Almanac as U.T. in the columns referring to the sun. It is the mean time of the meridian of Greenwich and is obtained by taking the average time of rotation of the earth as 24 hours.

When the actual sun is at upper transit at Greenwich, we have Greenwich Apparent Noon (G.A.N.), and similarly, Greenwich Mean Noon (G.M.N.), or G.M.T. 12.00 hr obtains when the mean sun crosses that meridian at upper transit. At Greenwich mean midnight (G.M.M.) Greenwich Mean Time is 00.00 hr.

Time is reckoned from midnight (00.00 hr) and thus apparent solar time will equal the hour angle of the sun plus 12 hr between upper and lower transits. If the sun's hour angle is greater than 12 hr, then 12 hr

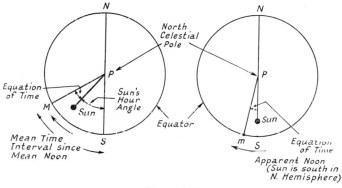

Fig. 12.15

is deducted from the hour angle to obtain apparent time. Similarly G.M.T. is equal to the hour angle of the mean sun at Greenwich plus or minus 12 hr depending on its position in relation to upper transit.

Both the actual and mean suns will have a right ascension measured from the First Point of Aries, and the difference between the two is termed the *equation of time* (Fig. 12.15 illustrates this although ♈ is not shown in the diagram). We thus have the relationship—

Mean time = apparent time ± equation of time.

The equation of time, which is a variable, may be obtained from data in the Star Almanac. The hour angle of the sun at Greenwich may be computed from—

G.H.A. of sun = E + U.T.

both of which can be obtained from the Star Almanac. (E has not the same value as the equation of time referred to in the previous expression but is related to it.)

365·2422 mean solar days = 366·2422 sidereal days

then   1 mean solar day   = 24 hr 3 min 56·56 sec sidereal time

1 mean solar hour   = 1 sidereal hr + 9·857 sidereal sec

1 sidereal day   = 23 hr 56 min 4·09 sec mean time

1 sidereal hour   = 1 mean solar hr − 9·830 solar sec

The time at which ♈ is at upper transit at Greenwich gives Greenwich Sidereal 0 hr or G.S.T. 0 hr. The G.S.T. for any instant of G.M.T. is given by adding corresponding values in the Star Almanac headed UT and R.

At a place whose longitude is east of Greenwich, the sun, or a star, will cross the meridian before it will cross the meridian at Greenwich. Similarly, if the place is west of Greenwich, the body transits later. Times are then related as follows—

$$\textit{Local Time} \qquad\qquad \textit{Greenwich Time}$$

Mean $\quad$ L.M.T. $=$ G.M.T. $\mp$ Longitude in hours $\dfrac{\text{West}}{\text{East}}$

Sidereal $\quad$ L.S.T. $=$ G.S.T. $\mp$ Longitude in hours $\dfrac{\text{West}}{\text{East}}$

Apparent $\quad$ L.A.T. $=$ G.A.T. $\mp$ Longitude in hours $\dfrac{\text{West}}{\text{East}}$

$1°$ Longitude $=$ 4 minutes time

$1'$ Longitude $=$ 4 seconds time

$1''$ Longitude $=$ $\frac{1}{15}$ second time.

Care must be taken not to confuse the formulae. Local mean time must not be related to local sidereal time simply by making the longitude correction, but a further set of relationships may be used—

Local Sidereal Time of Local Mean Noon

$=$ Greenwich Sidereal Time of Greenwich Mean Noon

$\pm$ 9·857 sec/hr of

Longitude $\dfrac{\text{West}}{\text{East}}$

If required, midnight may be substituted in lieu of noon in the above expression.

EXAMPLE. To find G.S.T. at G.M.M. if G.S.T. at G.M.N. on the same day is
(a) 20 hr 53 min 28·3 sec, i.e. $R = 08$ hr 53 min 28·3 sec;
(b) 2 hr 00 min 2·4 sec, i.e. $R = 14$ hr 00 min 02·4 sec.

(a) Mean Time interval between mid-
night and noon $\qquad\qquad = \quad$ 12 hr 00 min 00 sec
For equivalent sidereal interval add
$12 \times 9·857 = 118·28$ sec $\qquad = \qquad$ 01 min 58·3 sec

$\qquad\qquad\qquad\qquad\qquad\qquad\qquad$ 12 hr 01 min 58·3 sec

G.S.T. at G.M.N. $=\quad$ 20 $\quad$ 53 $\qquad$ 28·3
Deduct $\qquad\qquad\qquad$ 12 $\quad$ 01 $\qquad$ 58·3

G.S.T. at G.M.M. $=\quad$ 8 hr 51 min 30·0 sec

(b)

G.S.T. at G.M.N. =	2	00	02·4
Add	24	00	00
	26	00	02·4
Deduct	12	01	58·3

∴   G.S.T. at G.M.M. =   13 hr 58 min 04·1 sec

Note: G.S.T. at G.M.M. *on the day following* will be 2 hr 00 min 02·4 sec + 12 hr 1 min 58·3 sec = 14 hr 2 min 00·7 sec, and it will be observed that there has been an *increase* of 3 min 56·6 sec in sidereal time in a 24 hr mean time interval. In this type of problem multiples of 24 hr may be omitted or introduced. It will be seen in (b) that the sidereal interval is longer than the value from which it is to be deducted; in such a case, 24 hr is first added.

EXAMPLE. To find L.S.T. at L.M.N. (12.00hr) in longitude 90° W if G.S.T. at G.M.N. is 2 hr 00 min 02·4 sec.

G.S.T. at G.M.N.	= 2 hr 00 min 02·4 sec
90° W = 6 hr W	
Add 6 × 9·857 = 59·142 sec	59·1

∴   L.S.T. at L.M.N.   = 2 hr 01 min 01·5 sec

EXAMPLE. To find L.S.T. at 10 hr 00 min a.m. (Local Mean Time) in longitude 90° W if G.S.T. at G.M.M. is 13 hr 58 min 04·1 sec.

From above

G.S.T. at G.M.M. Now for	13 hr 58 min 04·1 sec
L.S.T. at L.M.M. add 59·1 as above	59·1
	13 hr 59 min 03·2 sec
Now 10.00 hr (mean) converted to sidereal time requires a correction of 9·857 sec/hr = 98·57 sec added	10 hr 1 min 38·6 sec

∴   L.S.T. at 10 hr L.M.N.   = 24 hr 00 min 41·8 sec

i.e. L.S.T. at 10 hr L.M.T.   = 00 hr 00 min 41·8 sec

Alternatively, if G.S.T. at G.M.N. = 02 hr 00 min 02·4 sec
From previous example
L.S.T. at L.M.N.   = 02 hr 01 min 01·5 sec
We require L.S.T. at 2 hr mean time before L.M.N. (i.e. 10.00 a.m.)
2 hr mean = 2 hr sidereal + 2 × 9·857 sec
= 2 hr 00 min 19·7 sec
∴ L.S.T. at 10 hr L.M.T. = L.S.T. at L.M.N. − 2 hr 0 min 19·7 sec
= 00 hr 00 min 41·8 sec

EXAMPLE. To find L.M.T. of upper transit of the star Capella (R.A. 5 hr 13 min 15·4 sec) at a place in longitude 2° 30′ W when G.S.T. at G.M.N. = 20 hr 53 min 28·3 sec.

G.S.T. at G.M.N.   = 20 hr 53 min 28·3 sec

L.S.T. at L.M.N. 2° 30′ W = 10 min W

Add $\dfrac{10}{60} \times 9{\cdot}857 = 1{\cdot}64$ 01·64

L.S.T. at L.M.N.   = 20 hr 53 min 29·94 sec

which is the interval between ♈ being at upper transit and the mean sum being at upper transit at the place, i.e. local mean noon.

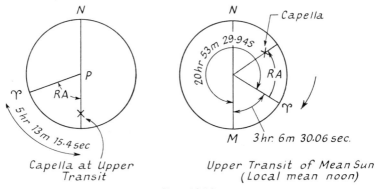

FIG. 12.16

It will be seen from Fig. 12.16 that Capella will transit in

3 hr 06 min 30·06 sec + 5 hr 13 min 15·40 sec

= 8 hr 19 min 45·46 sec (sidereal) after noon.

Convert the sidereal interval into mean time by deducting 9·83 sec per sidereal hour,

$$(8 \times 9{\cdot}83) + \left(\frac{19}{60} \times 9{\cdot}83\right) + \left(\frac{45}{3{,}600} \times 9{\cdot}83\right) = 81{\cdot}77 \text{ sec}$$

∴ Mean Time interval since L.M. Noon

= 8 hr 19 min 45·46 sec − 1 min 21·77 sec = 8 hr 18 min 23·69 sec

∴   L.M.T. = 8 hr 18 min 23·69 sec p.m., or 20 hr 18 min 23·69 sec

Note :   G.M.T. = 8 hr 28 min 23·69 sec

EXAMPLE. Determine L.A.T. of an observation at a place in longitude 2° 30′ W if L.M.T. is 2.20 p.m.: the equation of time at G.M.N. is 5 min 58·7 sec additive to apparent time and increasing at 0·22 sec/hr.

L.M.T. of observation   = 14 hr 20 min 00 sec

Add longitude (10 min)   10   00

∴   G.M.T. of observation = 14 hr 30 min 00 sec

which is 2·5 hr after noon.

Equation of time at G.M.N. = 0 hr 05 min 58·7 sec
Add 2·5 × 0·22 for increase   0·55
Equation of time   = 0 hr 05 min 59·25 sec

Now Mean Time = Apparent Time + Equation of time

∴   G.A.T. of observation = G.M.T. − Equation of time

∴   G.A.T.   = 14 hr 24 min 0·75 sec

       Deduct for longitude   10 min 0·00 sec
       L.A.T.   = 14 hr 14 min 0·75 sec

EXAMPLE. Find the Greenwich hour angle of the sun on 2nd Jan., 1954, at G.M.N.

From the Star Almanac

    If UT = 1200 then E =   11 hr 56 min 02·7 sec at this instant
∴   G.H.A. =   12   00   00
             +   11   56   02·7
                23 hr 56 min 02·7 sec

i.e. the actual sun is 3 min 57·3 sec behind the mean sun. This is the equation of time at this instant and is added to apparent time. Thus the sun transits at Greenwich at a mean time of

12 hr 00 min 00 sec + 3 min   57·3 sec = 12 hr 03 min 57·3 sec
                                    = 24 hr − E

EXAMPLE. Find G.M.T. for the instant at which G.H.A. of sun is 07 hr 43 min 59·1 sec on 20th Aug., 1961. At 1800 G.M.T., E is 11 hr 56 min 41·6 sec, increasing at 0·6 sec/hr.

       G.H.A. of sun   = 07 hr 43 min 59·1 sec
       Approx. E   = 11   56   41·6
       G.M.T.   = 19   47   17·5
       Correction to E =   1·0
∴   E (corrected)   = 11   56   42·6
and   G.M.T.   = 19   47   16·5

The value of E is listed against U.T. not G.A.T., so that some form of successive approximation is adopted to determine a suitable value of E.

Corrections to Observations of Altitude

The altitude of the sun or stars is most accurately measured with the theodolite, and in this case, in addition to the correction which will be required if the altitude bubble is not central when reading the vertical angle, other corrections may be required as follows.

Correction for Refraction is required for all observations. A ray of light from the body is refracted as it passes through the atmosphere such that the body appears at a greater altitude than it actually is. An approximate value for this correction is − 58″ cot α, where α is the apparent altitude, and should be over 20° at least for this correction to be used. Accurate values call for readings of atmospheric pressure and air temperature and recourse to the *Refraction Tables of the Star Almanac*.

Correction for Parallax is not required for observations on the fixed stars. A measurable angle is subtended at the sun by the earth's radius, however, and the observed altitude of the sun from the earth's surface will be slightly smaller than the altitude which would be observed from the earth's centre.

The correction required is $+ 8 \cdot 80''$ cos α, where α is the altitude corrected for refraction.

Correction for Semi-diameter is also required for solar observations. It is difficult to make an observation on the centre of the sun, and

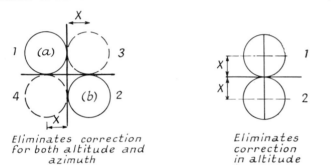

Eliminates correction
for both altitude and
azimuth

Eliminates
correction
in altitude

Fig. 12.17

observations are usually made instead on the upper or lower limbs (or edges) of the sun for altitudes, and on the east or west limbs when measuring azimuthal angles (*see* Fig. 12.17). A correction is then made to these readings reducing them to the sun's centre, called the correction for *semi-diameter*, and its value is given in the Nautical Almanac and Star Almanac. It is usual, however, to take one reading (*a*) on one limb, followed immediately by a further reading (*b*) on the other limb and a mean value taken so as to avoid this correction. If α be the altitude of the centre of the sun, the difference in azimuth between the centre and either left or right limb is X sec α.

A further *correction for dip* must be applied at sea when sighting either a sun or a star. On land the theodolite correctly set up has its vertical axis truly vertical and the plates in a horizontal plane, so that vertical angles are referred to the horizontal; at sea, however, the altitude is measured by sextant from the *visible horizon* and a correction must be deducted from the observed altitude. This correction depends upon the height of the observer above sea level, and special tables are available which give the value of dip for various heights above sea level.

Determination of Azimuth—True Bearing

Several methods are available for the determination of the horizontal angle between the meridian at a station and a survey line at the station. Sufficient observations are taken on a star (or on the sun) to determine its azimuth at that instant, and the horizontal angle between the star and the survey line is also measured.

A reference mark or referring object (R.M.) is established either (*a*) at the station at the other end of the line, or (*b*) at some other position, not necessarily on the line, such that the angle between the line and the R.M. can be measured. The reference mark should be (i) clearly visible, (ii) high enough from the ground to avoid refraction problems, and (iii) sufficiently distant from the instrument that the focusing of the

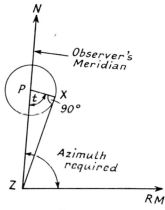

FIG. 12.18

telescope need not be readjusted when the pointing from the R.M. to the star (or sun) is changed.

A star appears as a point of light in the telescope and gives a good bisection, and a particular star (or stars) may be chosen to suit the task in hand out of the many available, but the sun allows the advantage of daylight observation. A few methods will now be described.

1. **Observation on a Circumpolar Star at Elongation.** The azimuth $P\hat{Z}X$ of such a star has its maximum value at eastern and western elongation, at which times $P\hat{X}Z$ is a right angle (*see* Fig. 12.18). At a short time before elongation there is little apparent motion of the star in azimuth, and when observed at this time the star appears to rise up the vertical hair at eastern elongation, and to descend at western elongation. In either case it is possible to change face and take a further observation without perceptible change in azimuth of the star. The time of elongation is calculated in advance, and the instrument is set up at the station and levelled about ten minutes before this time. Readings of the horizontal circle are then noted for the following successive pointings,

    (i) on the reference mark,
    (ii) follow the star with the vertical hair, starting near intersection of cross-hairs, until it appears to rise or fall on the vertical hair,

(iii) change face and reobserve the star,

(iv) on the reference mark.

A typical example of booking and computation now follows.

EXAMPLE. Observations were taken on a star (Dec. 86° 36′ 16″ N, R.A. 47 min 06 sec) at eastern elongation at a place in lat. 53° 30′ N, and the following data obtained. Find the azimuth of the reference mark.

Object	Face	Horizontal Circle Verniers		Mean	Angle between R.M. and star
R.M.	R	241° 20′ 20″	61° 20′ 40″	241° 20′ 30″	} 78° 40′ 00″
Star	R	162° 40′ 20″	342° 40′ 40″	162° 40′ 30″	
Star	L	342° 40′ 00″	162° 40′ 20″	342° 40′ 10″	} 78° 40′ 20″
R.M.	L	61° 20′ 40″	241° 20′ 20″	61° 20′ 30″	

Mean 78° 40′ 10″

Now
$$\sin Z = \frac{\cos \delta}{\cos \phi} = \frac{\cos 86° 36′ 16″}{\cos 53° 30′ 0″}$$

Therefore
$$Z = 5° 42′ 53″$$

Since the readings increase towards R.M. and the instrument reads clockwise, then R.M. is east of the star.

Azimuth of R.M. = 5° 42′ 53″ + 78° 40′ 10″

= 84° 23′ 03″ clockwise from N.

If G.S.T. at G.M.M. on the following day was 15 hr 20 min 51·8 sec and the place was in longitude 2° 30′ W, the G.M.T. of elongation would be computed as follows.

G.S.T. at G.M.M.   15 hr 20 min 51·8 sec

Add for 2° 30′ W

$$\left[\frac{10}{60} \times 9{\cdot}857\right]$$ 1·6 sec

L.S.T. at L.M.M.   15 hr 20 min 53·4 sec

Now $\cos t = \tan \phi/\tan \delta$

$$= \frac{\tan 53° 30′ 00″}{\tan 86° 36′ 16″}$$

Therefore $t = 85° 24′ 03″$ (or 5 hr 41 min 36·2 sec)

Referring to Fig. 12.19, the position of ♈ must be located to determine the L.S.T. of elongation

$$\begin{aligned}
\text{H.A. of star} &= \text{18 hr 18 min 23·8 sec} \\
\text{R.A. of star} &= \text{17 hr 47 min 06·0 sec} \\
\hline
&\text{36 hr 05 min 29·8 sec} \\
\hline
\end{aligned}$$

$$\text{L.S.T. of elongation} = \text{12 hr 05 min 29·8 sec}$$

i.e. it occurs at a sidereal period of 3 hr 15 min 23·6 sec before midnight.

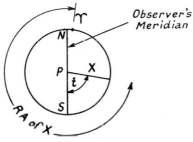

F<small>IG</small>. 12.19

This is now converted to mean time by deducting 9·83 sec/hr to give a mean time interval of 3 hr 14 min 51·6 sec.

Therefore
$$\text{L.M.T.} = \text{20 hr 45 min 8·4 sec}$$
$$\text{G.M.T.} = \text{20 hr 55 min 8·4 sec}$$

2. Elongation of Two Circumpolar Stars. It is obvious from the previous method that the latitude of the place must be known accurately. If it is not so known, then observations may be taken on two stars which reach elongation within a short time of each other. They may be at the same or opposite elongations as in Fig. 12.20.

Now
$$\sin Z_1 = \frac{\cos \delta_1}{\cos \phi}$$

and
$$\sin Z_2 = \frac{\cos \delta_2}{\cos \phi}$$

$\therefore$
$$\frac{\sin Z_1}{\sin Z_2} = \frac{\cos \delta_1}{\cos \delta_2} = B \quad \text{(a known constant since } \delta_1 \text{ and } \delta_2 \text{ are known).}$$

At opposite elongations, as in Fig. 12.20,

Let
$$Z_1 + Z_2 = X^1$$
$$\begin{aligned}
\sin Z_1 &= \sin (X^1 - Z_2) \\
&= \sin X^1 \cos Z_2 - \cos X^1 \sin Z_2 \\
&= B \sin Z_2
\end{aligned}$$

$\therefore$
$$\sin X^1 \cot Z_2 - \cos X^1 = B$$

$\therefore$
$$\cot Z_2 = \frac{B + \cos X^1}{\sin X^1}$$

In the case of the stars being at the same elongation (Fig. 12.21), let

$$Z_2 - Z_1 = X^1$$

then
$$Z_1 = Z_2 - X^1$$
$$\sin Z_1 = \sin Z_2 \cos X^1 - \cos Z_2 \sin X$$
$$= B \sin Z_2$$

$\therefore$
$$\cos X^1 - \cot Z_2 \sin X^1 = B$$

$\therefore$
$$\cot Z_2 = \frac{\cos X^1 - B}{\sin X^1}$$

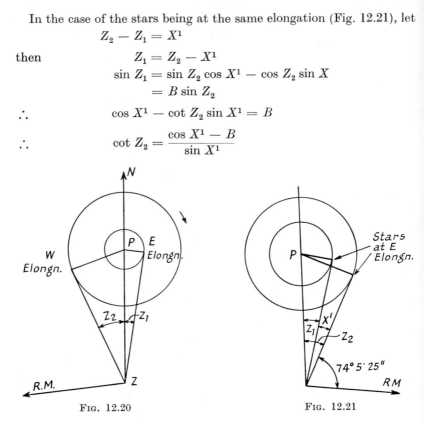

FIG. 12.20   FIG. 12.21

X^1 is the difference between the two horizontal angles from R.M. to the stars.

EXAMPLE. A star of declination $82° 06' 44''$ N is observed at E elongation when the anti-clockwise angle from a reference mark was $104° 23' 00''$. Immediately afterwards a star of declination $65° 46' 4''$ N was observed at E elongation, and the anti-clockwise angle noted was $74° 5' 25''$. Determine the azimuth of R.M.

$$\delta_1 = 82° 06' 44''$$
$$\delta_2 = 65° 46' 04''$$

$\therefore$
$$B = \frac{\cos \delta_1}{\cos \delta_2} = \frac{\cos 82° 06' 44''}{\cos 65° 46' 04''}$$
$$= 0 \cdot 3343595$$

Now, in Fig. 12.21
$$X^1 = 104° 23' 00'' - 74° 5' 25''$$
$$= 30° 17' 35''$$

$$\cot Z_2 = \frac{\cos 30° 17' 35'' - 0 \cdot 3343595}{\sin 30° 17' 35''}$$

$$\therefore \qquad Z_2 = 43° 37' 57''$$

$$\text{Azimuth of R.M.} = 43° 37' 57'' + 74° 5' 25''$$

$$= 117° 43' 22'' \text{ clockwise from North.}$$

3. Extra Meridian Altitude of (*a*) **a Star,** (*b*) **the Sun.** (*a*) This is a convenient method in that it can be carried out when stars are visible, using many repeated observations on different stars to give a mean result. The stars should be observed when slowly changing azimuth, i.e. near the observer's prime vertical, when angle PZX is therefore approximately $90°$. The readings are taken as follows—

(i) Bisect the reference mark and read the horizontal circle.

(ii) Set the horizontal cross-hair ahead of star, bisect with the vertical hair, and follow the star with the horizontal slow motion screw until the star is at the intersection of the cross-hairs. Read horizontal and vertical circles and the altitude level bubble.

(iii) Change face and repeat observation (ii) forthwith.

(iv) Direct back to reference mark reading the horizontal circle.

Accuracy will be increased if at least one other set of readings be taken on that star. To reduce errors due to refraction, it is good practice to pair, i.e. to repeat all observations, on another star placed similarly to the first star but on the other side of the observer's meridian. If possible, barometer and thermometer readings should be taken and the refraction correction computed from Bessel's Tables; otherwise the expression — $58'' \cot \alpha$ is used, as in the example given later.

(*b*) If the sun be observed, it is not so easy to bisect as a star, and it is usual to observe it in two opposite quadrants, 1, 2, of the cross-hairs, as indicated in Fig. 12.17. In this way the sun's semi-diameter in both altitude and azimuth will be eliminated. A second set, 3, 4, can then be taken immediately afterwards.

The procedure is similar to that described for the star observation, the horizontal hair being set in advance and the vertical hair kept in contact with the chosen limb of the sun, until both hairs are in contact. Apart from reading both circles and the altitude bubble, the time of contact is required since the sun's declination varies. The parallax correction must be applied to the altitude to reduce the observation to earth's centre. Pairing is of course only possible when the sun has reached a similar position on the other side of the observer's meridian, by which time refraction effects may have changed.

EXAMPLE. Observations taken on the sun at a place in N $53° 29' 19''$ gave the tabulated data. (Dec. $3° 25' 6''$ S at G.M.N. decreasing at $59''$ per hour.) One division of altitude bubble = $15''$. Determine the azimuth of R.M.

$$\text{Mean observed altitude} = \frac{22° 56' 10'' + 22° 12' 50''}{2}$$

$$= 22° 34' 30''$$

$$\text{Level correction} = \tfrac{1}{4}[3 \cdot 5 - 2 \cdot 5 + 4 \cdot 5 - 1 \cdot 5] \times 15''$$

$$= + 15''$$

Object	Face	Readings				Alt. O	Level E	G.M. Time (p.m.)
		Horizontal	Verniers	Vertical	Verniers			
RM	R	60° 00′ 00″	240° 00′ 00″					
Sun ☉⊢	R	191° 38′ 00″	11° 38′ 00″	22° 56′ 20″	22° 56′ 00″	3·5	2·5	3 hr 12 min 3 sec
Sun ⊢☉	L	12° 39′ 20″	192° 39′ 40″	22° 12′ 40″	22° 13′ 00″	4·5	1·5	3 hr 13 min 51 sec
RM	L	240° 00′ 00″	60° 00′ 00″					

Refraction correction $= -58''\cot 22° 34' 45''$
$$= -2' 19''$$

$$\text{Parallax} = +8\cdot8'' \cos \alpha$$
$$= +8\cdot8'' \cos 22° 32' 26''$$
$$= +8\cdot0''$$

$\therefore$ Corrected altitude $(\alpha_1) = 22° 32' 34''$

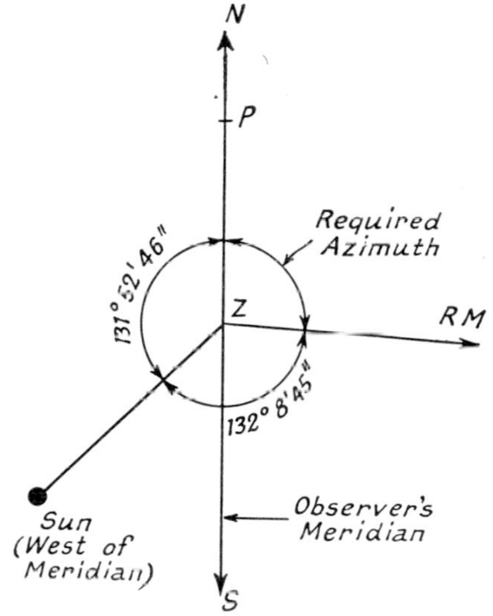

Fig. 12.22

Mean time of observation $= \frac{1}{2}[3 \text{ hr } 12 \text{ min } 3 \text{ sec} + 3 \text{ hr } 13 \text{ min } 51 \text{ sec}]$
$$= 3 \text{ hr } 12 \text{ min } 57 \text{ sec}$$
$$= 3\cdot216 \text{ hr}$$

$\therefore$ Declination of sun $(\delta) = 3° 25' 6'' - 3\cdot216 \times 59''$
$$= 3° 21' 56'' \text{ S}$$

Latitude of place $= 53° 29' 19'' \text{ N}$

$PZ = 90 - \phi = 36° 30' 41''$	$S - PZ = 62° 09' 20''$
$ZX = 90 - \alpha_1 = 67° 27' 26''$	$S - ZX = 31° 12' 35''$
$PX = 90 + \delta = 93° 21' 56''$	$S - PX = 5° 18' 05''$

$\therefore$
$$2S = 197° 20' 03''$$
$$S = 98° 40' 01''$$

Now
$$\tan \frac{Z}{2} = \sqrt{\frac{\sin(S - ZX)\sin(S - PZ)}{\sin S \sin(S - PX)}}$$

whence $\qquad\qquad Z = 131° \; 52' \; 46''$

Now mean horizontal angle between R.M. and sun

$$= \tfrac{1}{2}[131° \; 38' \; 00'' + 132° \; 39' \; 30'']$$
$$= 132° \; 08' \; 45''$$

∴ $\qquad$ *Azimuth of R.M.* $= 360° - 132° \; 08' \; 45'' - 131° \; 52' \; 46''$
$$\qquad\qquad\qquad = 95° \; 58' \; 29'' \text{ clockwise from north.}$$

Note : (i) Sun's southerly declination gives a co-declination o
$(90 + 3° \; 21' \; 56'') = 93° \; 21' \; 56''$; (ii) the observations, on a clock
wise reading instrument, place the sun "ahead" of R.M.; then the R.M
is placed relative to the sun as in Fig. 12.22; (iii) since observation
are taken after the sun's upper transit then the sun will be west of th
observer's meridian.

4. Extra Meridian Observation by Hour Angle. If the time b
accurately known and the instrument be in adjustment, azimuth ca
be determined from a knowledge of the hour angle of the body observed
When the observation is being carried out, time and horizontal circl
readings are noted at the instant of contact. Altitude is not require
if the instrument is in correct adjustment and so the vertical hair ca
be set in advance of the body: instrument face is changed as required
Various methods of solution are possible since in triangle PZX (Fig
12.13), angle P and sides PZ and PX are known, and in the example
given ZX is determined and in turn angle Z deduced.

The data given in the previous example are used again to determin
azimuth, but neglecting the altitude values. The two final results d
not give very close agreement, but this particular set of reading
formed part of the coursework of one group of students and the erro
is probably in the altitude measurement. The value for E must b
determined from the Star Almanac.

G.M.T. of observation	15 hr	12 min	57 sec
E	11 hr	50 min	06 sec
	27 hr	03 min	03 sec
Deduct	24 hr	00 min	00 sec
G.H.A. Sun	03 hr	03 min	03 sec
Deduct for Long. W		09 min	05 sec
∴ L.H.A. Sun	02 hr	53 min	58 sec

Whence $\qquad\qquad Z\hat{P}X = 43° \qquad 29' \qquad 30''$

Now $\qquad\quad \cos P = \dfrac{\cos ZX - \cos PX \cos PZ}{\sin PX \sin PZ}$

∴ $\qquad \cos ZX = \cos PX \cos PZ + \sin PX \sin PZ \cos P$

Now $PX = 93° 21' 56''$ and $PZ = 36° 30' 41''$

whence $ZX = 67° 26' 09''$

But $$\frac{\sin Z}{\sin PX} = \frac{\sin P}{\sin ZX}$$

$\therefore$ $$\sin Z = \frac{\sin 93° 21' 56''}{\sin 67° 26' 09''} \sin 43° 29' 30''$$

and so $Z = 131° 55' 31''$,

whereas by the previous method its value was $131° 52' 46''$.

Effects of Errors on Determination of Azimuth

In the first method described a circumpolar star was to be observed at elongation and the latitude of the station must be known. The azimuth of the star is determined from the expression

$$\sin Z = \frac{\cos \delta}{\cos \phi}$$

δ will be found from the Star Almanac and the error, dZ, which may occur in Z will be due to an error $d\phi$ in latitude. Then, on differentiating,

$$\cos Z \, dZ = - \frac{\cos \delta(- \sin \phi)}{\cos^2 \phi} \, d\phi$$

$\therefore$ $$dZ = \frac{\cos \delta \sin \phi}{\cos Z \cos^2 \phi} \, d\phi$$

$$= \frac{\sin Z \sin \phi}{\cos Z \cos \phi} \, d\phi$$

$$= \tan Z \tan \phi \, d\phi$$

Thus the greater Z or ϕ become, the larger is dZ for a given value of $d\phi$. The closer the circumpolar star to the celestial pole, the smaller is this maximum angle Z (see Fig. 12.20) and the smaller is dZ. On the other hand as the latitude of the station increases so dZ increases.

In the third method the latitude of the station together with the declination and altitude of the body are required.

(a) Let there be an error $d\phi$ in latitude.

Now $$\cos Z = \frac{\cos PX - \cos PZ \cos XZ}{\sin PZ \sin XZ}$$

$$= \frac{\sin \delta - \sin \phi \sin \alpha}{\cos \phi \cos \alpha}$$

Whence on differentiating, we have $- \sin Z \, dZ$

$$= \frac{\cos \phi \cos \alpha(- \sin \alpha \cos \phi) - (\sin \delta - \sin \phi \sin \alpha)(- \cos \alpha \sin \phi)}{\cos^2 \phi \cos^2 \alpha} \, d\phi$$

$$= \frac{- \sin \alpha \cos^2 \phi + \sin \delta \sin \phi - \sin \alpha \sin^2 \phi}{\cos^2 \phi \cos \alpha} \, d\phi$$

$$= \frac{\sin \delta \sin \phi - \sin \alpha}{\cos^2 \phi \cos \alpha}\, d\phi$$

$\therefore$ $\qquad$ $dZ = \dfrac{\sin \alpha - \sin \delta \sin \phi}{\sin Z \cos \alpha \cos^2 \phi}\, d\phi$

But $\qquad$ $\cos P = \dfrac{\cos ZX - \cos PX \cos PZ}{\sin PX \sin PZ}$

$$= \frac{\sin \alpha - \sin \delta \sin \phi}{\cos \delta \cos \phi}$$

$\therefore$ $\qquad$ $dZ = \dfrac{\cos P \cos \delta \cos \phi}{\sin Z \cos \alpha \cos^2 \phi}\, d\phi$

$$= \frac{\cos P \cos \delta}{\sin Z \cos \alpha \cos \phi}\, d\phi$$

Also since $\qquad$ $\dfrac{\sin Z}{\sin PX} = \dfrac{\sin P}{\sin ZX}$

Then $\qquad$ $\sin Z \cos \alpha = \sin P \cos \delta$

$\therefore$ $\qquad$ $dZ = \cot P \sec \phi\, d\phi$

(b) Let there be an error $d\delta$ in declination. This will arise in a sun observation if the recorded times are in error since the declination of the sun varies with time.

Again $\qquad$ $\cos Z = \dfrac{\sin \delta - \sin \phi \sin \alpha}{\cos \phi \cos \alpha}$

$$- \sin Z\, dZ = \frac{\cos \delta}{\cos \phi \cos \alpha}\, d\delta$$

$$= \frac{\sin Z}{\sin P \cos \phi}\, d\delta$$

$\therefore$ $\qquad$ $dZ = - \operatorname{cosec} P \sec \phi\, d\delta$

It will be observed that in case (a) or (b) for a given error $d\phi$ or $d\delta$, dZ increases as P decreases or as ϕ increases.

(c) Let there be an error $d\alpha$ in altitude.

Again $\qquad$ $\cos Z = \dfrac{\sin \delta - \sin \phi \sin \alpha}{\cos \phi \cos \alpha}$

$\therefore$ $\quad - \sin Z\, dZ$

$$= \frac{\cos \phi \cos \alpha\,(- \sin \phi \cos \alpha) - (\sin \delta - \sin \phi \sin \alpha)(- \cos \phi \sin \alpha)}{\cos^2 \phi \cos^2 \alpha}\, d\alpha$$

$$= \frac{- \sin \phi \cos^2 \alpha + \sin \delta \sin \alpha - \sin \phi \sin^2 \alpha}{\cos \phi \cos^2 \alpha}\, d\alpha$$

$$= \frac{\sin \delta \sin \alpha - \sin \phi}{\cos \phi \cos^2 \alpha}\, d\alpha$$

$$\therefore \quad dZ = \frac{\sin \phi - \sin \delta \sin \alpha}{\sin Z \cos \phi \cos^2 \alpha} \, d\alpha$$

But

$$(\sin \phi - \sin \delta \sin \alpha) = \cos \delta \cos \alpha \cos X$$

$$\therefore \quad dZ = \frac{\cos \delta \cos \alpha \cos X}{\sin Z \cos \phi \cos^2 \alpha} \, d\alpha$$

and since

$$\frac{\sin Z}{\sin PX} = \frac{\sin X}{\sin PZ}$$

then

$$\sin Z \cos \phi = \sin X \cos \delta$$

$$\therefore \quad dZ = \frac{\cos \delta \cos \alpha \cos X}{\sin X \cos \delta \cos^2 \alpha} \, d\alpha$$

$$= \cot X \sec \alpha \, d\alpha$$

dZ then increases as X decreases or as α increases. The body should thus be observed when away from the meridian for X to have as large a value as possible. α should not be lower than $20°$ when refraction correction values become difficult to assess.

Solar Prism. This device, designed by Professor Roelofs, is manufactured by Wild, and by means of an adaptor can be fitted to any make of theodolite.

Fig. 12.17 has shown the procedure adopted to eliminate the use of the correction for semi-diameter of the sun, but this device fitted at the objective end of the theodolite presents four overlapping images of the sun by means of semi-circular glass plates, which can be seen on looking into the solar prism. A small square is formed at the centre of the field of view which can be adjusted to the intersection of the horizontal and vertical lines. In this way the telescope can be pointed directly to the sun's centre and a more accurate sighting is possible.

Determination of the Meridian by the Gyroscope

The gyroscope has been used in navigation as a north-seeking device for a considerable period of time and certain manufacturers of surveying instruments are now producing units which allow the direct establishment of the meridian by theodolite without the need for calculations based on astronomical observations.

The development of the gyroscope for the precise transfer of bearings underground in mining surveying dates from about the beginning of the First World War, but an instrument capable of registering bearings to within one minute of arc did not appear until after the end of the Second World War. Even then it was of considerable weight, but advances in the design of gyroscopes for airborne inertial navigation systems have allowed much more compact and portable units. The total mass of the Precision Indicator of the Meridian Mk II manufactured by the British Aircraft Corporation is about 38 kg and an accuracy better than 10 seconds of arc is quoted by the manufacturer. Thomas, in an article in the *Chartered Surveyor* dated March 1965, has said that this accuracy can be improved in the observation programme. The Wild GAK1, a lighter and less sophisticated device, has a total mass of about 13·6 kg, and although its accuracy is not of the order of that of the P.I.M. MK II it is of the same order as that obtained in sun observations.

In each of the two instruments a rotor or spinner is driven at speeds in excess of 20,000 rev/min, the axis of spin being in a horizontal plane.

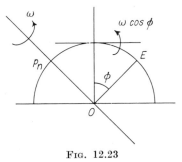

Fig. 12.23

Fig. 12.23 shows the earth rotating about its axis with an angular velocity of ω; at a place in latitude ϕ an angular velocity of $\omega \cos \phi$ obtains about the meridian. The spinner of a gyroscope set up at this point will try to maintain its initial spatial position provided that its angular momentum, I_ω is large, but the rotation of the earth itself pulls the spinner out of this plane. There is a consequent reaction in the form of a rotation, or precession, about the vertical (or output) axis of the spinner which holds until the spin axis lies in the meridian at the place. Fig. 12.24a shows the three axes of the spinner and Fig. 12.24b shows the spin axis at an angle of α to the meridian. The earth's rotation causes an interference torque (T) equal to $I_\omega \omega \cos \phi$, and a consequent precessional couple of $T \sin \alpha$ is induced which forces the spinner into the meridian with an angular rate of $\omega \cos \phi \sin \alpha$. When the spinner is suspended (as it is in the GAK 1 attachment) it will be seen that the maximum precessional couple occurs when ϕ is zero, i.e. at the equator, and that the gyro will float freely at the poles where $\cos \phi$ is zero. The precessing gyro does not align immediately on the meridian but oscillates about it.

GAK1 Attachment (Fig. 12.25a). This can be placed on instruments such as Wild T1A, T16 or T2 theodolites which must be suitably modified, and it is then connected to a control unit or converter. The oscillating system, containing the spinner and optical system with an index mark is carried by a tape suspended vertically within the "chimney" of the supporting case and fastened at the top. The converter contains a nickel–cadmium battery and an electronic unit;

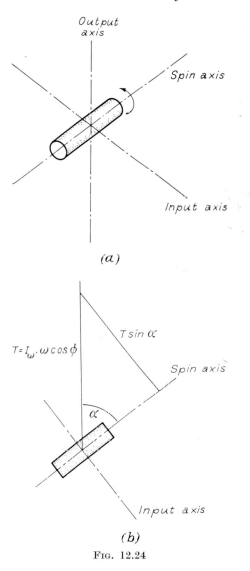

(a)

(b)

Fig. 12.24

the battery is connected by leads to a six-way adaptor on the supporting case, and is controlled by a selector switch. A further switch indicating "run" and "brake" is provided with two indicators, one of which, "measure," shows a green colour when the spinner is revolving at its working speed. When setting up, the theodolite telescope should be roughly pointed towards north by means of, say, prismatic compass or

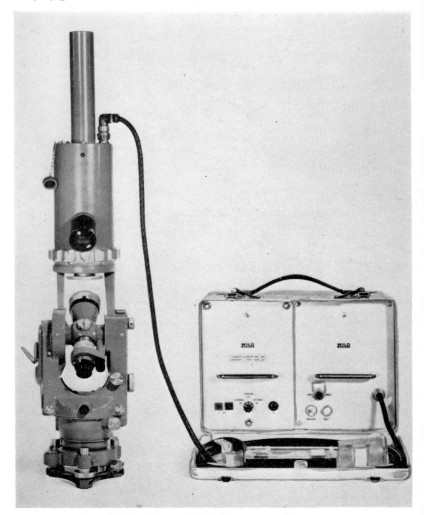

FIG. 12.25a. WILD GAK1 GYROSCOPE ATTACHMENT
(Courtesy of Wild Heerbrugg, Switzerland)

tubular compass observations, and the gyro attachment can then be fastened on the theodolite with the telescope *in the face-left position.* The cable from the converter unit is attached and the observer checks that the gyro is clamped by means of a device underneath the supporting case. The battery switch is set to "on" and when the other switch is set to "run" the "wait" indicator remains red for a short period of time before the green colour shows at "measure," indicating that the spinner is now running at its correct operational speed and that it can be unclamped. As the spinner oscillates about the meridian the gyro index mark can be seen moving across an auxiliary scale in an observation tube, and when the spin axis and the line of sight of the theodolite are in the same vertical plane that mark should be centred in the middle of the scale, which is also marked by a V-shaped index (Fig. 12.25*b*).

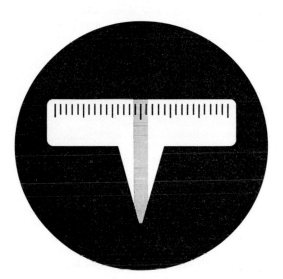

Fig. 12.25*b*. GAK1: Image of the Index and
Auxiliary Scale in the Observation Tube
(Courtesy of Wild Heerbrugg, Switzerland)

Thus if the mid-position of the oscillations is established, the telescope line of sight at the station is oriented towards the north, and the purpose of all observations is to determine that position. A calibration constant (E), which is the horizontal angle between the direction of the line of sight in the mid-position of oscillation and the meridian at a station, must however be determined for each instrument at frequent intervals. The calibration constants of gyrotheodolites can show a systematic drift caused by instrumental changes: Chrzanowski has suggested in an article "New Techniques in Mine Orientation Surveys", *Canadian Surveyor* 24 (I) 1970, that this be done immediately before and after the underground work and on the same day.

The form of oscillation is sinusoidal and in the middle of an oscillation the index mark is moving at maximum speed in the relevant field of view but it slows down quite markedly as the turning points are approached and for a very short period is stationary at such points. If the gyro index mark has been kept centered within the V-shaped index by using the horizontal motion of the theodolite alidade the corresponding horizontal circle reading of the theodolite can be obtained for this turning point.

Various methods of orientation are available including quick methods for pre-orientation before the adoption of a more precise method. In one of the quick methods the line of sight should be set within 15° of the estimated meridian and by following up the gyro index mark, with the V-shaped index in coincidence, horizontal circle readings are obtained for two successive turning points on opposite sides of the meridian. The mean horizontal circle reading gives the approximate meridian and the line of sight of the telescope can therefore be so established. It is essential that the two index marks be kept in coincidence so that the suspension tape is held free from torsion. Some initial practice is required for this and to facilitate the work some instruments are provided with a wider-range horizontal slow-motion screw. An accuracy of the order of $\pm$ 3' is quoted by Messrs. Wild for this method.

Two precise methods are available, one of which is an extension of the above when several successive turning points are observed, ensuring that the coincidence previously mentioned is maintained. The corresponding horizontal circle readings have to be obtained very quickly since the gyro index mark, relating to the spin axis, is only stationary for a very short period when it also has to be centred accurately on the zero division of the scale. Knowing the circle readings for pointings on the referring object (M, mean of readings on two faces) and the meridian (N), the two are related to give the geographical azimuth as ($M - N + E$). Four to six turning points are observed for highest accuracy.

EXAMPLE 1. Horizontal circle readings of 34° 20·8' and 214° 20·4' were measured face-left and face-right respectively by pointings on a reference mark from a station. Turning-point readings were given by gyro observation as follows—

Turning Point Left	Turning Point Right
353° 44·0'	357° 52·8'
353° 46·6'	357° 51·6'
353° 47·0'	357° 50·8'

If the calibration constant of the instrument is $+$ 2·1' determine the azimuth of the reference mark from the station.

The circle value for the mean of the oscillations can be derived by an approximate method known as the Schuler Mean. Taking three successive turning-point readings r_1, r_2, and r_3, this mean (r_0) is given by the expression

$$r_0 = \frac{r_1 + 2r_2 + r_3}{4}$$

Hence the following obtain—

TP Left	TP Right	Schuler Mean
353° 44·0′		
	357° 52·8′	355° 49·0′
353° 46·6′		355° 49·4′
	357° 51·6′	355° 49·2′
353° 47·0′		355° 49·1′
	357° 50·8′	————

Mean 355° 49·2′

Azimuth of R.M. = 34° 20·6′ + (360° 00·0′) − 355° 49·2′ + 2·1′
= 38° 33·5′ clockwise from north

Thomas, in the paper previously mentioned, suggests the use the four-point mean of

$$r_0 = \frac{r_1 + 3r_2 + 3r_3 + r_4}{8}$$

The second method (Fig. 12.26) which can be adopted is that of timing the transits of the gyro index mark through the middle of the scale by means of a stop-watch; at this instant the light slot on the

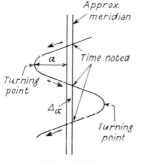

Fig. 12.26

gyro index mark is centred on the zero division of the scale. Now the gyro oscillations must be clamped to ensure that amplitudes (*a*) do not exceed the range of the auxiliary scale (15 divisions). To obtain an accuracy of, say, ± 15″ four or five transits are timed but the two index marks do not have to be kept in coincidence, the oscillation rhythm is not disturbed and the method is less tiring than the first method. In fact once the approximate meridian is established (and this must be to within ± 15′) the telescope remains clamped in this position and only the gyro is moving. Differences between successive transit times (Δ*t*) are evaluated and these are related to the angle (Δ*N*) between the approximate and actual meridians by means of a proportionality factor (c), which has to be obtained. Once established however, c can be considered an instrumental constant within a range of 1,000 km even though it actually depends upon latitude. Δt and ΔN are related by the expressions $\Delta N = c \cdot a' \cdot \Delta t$, providing that Δt is not greater than about 40 seconds. This ensures that the linear range of the sine curve obtains for the correction ΔN: a' is the mean of the amplitudes measured, left and right. Since the horizontal motion of the theodolite remains clamped, the upper clamp of the suspension tape of the gyro remains position-fixed during the measurements. The swing time is therefore influenced by the tape torsion, and the centre-line position of the oscillations is identified with the resultant of the earth's rotation torque and the tape torsion torque. This latter forces the centre-line position (a from turning point, Fig. 12.26) away from the true meridian.

EXAMPLE 2. Horizontal circle readings of a reference mark were determined at a station as 326° 34' 00" FL and 146° 34' 00" FR respectively and an approximate gyro orientation of 184° 12·5' was then established. Successive transit times were then observed as 00 min 00 sec, 03 min 38·4 sec (swinging right), 07 min 27·7 sec (swinging left), 11 min 06·2 sec (swinging right) and 14 min 55·7 sec (swinging left). If a mean amplitude of 9·1 scale divisions was measured during these observations, find the azimuth of the reference mark. Take the proportionality factor (c) as 0·047'/sec and the instrument calibration value (E) as $-$ 2·1'.

Time of Transit	Swing Time (+ ve left)	Time Difference Δt
00 min 00·0 sec		
	$-$ 03 min 38·4 sec	
03 min 38·4 sec		$+$ 10·9 sec
	$+$ 03 min 49·3 sec	
07 min 27·7 sec		$+$ 10·8 sec
	$-$ 03 min 38·5 sec	
11 min 06·2 sec		$+$ 11·2 sec
	$+$ 03 min 49·5 sec	
14 min 55·7 sec		

Taking $\Delta t = 11\cdot0$ sec, then $\Delta N = + \ 0\cdot047 \times 9\cdot1 \times 11\cdot0 = + \ 4\cdot7'$

It will be noted that the signs of Δt are related to those of the swing times, that ΔN takes the same sign as Δt, and that in so far as these readings are concerned, the approximate and correct meridians are in the reverse position to those on Fig. 12.26.

The corrected azimuth of the R.M. is then given as $M - N + E$

$$= 326° \ 34' \ 00" - (184° \ 12\cdot5' + 4\cdot7') - 2\cdot1'$$
$$= 142° \ 14\cdot7'$$

Messrs. Fennel of Kassel also manufacture a Gyrotheodolite KT3 (Fig. 12.27), and a Gyrotheodolite KT4 which can be adapted to fit any theodolite. As in the GAK1, the gyro is suspended in the horizontal plane by a tape and its movements are similarly observed when establishing the meridian. The KT3 instrument, consisting of the upper

FIG. 12.27. FENNEL GYROTHEODOLITE KT3
(Courtesy of Otto Fennel G.m.b.H.)

gyro part, rotatable horizontal circle read by a scale microscope, and a periscope telescope is mounted on a levelling head; by interchanging with targets in the levelling heads the KT3 instrument can be used in the three-tripod system of traversing if required.

PRECISION INDICATOR OF THE MERIDIAN MK II. The spinner in this instrument is a synchronous hysteresis-type motor contained in a sealed cylindrical gimbal, pivoted inside a hermetically sealed case (Fig. 12.28). A special oil, maintained at 71°C, is contained inside the

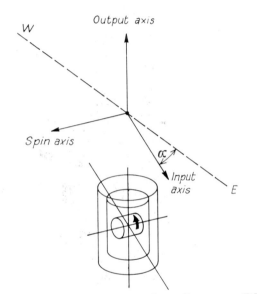

FIG. 12.28. PRINCIPLE OF THE GYRO SYSTEM OF P.I.M.

narrow gap between the gimbal and the case: this floats the gimbal at neutral buoyancy, provides damping and acts as a shock protector. The unit is mounted underneath an adapted theodolite (which can be a Hilger and Watts Microptic No. 1 instrument) so that the vertical axis of the gimbal, and the output axis of the spinner, lie in the vertical axis of the theodolite and the input axis lies parallel to the telescope longitudinal axis.

The gimbal is linked electromagnetically to the outer case by means of a pick-off and a torque motor so that if the gimbal rotates about its vertical axis away from its position of electrical equilibrium, the pick-off transmits a signal which is amplified and fed back to the torque meter, thereby ensuring a balance and preventing further rotation of the gimbal. For a fuller discussion of the behaviour of the gimbal, the student is referred to a paper by Thomas in the *Proceedings of the Third South African National Survey Conference*, 1967. Providing that the theodolite has been carefully set up the balancing torque is equal to $I_\omega \omega \cos \phi \sin \alpha$ and this can be suitably registered on a meter in the control unit established nearby.

After setting up, the instrument is switched on so that the working temperature is achieved and in this period the reference mark could be sighted. Then the telescope may be directed to within a few degrees of north, using perhaps a compass, or alternatively the meter reading. In this latter case the reading will be proportional to $\cos \phi \sin \alpha$, and since the spin axis is at right angles to the line of sight then α will be about 90°, $\sin \alpha$ will be virtually unity and the meter reading will therefore be a maximum when the telescope is roughly so aligned. The meter can then in fact be set to read its full scale deflection to give $\sin \alpha$ on a scale reading from $- 25$ divisions through zero to $+ 25$ divisions, i.e. $\sin \alpha/25$ per division. This calibration which includes for latitude, is effected by means of a "scale factor switch" which must always be turned to "Adjust" when the theodolite is being rotated in azimuth. The calibration is carried out in the second position of the switch and two other positions allow "finer" readings later, each division then being equivalent to 10 minutes of arc and 1 minute of arc respectively. The theodolite is now turned counter-clockwise to point on a West heading so that the spin axis is headed towards North, and, by actuating the scale factor switch so as to introduce the two other ranges of meter readings, a correction can be applied to the circle reading of the theodolite. Further counter-clockwsie rotation puts the line of sight on an east heading and fine-reading corrections are again established. This forms a "pair," thereby allowing the elimination of systematic bias of the gyroscope, which causes a null meter-reading to be given when the spin axis is not actually in the meridian. It is good practice to repeat the above but rotating in a clockwise direction to obtain a quartet. With repetitions of quartets, British Aircraft Corporation, who manufacture the PIM state that azimuth can be established to within 10 seconds of arc.

Astronomical and Geodetic Azimuth. In this chapter, and elsewhere, the earth has been considered as a sphere whereas it actually approximates to a spheroid, i.e. the solid produced when an ellipse is rotated about its minor axis.

The *geoid*, or mean sea level surface (assumed continued through land masses) is very nearly a spheroid; the ocean surface when considered motionless gives an equipotential surface of gravitation, a plumb-line being perpendicular to it at any point. As mentioned in Chapter 3, heights can be determined relative to the geoid by levelling. However, since it is not exactly a spheroid, the geoid is not taken to be suitable for geodetic computation, and a spheroid of reference is adopted.

Observations have been made at various times in various parts on the surface of the earth, and the form of the spheroid of reference indicated by these observations varies somewhat, though not greatly. As a result a spheroid of reference may be chosen to suit the area under consideration. Spheroidal or geodetic co-ordinates of latitude and longitude may be related to an origin as on the celestial sphere.

When the astronomical azimuth is being found, the instrument is set up with its vertical axis in the direction of gravity, and this axis in general does not coincide with the normal to the spheroid of reference. The angle which the normal makes with the plane of the equator gives the geodetic latitude. Geodetic azimuth can be deduced from astronomical azimuth by means of the Laplace equation which states that the difference between astronomical and geodetic azimuth is equal to the product of the sine of the geodetic latitude and the difference between astronomical and geodetic longitudes.

A control of azimuth errors is thus possible in a large triangulation survey, and the stations at which this control occurs are Laplace stations.

Over long distances the azimuth of a great-circle line AB will change from α to $\alpha + \delta\alpha$ if A and B are at different latitudes, since the meridians at A and B converge at the poles; it can be shown that $\delta\alpha = \delta\theta \sin\dfrac{\phi_A + \phi_B}{2}$ where $\delta\theta$ is the difference in longitude between A and B. Fig. 12.29a refers.

In so far as relatively short lines (up to 40 km, say) are concerned, lengths and azimuths in respect of points of known latitude and longitude can be estimated by considering the meridians and parallels of latitude through A and B to give rectangular axes. The separation distances between the axes are then denoted by $\lambda \cdot (\phi_B - \phi_A)$ and $\mu \cdot \delta\theta$ respectively, where λ and μ are the lengths of $1''$ of latitude and $1''$ of longitude at the mean latitude of $\dfrac{\phi_A + \phi_B}{2}$.

Thus in Fig. 12.29b we have

$$\lambda\delta\phi = l \cos\left(\alpha + \frac{\delta\alpha}{2}\right)$$

and

$$\mu\delta\theta = l \sin\left(\alpha + \frac{\delta\alpha}{2}\right)$$

These formulae are reasonably valid for latitudes not exceeding 60°. If the latitudes and longitudes of the terminal stations are given then $\delta\alpha$, α, and l can be determined directly. In other cases an indirect approach may be needed as shown in the following example.

EXAMPLE. A line AB of length 10,854 m has an azimuth of 46° 21′ 20″ at A in latitude 52° 20′ 15″ N and longitude 24° 28′ 50″ E. Estimate the reverse azimuth at B.

Latitude	Length of 1″ of Latitude	Length of 1″ of Longitude
52° 20′	30·9022 m	18·9364 m
52° 25′	30·9107 m	18·9008 m

It will be noted that $\dfrac{\phi_A + \phi_B}{2}$ is not known and therefore some approximations must be made.

In the first instance we can assume that the latitude of A is nearly equal to the mid-latitude of AB, i.e. 52° 20′ 15″ N.

Thus $\quad\quad\quad \lambda = 30 \cdot 9022 + \dfrac{15}{300} \times 0 \cdot 0085$

$\quad\quad\quad\quad\quad = 30 \cdot 9026 \text{ m}$

Then the approximate latitude difference

$$= \frac{l \cos \alpha}{\lambda} = \frac{10{,}854 \cos 46° \ 21′ \ 20″}{30 \cdot 9026}$$

$$= 242 \cdot 4″, \text{ neglecting } \delta\alpha.$$

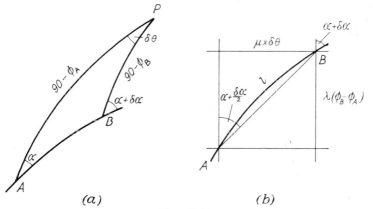

(a) (b)

Fig. 12.29

The approximate mid-latitude therefore

$\quad\quad\quad\quad = 52° \ 20′ \ 15″ + 121″$

$\quad\quad\quad\quad = 52° \ 22′ \ 16″ \text{ to the nearest second.}$

For this value

$$\mu - 18 \cdot 9364 - \frac{136}{300} \times 0 \cdot 0356$$

$$= 18 \cdot 9203 \text{ m}$$

and $\quad\quad\quad \delta\theta = \dfrac{10{,}854 \sin 46° \ 21′ \ 20″}{18 \cdot 9023}$

$\quad\quad\quad\quad\quad = 415 \cdot 1″, \text{ neglecting } \delta\alpha.$

This allows an assessment of $\delta\alpha$ to now be made since

$$\delta\alpha = \delta\theta \ . \ \sin \frac{\phi_A + \phi_B}{2}$$

$$= 415 \cdot 1″ \times \sin 52° \ 22′ \ 16″$$

$$= 328 \cdot 8″$$

$$\therefore \qquad \frac{\delta\alpha}{2} = 164\cdot4'', \text{ or } 164'' \text{ to the nearest second.}$$

Now
$$\delta\theta = \frac{10{,}854 \sin (46°\ 21'\ 20'' + 02'\ 44'')}{18\cdot9203}$$

$$= 415\cdot4'', \text{ as a further estimation,}$$

with
$$\delta\alpha = 415\cdot4'' \times \sin 52°\ 22'\ 16'' = 329''$$

A check can be made at this stage to revise the first approximation of latitude difference, and the student can check that $\delta\phi$ is now about $242\cdot2''$ which still implies a mid-latitude of $52°\ 22'\ 16''$ to the nearest second. Therefore the value of $\delta\alpha = 05'\ 29''$ can be accepted, to give the azimuth of AB at B equal to $46°\ 26'\ 49''$ and the reverse bearing (or the azimuth of BA) at B as $226°\ 26'\ 49''$.

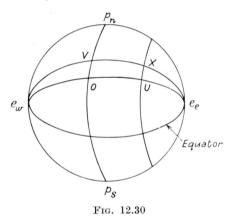

Fig. 12.30

PROJECTION AND NATIONAL GRID

The position of a point can be defined by its latitude and longitude, the "axes" being the equator and the Greenwich meridian respectively. However, if desired, the meridian could pass through another point, and another origin, O, could be established on that meridian. Thus point X (Fig. 12.30) could be located by its perpendicular distance VX along the great circle e_wVe_e from p_sOp_n and its distance OV from O along that meridian. VX can be defined as the easting of X, and OV as the northing of X in this system. All points having the same easting as X would lie on the small circle XU and all points having the same northing would lie on e_wVe_e, but it will be apparent that the latitudes of points on e_wVe_e will vary. Thus northing co-ordinates measured from e_wOe_e along the relevant small circles, e.g. UX, would be smaller than distances measured along p_sOp_n. It will be seen therefore that from this aspect alone, if p_sOp_n and e_wOe_e were laid down as straight lines mutually at right angles and in a plane, X would have to be displaced so that OV was its actual northing co-ordinate, and not UX.

Plane rectangular axes are however very convenient for recording the positions of points by rectangular co-ordinates, and the National Grid reference system of the Ordnance Survey consists of a series of lines parallel to the adopted central meridian with a series of lines at right angles thereto, to give a square grid. The positions of the tri- angulation stations in Great Britain are defined by their National Grid co-ordinates to 0·001 m, and are published in list form by the Ordnance Survey. This grid was obtained by a Transverse Mercator Projection, the origin being at 49° N on the central meridian through 2° W: the Airy spheroid of reference was taken to represent the earth. In the basic projection the spheroid would be represented on a cylinder touching it along the central meridian, and the scale would be constant along that meridian but would increase to either side. However, in the actual projection some adjustments were made, resulting in the scale being correct on lines nearly parallel to the central meridian, and about two-thirds towards the edges of the projection.

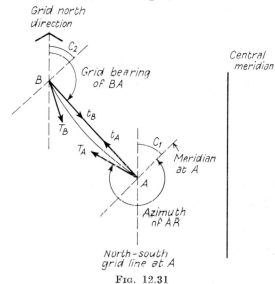

Fig. 12.31

The north–south lines on the grid are parallel to the central meridian throughout, and therefore grid north at any point off the meridian will not be the same as true north at that point. Also since the grid is square (being based on 10 km sides) the scales of distance along the lines perpendicular to the central meridian had to be increased to agree with the scales parallel to that meridian. The scale is therefore constant in all directions at any one point, although it changes slightly with distance from the central meridian, and the projection is said to be orthomorphic. To convert true distances (S) measured along the spheroid to grid distances (s) derived from the grid co-ordinates a scale factor (F) must be applied in accordance with the expression

$s = S \times F$. Values of F are listed in *Constants, Formulae and Methods used in Transverse Mercator Projection* published by H.M.S.O., F being 1·00000 at easting distances of 180 km from the central meridian and 0·99960 on the meridian. It must be pointed out that the National Grid has in fact been based for convenience on a false origin 400 km to the west and 100 km to the north of the 49° N, 2° W origin so that all eastings will in fact be positive (at 2° W they are 400 km) and no northing will exceed 1,000 km.

Furthermore, to convert grid bearings to true azimuths a convergence factor C must be introduced to allow for the angle between the north–south grid line and the meridian at a point. In addition, for very long sighting distances and in very accurate work, an extra correction known as the $(t - T)$ correction is to be applied, to allow for the fact that the line of sight between two points will be effectively curved in the projection as in Fig. 12.31. Data for these corrections can be found in *Projection Tables for the Transverse Mercator Projection of Great Britain* published by H.M.S.O.

EXERCISES 12

1. Find the altitude of the sun (declination N 23° 15′) at upper transit at a place in latitude 53° 30′ N. At what latitude in the southern hemisphere will the sun have the same meridian altitude at that instant?
Answer: 59° 45′ : 7°S.

2. Find the latitude of a place in the northern hemisphere at which a star of declination N 6° 18′ 34″ will have an altitude at upper transit of 45° 3′ 4″. Will this star set?
Answer: 51° 15′ 30″ N; Yes.

3. At what altitudes will Capella (declination N 45° 57′ 22″) cross the meridian of an observer in latitude 54° 15′ 20″ N?
Answer: 81° 42′ 02″; 10° 12′ 42″.

4. Find the G.S.T. at 0600 hr G.M.T. if G.S.T. at G.M.N. on that day was 14 hr 29 min 7·9 sec.
Answer: 8 hr 28 min 08·8 sec.

5. Find L.S.T. at L.M.N. in longitude 60° E if G.S.T. at G.M.N. is 14 hr 29 min 7·9 sec.
Answer: 14 hr 28 min 28·5 sec.

6. If L.S.T. at L.M.N. at a place in 30° W is 9 hr 25 min 53·0 sec, find L.S.T. at L.M.T. 1,600 hr at a place 45° E.
Answer: 13 hr 25 min 43·2 sec.

7. Find G.M.T. of upper transit of Betelgeuse (RA 5 hr 52 min 43·2 sec) at a place in longitude 3° 30′ W if G.S.T. at G.M.M. on that day was 6 hr 40 min 26·5 sec.
Answer: 23 hr 22 min 25·9 sec.

8. Compute longitude and latitude from the following data—

   (a) observed altitude of lower limb of sun when crossing south meridian = 43° 25′ 20″,

   (b) vernier arm bubble readings 0, 8; E, 6; angular value of one division, 12″,
   (c) time of observation, 9 hr 14 min 24 sec a.m. G.M.T.,
   (d) declination of sun at G.M.N. is 14° 11′ 18″ N, increasing 35″ per hour,
   (e) equation of time at G.M.N., 4 min 10 sec (add to A.T. to give M.T.), decreasing numerically 0·42 sec/hr,

(*f*) sun's semi-diameter 15′ 45″. Horizontal parallax 8·9″. Refraction 58″ cot α. (*L.U., B.Sc.*)
Answer: Long. 42° 26′ 48″ E. Lat. 60° 29′ 20″ N.

9. Explain, using sketches where necessary, the Declination, Right Ascension, and Hour Angle of a star.
A star was observed to reach an altitude of 50° 20′ 30″ at 9 hr 46 min 16 sec G.S.T., and to return to the same altitude at 10 hr 58 min 20 sec, G.S.T. The R.A. of the star was 10 hr 6 min 6 sec. Determine the longitude of the observer. (*L.U., B.Sc.*)
Answer: 04° 03′ W.

10. In order to determine the azimuth of a survey line *XY*, a theodolite was set up at *X*, west of Greenwich, and an afternoon extra-meridian observation made on the Sun. Using the following data, determine the azimuth of the line—
Corrected altitude of Sun's centre, 17° 38′ 11″.
G.M.T. of observation, 15 hr 18 min 14 sec.
Sun's declination at previous G.M.M., 9° 46′ 47·8″ S, decreasing by 1,327″ per day.
Latitude of station *X*, 51° 32′ 25″ N.
Horizontal angle from Sun to station *Y*, 55° 26′ 28″ (measured clockwise).

$$\tan^2 \frac{A}{2} = \frac{\sin (s - b) \,.\, \sin (s - c)}{\sin s \,.\, \sin (s - a)}$$

where $2s = a + b + c$

(*L.U., B.Sc.*)

Answer: 282° 35′ 48″.

11. A star (R.A. 15 hr 22 min 6 sec, Dec. 16° 19′ 14″ N) is to be observed in the east from a station in latitude 50° 51′ 00″ N. Find the azimuthal angle from the meridian plane when the star is at an altitude of 34°. (*L.U., B.Sc.*)
Answer: 106° 57′ 20″.

12. A star (R.A. 10 hr 5 min 5 sec, Dec. 12° 20′ 02″ N) is to be observed in the east from a station in latitude 42° 29′ 45″ N and longitude 4° 48′ 30″ W. Determine G.S.T., when the star is at a true altitude of 43°. (*L.U., B.Sc.*)
Answer: 7 hr 37 min 27·8 sec.

13. Explain what you understand by (*a*) Apparent Time; (*b*) Local Mean Time; (*c*) Sidereal Time; (*d*) Equation of Time. Find the Local Sidereal Time of western elongation of η Draconis on 10th June, 1945, at a place in latitude 55° 01′ 20″ N, given that the R.A. of the star was 16 hr 23 min 17 sec, and the Declination was 61° 38′ 23″ N on that date. (*L.U., B.Sc.*)
Answer: 19 hr 01 min 18·5 sec.

CHAPTER 13

THEORY OF ERRORS

IN CHAPTER 2, the differences between mistakes, systematic errors and accidental errors were discussed in general terms. It was pointed out that surveying work should be free from mistakes, and that systematic errors, where known, should be corrected for; this chapter concerns itself, therefore, with small random accidental errors. These are equally likely to be positive or negative, considered with respect to the "true" value of the quantity being measured. In precise surveying, knowledge of such errors and of their adjustment becomes important.

Distributions

If a large number of measurements are made of an angle or line all with equal precision, a range of results will be obtained. Consider as an example the following measurements of an angle which were taken in a test on a Watts Microptic Theodolite No. 2 (ST 200).*

Mean Angle	Mean Angle	Mean Angle	Mean Angle
94° 4′ 19·00″	94° 4′ 20·00″	94° 4′ 22·25″	94° 4′ 21·00″
94° 4′ 21·57″	94° 4′ 20·00″	94° 4′ 19·50″	94° 4′ 20·50″
94° 4′ 22·50″	94° 4′ 21·25″	94° 4′ 19·75″	94° 4′ 20·50″
94° 4′ 21·25″	94° 4′ 21·25″	94° 4′ 20·50″	94° 4′ 20·50″
94° 4′ 20·25″	94° 4′ 22·00″	94° 4′ 18·75″	94° 4′ 20·25″
94° 4′ 21·50″	94° 4′ 21·50″	94° 4′ 18·75″	94° 4′ 20·50″
			94° 4′ 19·00″

If we divide the range of variation into a number of groups and count the number of observations falling in each group, the results can be plotted as a *histogram,* as in Fig. 13.1. The horizontal axis is divided into segments corresponding to the ranges of the groups, and on each segment is constructed a rectangle whose area is proportional to the group frequency. In the example, since the range is the same for all the groups, the heights of the rectangles are proportional to the frequencies. It is a reasonable supposition, and found to be so in practice, that if the number of observations was made very much larger and the groups much narrower, the irregular shape would become a smooth curve as shown dotted in Fig. 13.1. Such a curve is referred to as a frequency curve.

If the results are expressed as proportional frequencies as in Fig. 13.2, the curve can then be referred to as the probability distribution. The probability of the occurrence of a given observation is defined as the proportional frequency with which it occurs in a large number of observations.

In most cases, the observed results are the only direct information available, and the problem is to deduce the probability distribution.

* Hilger & Watts Technical Report BS 7.

This is only possible if a very large number of observations have been made, whereas in most cases, relatively few are available. It is often possible, however, to make an assumption as to the general form of the probability distribution with a fair degree of confidence, using as a basis either past experience or knowledge of similar situations. In

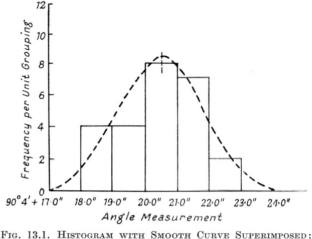

Fig. 13.1. Histogram with Smooth Curve Superimposed: Tests on Watts Microptic No. 2 Theodolite

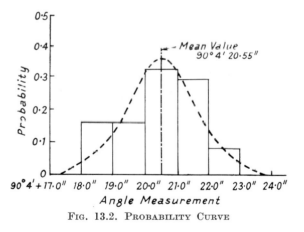

Fig. 13.2. Probability Curve

surveying where we are concerned with direct measurement, the distribution can be assumed to be of the normal (or Gaussian) type.

This is in fact the most commonly used of all the theoretical distributions; it gives a bell-shaped probability curve, and is expressed mathematically as

$$dp = \frac{1}{\sigma\sqrt{2\pi}} e^{-(x_1-\mu)^2/2\sigma^2} dx . \qquad . \qquad . \quad (13.1)$$

For the derivation of this expression, the student is referred to any book on statistics (*see*, for example, *Statistical Methods for Technologists*, by Paradine and Rivett, E.U.P.). If the data are normally distributed, the probability that x will assume a value between x_1 and $(x_1 + dx)$ is then given by eqn. (13.1) in which μ is the true mean of the Universe or population and σ is the Universe standard deviation.

The distribution is thus defined by μ and σ, where μ is the central value and σ is a measure of the spread. The population (or Universe, both terms being used by statisticians) is the complete set of values of a given variate. The number of individual values is generally very large or even infinite.

Arithmetic Mean

In practice, μ and σ are now known and have to be estimated from the limited observations taken. Instead of the Universe mean μ, we use the arithmetic mean of the observations—

$$\bar{x} = \frac{\Sigma x}{n} \qquad . \qquad . \qquad . \qquad . \qquad (13.2)$$

where n is the sample size
and where $\bar{x} \to \mu$ as $n \to \infty$.

Eqn. (13.2) can be written in the form

$$\bar{x} = x_0 + \frac{\Sigma(x - x_0)}{n} \qquad . \qquad . \qquad . \qquad (13.3)$$

where x_0 is a datum quantity lower in value than the smallest observation.

This method of calculating mean values is particularly useful when dealing with angle measurement. Thus the mean of the angles given earlier in this chapter may be obtained, by putting $x_0 = 90° 04' 18''$, as

$$90° 04' 18'' + [(1 + 3\cdot57 + 4\cdot5 + 3\cdot25 + 2\cdot25 + 3\cdot5$$
$$+ 2 + 2 + 3\cdot25 + 3\cdot25 + 4 + 3\cdot5$$
$$+ 4\cdot25 + 1\cdot5 + 1\cdot75 + 2\cdot5 + 0\cdot75 + 0\cdot75$$
$$+ 3 + 2\cdot5 + 2\cdot5 + 2\cdot5 + 2\cdot25 + 2\cdot5 + 1)$$
$$\div 25]$$
$$= 90° 04' 18'' + [63\cdot82 \div 25] = 90° 04' 18'' + 2\cdot55''$$
$$= 90° 04' 20\cdot55''$$

Standard Deviation

The standard deviation of the Universe is given by

$$\sigma = \sqrt{\frac{\Sigma(x - \mu)^2}{n}} \qquad . \qquad . \qquad . \qquad (13.4)$$

Since μ is not known, the arithmetic mean, $\bar{x}$, of the sample must be used. It can be shown that $\Sigma(x - \bar{x})^2$ is less than $\Sigma(x - \mu)^2$, and to correct for the consequent underestimate of the standard deviation,

$(n - 1)$ is used in place of n in the divisor. The estimate of the population standard deviation derived from the sample is denoted by s, where

$$s = \sqrt{\frac{\Sigma(x - \bar{x})^2}{n - 1}}. \qquad \cdot \qquad \cdot \qquad \cdot \qquad (13.5)$$

in which $(x - \bar{x})$ is termed a *residual* to distinguish it from the error of a reading, which is its deviation from the true (but unknown) mean, μ. Thus as

$$n \to \infty, \quad s \to \sigma$$

The square of the standard deviation is known as the variance, V, and is also used as a measure of dispersion or spread; but in surveying, the standard deviation is most commonly met with. It has the same dimensions as the variable and is thus easier to comprehend as a measure of the spread of the distribution.

The computation of the term $\Sigma(x - \bar{x})^2$ may give rise to awkward numbers and use should be made of the identity

$$\Sigma(x - \bar{x})^2 = \Sigma x^2 - \frac{(\Sigma x)^2}{n} \qquad \cdot \qquad \cdot \qquad \cdot \qquad (13.6)$$

Put $S = \Sigma(x)$; then

$$s = \sqrt{\frac{\Sigma x^2 - (S^2/n)}{n - 1}} \qquad \cdot \qquad \cdot \qquad \cdot \qquad (13.7)$$

$$= \sqrt{\frac{n\Sigma x^2 - S^2}{n(n - 1)}}$$

Since S has already been determined in the computation of $\bar{x}$, the arithmetic mean is reduced, and the device given in eqn. (13.3) can be used to reduce it further. As an example, the standard deviation of the observations given earlier is calculated.

Taking a datum value, x_0, of $90°\ 04'\ 18''$, the working values in seconds are as shown in table on p. 468.

Standard and Probable Error

If in eqn. (13.1) the substitution is made

$$u = \frac{(x - \mu)}{\sigma}$$

then

$$dp = \frac{1}{\sqrt{2\pi}}\ e^{-u^2/2}\ du \qquad \cdot \qquad \cdot \qquad \cdot \qquad (13.8)$$

This is known as the standardized form of the normal equation; if the value of u is known, dp/du is determined uniquely. Eqn. (13.8) represents the probability of occurrence of observations for which u lies between u and $u + du$, and by integration gives the total probability of observations having u less than a given value. The definite integral has been evaluated for a wide range of values of u. The normal curve (Fig. 13.3) is symmetrical about $u = 0$, the areas on each side being equal to $\frac{1}{2}$ (the total area under the curve must equal 1). It is thus only necessary to consider positive values of u, and in Fig. 13.3 the shaded area α to

x_0	$(x - x_0) = X$	$(x - x_0)^2 = X^2$
90° 04′ 18″	7·00	1·00
	3·57	12·74
	4·50	20·25
	3·25	10·56
	2·25	5·06
	3·50	12·60
	2·00	4·00
	2·00	4·00
	3·25	10·56
	3·25	10·56
	4·00	16·00
	3·50	12·60
	4·25	18·06
	1·50	2·25
	1·75	3·06
	2·50	6·25
	0·75	0·56
	0·75	0·56
	3·00	9·00
	2·50	6·25
	2·50	6·25
	2·50	6·25
	2·25	5·06
	2·50	6·25
	1·00	1·00

$$S = 63\cdot82 \qquad \Sigma X^2 = 190\cdot73$$

$$S^2 = 63\cdot82^2 = 4{,}072\cdot99, \text{ and } n\Sigma X^2 = 4{,}768\cdot25$$

$$\therefore \qquad s = \sqrt{\frac{n\Sigma X^2 - S^2}{n(n-1)}}$$

$$= \sqrt{\frac{695\cdot26}{600}}$$

$$= 1\cdot07''$$

Note that the mean can be calculated from the same computation.

---

the right of $+ u_1$ is the probability that u will be greater than or equal to u_1. It is also the probability that u will be less than or equal to $- u_1$ and the total probability that u will lie outside the range $+ u_1$ to $- u_1$ is 2α. For $u = 3\sigma$, the probability that a value will lie outside the range 3σ to $- 3\sigma$ is about 0·003 or 1 chance in 400, and for $u = 2\sigma$ the probability is about 0·045 or about 1 chance in 20. The extent to which a single observation may depart from its true value, therefore, is measured by the standard deviation.

In addition it can be shown that if the population is normal, with mean μ and standard deviation σ, then the mean of a random sample (e.g. a set of independent observations made under uniform conditions) is also normally distributed with mean μ and standard deviation $\sigma/\sqrt{n}$. This is an indication of how the reliability of the mean is related to sample size.

The standard deviation of the mean is usually referred to as its *standard error* and this is inversely proportional to the square root of n.

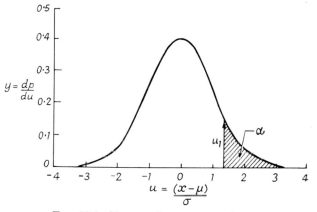

FIG. 13.3. NORMAL PROBABILITY CURVE

If the sample is increased 4 times, the standard error of the mean is halved and this in turn is a measure of the precision of the estimate. Thus from the equations already given,

S.E.$_{\bar{x}}$ = standard error of the mean

$$= \sqrt{\frac{\Sigma(x - \bar{x})^2}{n(n - 1)}} \qquad . \qquad . \qquad . \qquad . \qquad . \qquad (13.9)$$

S.E.$_x$ = standard error of a single observation

$$= \sqrt{\frac{\Sigma(x - \bar{x})^2}{n - 1}} \qquad . \qquad . \qquad n \qquad . \qquad . \qquad (13.10)$$

$$= \text{S.E.}_{\bar{x}} \times \sqrt{n}$$

It will be observed that standard deviation and standard error of a single observation are similarly defined by equations (13.5) and (13.10), and in fact the two terms seem to be used synonymously by surveyors.

For estimates of a population mean derived from samples of the same size, the standard error of the estimates is that error which is numerically exceeded with a relative frequency of about one in three in a large number of trials. In surveying it has been the convention for many years to estimate the value of the *probable error*, E, such that the probability of an error greater than E is $\frac{1}{2}$. From a study of tabulated values of eqn. (13.8) it can be seen that E will have a value of 0·6745 times the standard deviation—

$$E_{\bar{x}} = \text{probable error of the mean}$$

$$= 0·6745 \sqrt{\frac{\Sigma(x - \bar{x})^2}{n(n - 1)}} \qquad . \qquad . \qquad . \qquad (13.11)$$

$$E_x = \text{probable error of a single observation}$$

$$= 0·6745 \sqrt{\frac{\Sigma(x - \bar{x})^2}{n - 1}} \qquad . \qquad . \qquad . \qquad (13.12)$$

It is convenient to replace 0·6745 by $\frac{2}{3}$.

Thus in the example given previously, we have

$$\text{S.E.}_{\bar{x}} = \frac{1·07''}{\sqrt{25}} = 0·21''$$
$$\text{S.E.}_x = 1·07''$$
$$E_{\bar{x}} = \tfrac{2}{3} \times 0·21'' = 0·14''$$
$$E_x = \tfrac{2}{3} \times 1·07'' = 0·72''$$

Thus the estimated mean might be given as $\bar{x} \pm E_{\bar{x}}$ (in the example, $90° \ 04' \ 20·55'' \pm 0·14''$), implying that we expect to be right in 50 per cent of cases in saying that the true mean lies within this range. The mean might also be given as $\bar{x} \pm 2$ (S.E. $_{\bar{x}}$), implying that we expect to be right in 95 per cent of cases. It might be mentioned here that statisticians nowadays use standard error, and tend to deplore the use of probable error.

Rejection of Observations. The assumptions made in the preceding paragraphs are that we are considering small random errors which are normally distributed, and from a consideration of the normal distribution it can be said that the chance of a result deviating from the mean by more than three times the standard deviation is about 1 in 400, and by more than four times the standard deviation, 1 in 10,000. Thus when the result deviates by a large amount from the mean, there is a possibility that unusual circumstances have operated and that the result does not belong to the same normal distribution. It is reasonable in such a case to reject the doubtful observation, and Wright and Hayford, in *Adjustment of Observations*, suggest that for a deviation of more than five times the estimated probable error, the observation should be rejected. In terms of standard error, this implies rejection where $|x - \bar{x}| > 3\frac{1}{2}$ (S.E.$_x$).

In the previous example, the student can confirm that the largest residual is 1·95''. The estimated probable error of a single observation is 0·72'', so that the largest deviation is 2·7 times the estimated probable error.

Combinations of Errors. Many observations in surveying are of a quantity which must be calculated from quantities whose observed or estimated values are liable to error, and it is necessary to estimate the precision of the required quanity. In theodolite readings, the error in angle measurement can be considered as a function of the error in angle reading and the error in signal bisection. In the precise spirit level (*see* Chapter 3), precision is influenced by error in registered reading error and bubble displacement.

If $z = f(x, y)$, where x and y are independent variables, then, to the first order,

$$\delta z = \frac{\partial f}{\partial x}\, \delta x + \frac{\partial f}{\partial y}\, \delta y$$

Since δx and δy are deviations from their mean values, the partial derivatives are given their values at $(\bar{x}, \bar{y})$, and it can be shown that, for any number of independent variables,

$$\sigma_z^2 = \left(\frac{\partial f}{\partial x}\right)^2 \sigma_x^2 + \left(\frac{\partial f}{\partial y}\right)^2 \sigma_y^2 + \left(\frac{\partial f}{\partial a}\right)^2 \sigma_a^2$$

If $z = x + y + a + \ldots$,

$$\sigma_z^2 = \sigma_x^2 + \sigma_y^2 + \ldots \qquad \cdot \qquad \cdot \qquad \cdot \qquad (13.13)$$

and the standard deviation is

$$\sigma_z = \sqrt{\sigma_x^2 + \sigma_y^2 + \ldots}$$

For a fuller treatment, *see*, for example, *Statistical Methods for Technologists, op. cit.* (*see* p. 466).

EXAMPLE. A measuring tape of length 20 m has been calibrated to have a standard deviation of $\pm$ 1 mm. If it is used to measure a length of 180 m what is the standard deviation of this measurement? Assuming an accidental error of $\pm$ 2 mm occurred when measuring each bay, find the standard deviation of the whole measurement.

Standard deviation due to standardizing = $\pm$ 1 $\times$ 9   (since there are 9 bays)

= + 9 mm

Standard deviation due to measurement = $\pm$ 2$\sqrt{9}$

= $\pm$ 6 mm

Standard deviation of whole = $\pm$ $\sqrt{(9^2 + 6^2)}$

= $\pm$ 11 mm

or 1/16,400 say.

Tolerances. If prefabricated components are incorporated into structures without allowing for any deviation in their dimensions site adjustments, i.e. cutting or trimming, may well be needed. It is preferable that the designer should initially allow for the various inaccuracies which might rise, and this can be done to some extent by considering tolerances or permissible deviations. Tolerance is defined in B.S. 3626:1963 (*Recommendations for a System of Tolerances and Fits for Building*) as the difference between the limits within which a size, dimension or position must lie. An absolute value, i.e. 20 mm, is implied, whereas permissible deviation specifies limits of deviation (difference between a size or position and a specified size or position), i.e. $\pm$ 10 mm, within that value.

Not only might the components be delivered on the site not exactly to size, but also the initial survey and the subsequent setting out will have been subjected to errors, as will the erection processes. Suggested values of permissible deviation are listed in P.D. 6440, Part 2* together with a discussion of their combination: this follows lines laid down in this chapter. For instance, if the permissible deviations of $\pm$ 10 mm and $\pm$ 20 mm respectively be assigned to plan position measured from a grid line and verticality for a certain column then the combined erection tolerance can be taken as $\sqrt{(20^2 + 40^2)}$, or 45 mm, say.

Weighted Observations. It may be necessary to assess the best value of a quantity for which the observed values are of different precision. As examples, readings might be made by different observers, or with different instruments; in lines of levels between two points, different distances might be covered.

If the true value of a quantity is x and the observed values are $x_1\ x_2, \ldots x_n$, corresponding standard deviations being $\sigma_1, \sigma_2, \ldots, \sigma_n$, then, if the errors are distributed normally, the probabilities of the observed values are proportional to

$$e^{-(x-x_1)^2/2\sigma_1^2}, \ e^{-(x-x_2)^2/2\sigma_2^2}, \text{ etc.} \quad . \quad \text{(from eqn. (13.1))}$$

The total probability of the set of observations is thus proportional to

$$e^{-\Sigma(x-x_r)^2/2\sigma_r^2}$$

Choose the value of x which makes this a maximum, i.e.

$$\sum \frac{(x - x_r)^2}{2\sigma_r^2} \text{ is a minimum .} \qquad . \qquad . \quad (13.14)$$

Differentiating with respect to x and equating to zero,

$$\sum \frac{(x - x_r)}{\sigma_r^2} = 0$$

$$x = \sum \frac{x_r}{\sigma_r^2} \Big/ \sum \frac{1}{\sigma_r^2}$$

Put
$$w_r = \frac{1}{\sigma_r^2} \qquad . \qquad . \qquad . \qquad . \quad (13.15)$$

$\therefore$
$$x = \frac{\Sigma w_r x_r}{\Sigma w_r} \qquad . \qquad . \qquad . \quad (13.16)$$

---

* Draft for Development: *Accuracy in Building*, P.D. 6440, Part 2, November 1969 (British Standards Institution, London).

.e. the best value of x is given by a weighted mean in which the weights are inversely proportional to the squares of the standard deviations of the several observations (i.e. inversely proportional to the variances).

It will be realized by the reader that weights can also be assigned inversely to the squares of either the probable errors or the standard errors of the several observations.

The best value thus derived by eqn. (13.16) can also be assessed as to its precision. If the variance of the weighted mean is denoted by σ_m^2, it can be shown that

$$\sigma_m^2 = \frac{1}{(\Sigma w_r)^2}\{w_1^2\sigma_1^2 + w_2^2\sigma_2^2 + \dots\}$$

$$= \Sigma\frac{1}{\sigma_r^2}\bigg/\left(\Sigma\frac{1}{\sigma_r^2}\right)^2$$

$$\therefore \quad \frac{1}{\sigma_m^2} = \Sigma\frac{1}{\sigma_r^2}$$

$$\therefore \quad w_m = \Sigma w_r . \qquad . \qquad . \qquad . \qquad . \qquad . \qquad . \quad (13.17)$$

.e. the weight of the value of x given by eqn. (13.16) is the sum of the weights of the several observations.

EXAMPLE. In a traverse survey the following measurements of one of the angles were made by two different observers using the same instrument—

Readings by A			Readings by B		
°	′	″	°	′	″
43	51	10	43	50	50
	50	40		50	25
	50	30		50	40
	51	00		51	15
	50	40		50	30
	50	20		50	45
	50	30		51	20
	50	50		51	00

Compare the reliability of the two results and calulate the most probable value of the angle. (L.U., B.Sc.)

Taking a datum value of 43° 50′, the working values in seconds are

Readings by A		Readings by B	
X	X²	X	X²
70	4,900	50	2,500
40	1,600	25	625
30	900	40	1,600
60	3,600	75	5,625
40	1,600	30	900
20	400	45	2,025
30	900	80	6,400
50	2,500	60	3,600
340	16,400	405	23,275

A. $\Sigma X = 340;\quad \Sigma X^2 = 16,400$

$$s^2 = \frac{8(16,400) - (340)^2}{8(8-1)} = \frac{15,600}{56} = 278 \cdot 57$$

$\therefore$ $s_{\bar{x}_A} = \sqrt{\frac{278 \cdot 57}{8}} = \sqrt{34 \cdot 82}$

Also $\bar{X} = \frac{340}{8} = 42 \cdot 5''$

$\therefore$ $\bar{x}_A = 43° \ 50' \ 42 \cdot 5''$

B. $\Sigma X = 405;\quad \Sigma X^2 = 23,275$

$$s^2 = \frac{8(23,275) - 405^2}{8(8-1)} = \frac{22,175}{56}$$

$$= 395 \cdot 98$$

$\therefore$ $s_{\bar{x}_B} = \sqrt{\frac{395 \cdot 98}{8}} = \sqrt{49 \cdot 50}$

Also $\bar{X} = \frac{405}{8} = 50 \cdot 62''$

$\therefore$ $\bar{x}_B = 43° \ 50' \ 50 \cdot 62''$

From eqn. (13.15), $w_A = \frac{1}{s_{\bar{x}_A}^2} = \frac{1}{34 \cdot 82}$

$$w_B = \frac{1}{s_{\bar{x}_B}^2} = \frac{1}{49 \cdot 50}$$

i.e. the ratio of the weights is

$$w_A : w_B = 1 \cdot 42 : 1$$

which is a measure of the relative reliability of the two sets of readings.
From eqn. (13.16), the best value of the angle is given by

$$x = \frac{Aw_A + Bw_B}{w_A + w_B} = 43° \ 50' + \frac{(42 \cdot 5)(49 \cdot 50) + (50 \cdot 62)(34 \cdot 82)}{84 \cdot 32}$$

$$= 43° \ 50' + 45 \cdot 78''$$

$$= 43° \ 50' \ 45 \cdot 78''$$

From eqn. (13.17),

$$w_x = w_A + w_B = \frac{1}{34 \cdot 82} + \frac{1}{49 \cdot 50}$$

$$= \frac{84 \cdot 32}{(34 \cdot 82)(49 \cdot 50)}$$

$$s_x^2 = \frac{(34 \cdot 82)(49 \cdot 50)}{84 \cdot 32} = 20 \cdot 44$$

$$s_x = 4 \cdot 51''$$

This is the standard error of the derived best value, so that in terms of the probable error the result can be expressed as $43° \ 50' \ 46'' \pm$ p.e. of $3''$.

onfidence Limits

Another, and often convenient, manner of expressing the precision of n estimate is to give limits which, with a stated probability, include he true value. Such limits are termed *confidence limits*, that is to say hey are limits within which it can be said, with a given degree of onfidence, that the true value lies. Provided the distribution is nown—and, as stated previously, in surveying the assumption is made hat the original data are distributed normally—such limits can be alculated for any statistic such as the mean or the standard deviation.

For a normal population with mean μ, and standard deviation σ, the robability that the mean of n observations will lie within the range $- (3\sigma/\sqrt{n})$ to $\mu + (3\sigma/\sqrt{n})$ is 0·997, and conversely if the true mean is not known, the probability that μ lies within the range $- 3\sigma/\sqrt{n}$ $) + 3\sigma/\sqrt{n}$ about the estimated mean $\bar{x}$ is also 0·997. This implies nat in repeated measurements of the same sort, the assertion that the :ue value of μ lies within the range $\bar{x} \pm (3\sigma/\sqrt{n})$ would be right in 9·7 per cent of cases. These limits are thus the 99·7 per cent confidence mits for μ. It can be stated with 99·7 per cent confidence that lies somewhere in this range, its most probable value being $\bar{x}$. The egree of confidence is called the *confidence coefficient* and the interval etween the limits the *confidence interval*. Common intervals are the 5 per cent interval, which is roughly $\bar{x} - (2\sigma/\sqrt{n})$ to $\bar{x} + (2\sigma/\sqrt{n})$, nd the 50 per cent interval, which is roughly $\bar{x} - (\frac{2}{3}\sigma/\sqrt{n})$ to $\bar{x}$ $- (\frac{2}{3}\sigma/\sqrt{n})$. In general,

$$\text{Lower confidence limit} = \bar{x} - (u_\alpha\sigma/\sqrt{n})$$

nd

$$\text{Upper confidence limit} = \bar{x} + (u_\alpha\sigma/\sqrt{n})$$

here u_α corresponds to a stated value of α as shown in Fig. 13.3. The onfidence coefficient is then, for both limits, $100(1 - 2\alpha)$ per cent.

Student's "t". In the preceding paragraphs the assumption was made 1at σ was known exactly, whereas in most cases an estimate s has to e made using eqn. (13.5). This estimated standard deviation is also ibject to some uncertainty, and the confidence limits for μ the true 1ean will be farther apart than if σ were known. W. S. Gossett (who rote under the name of Student) allowed for this uncertainty by sing t in place of u, and the distribution of t was tabulated by R. A. isher (*see* table below). Where the sample is very large, the uncertainty s is very small and t becomes almost identical with u. Thus for alues of $\alpha = 0\cdot25$, $0\cdot10$, $0\cdot05$, $0\cdot025$, $0\cdot001$ and $0\cdot0005$ (confidence efficients of 50 per cent, 80 per cent, 90 per cent, 95 per cent, 99·8 per nt and 99·9 per cent respectively),

ϕ	t_{50}	t_{80}	t_{90}	t_{95}	$t_{99\cdot8}$	$t_{99\cdot9}$
∞	0·674	1·282	1·645	1·960	3·090	3·291 = u
25	0·684	1·316	1·708	2·060	3·450	3·725
10	0·700	1·372	1·812	2·228	4·144	4·587
5	0·727	1·476	2·015	2·571	5·893	6·869
1	1·000	3·078	6·314	12·706	318·31	636·62

The symbol ϕ is used to denote the number of degrees of freedom ar equals $n - 1$. Then

$$\text{Lower confidence limit} = \bar{x} - (t_\alpha s/\sqrt{n})$$

and $$\text{Upper confidence limit} = \bar{x} + (t_\alpha s/\sqrt{n})$$

where t_α is read off from the table using the appropriate number degrees of freedom. As examples, the 95 per cent confidence interva will be evaluated for (a) the mean from the tests on the Watts Micropt Theodolite No. 2 and (b) the mean values calculated in the examp given earlier.

(a) $n = 25$, therefore $\phi = 24$.

From tables,* $t = 2.06(\alpha = 0.025)$.

$$\bar{x} = 90° \ 04' \ 20.55''$$

$$s = 1.07'' \qquad \frac{s}{\sqrt{n}} = 0.21''$$

Therefore the 95 per cent confidence limits are $90° \ 04' \ 20.55'' \pm 0.4$

(b) | Readings by A | Readings by B |
|---|---|
| $n = 8, \phi = 7$ | $n = 8, \phi = 7$ |
| From tables,* $t = 2.36$ | From tables,* $t = 2.36$ |
| $\bar{x}_A = 43° \ 50' \ 42.5''$ | $\bar{x}_B = 43° \ 50' \ 50.6''$ |
| $s_{\bar{x}_A} = \sqrt{34.82} = 5.90''$ | $s_{\bar{x}_B} = \sqrt{49.50} = 7.04''$ |

Therefore the 95 per cent confidence limits are—

For *A:* $43° \ 50' \ 43'' \pm 14''$

For *B:* $43° \ 50' \ 51'' \pm 17''$ to the nearest second

As a further example, if an engineer accepts that 90 per cent of well-defined points should lie within 1 mm of their true position wh plotted to scale on a plan, then

$$1.645 \ \frac{s}{\sqrt{n}} = 1$$

i.e. $$\frac{s}{\sqrt{n}} = 0.6 \text{ mm, say}$$

This in fact implies that 68 per cent of the plotted points should within 0.6 mm of their true plotted position. Using the table it will seen that 50 per cent of all points should lie within 0.674×0.6 m $= 0.4$ mm of their true position whilst 99.8 per cent should lie with 1.9 mm of their true position.

* *Biometrika Tables for Statisticians*, Vol. 1 (Cambridge University Press).

Adjustment of Errors

In precise surveying, errors of the nature just discussed give rise to small discordances. In plane surveying involving angle measurement, such discrepancies can be detected by the normal rules of geometry: for example, (i) the angles closing the horizon, or a complete round at a point, should sum to 360°; (ii) the angles of a small triangle must sum to 180°; (iii) the sum of the angles of a polygon is fixed by the number of its sides. In triangulation, more difficult problems of adjustment can arise since, besides the exact angle sums, there are other exact relations to be satisfied due to the fact that a particular side may occur in more than one triangle; calculation of the spherical excess may also be required if the triangles are large (*see* Chapter 9). In levelling, the reduced levels of a number of points may be determined by direct measurement. If further observations are made of the differences in level between selected pairs of points, it is almost inevitable that some small inconsistency will be found between the two sets of observations.

These inconsistencies and discordances require to be adjusted and allocated, and the method of least squares allows a solution. Indirect mention has been made of this earlier in the chapter, where from statement (13.14) it will be seen that we have expressed the criterion of the best value in the form

$$\Sigma w_r(x - x_r)^2 \text{ is a minimum .} \qquad . \qquad . \quad (13.18)$$

$(x - x_r)$ is a residual, the true error not being known, and the statement (13.18) can be expressed in the form: the best value of an unknown quantity, for which discordant values of different weights have been observed, is that which makes the weighted sum of the squares of the residuals a minimum. This is the *principle of least squares*, and in surveying it can be used to incorporate exact relations by extension of the method used in the section dealing with weighted means. In the detailed working out, two approaches are possible—

(i) Reduction to a minimum number of unknowns.

(ii) Lagrange's method of undetermined multipliers, more usually known in surveying as the *method of correlates*.

These are perhaps best studied by application to particular problems, and examples of both are given.

EXAMPLE   Some levellings were carried out with the following results—

	Rise or Fall	Weight
P to Q	+ 4·32 m	1
Q to R	+ 3·17 m	1
R to S	+ 2·59 m	1
S to P	− 10·04 m	1
Q to S	+ 5·68 m	2

The reduced level of P is known to be 134·31 m above datum. Determine the probable levels of the other points.

Let the probable heights of Q, R and S above P be q, r, and s respectively. The observation or residual errors are then

$q - 4\cdot32$	$s - 10\cdot04$
$(r - q) - 3\cdot17$	$(s - q) - 5\cdot68$
$(s - r) - 2\cdot59$	

The sum (say E) of the weighted squares of the observation errors is to be a minimum and thus

$$\frac{\partial E}{\partial q} = 0; \quad \frac{\partial E}{\partial r} = 0; \quad \frac{\partial E}{\partial s} = 0.$$

$$E = 1[q - 4 \cdot 32]^2 + 1[(r - q) - 3 \cdot 17]^2 + 1[(s - r) - 2 \cdot 59]^2 + 1[s - 10 \cdot 04]^2 \\ + 2[(s - q) - 5 \cdot 68]^2$$

$$= 200 \cdot 74 + 20 \cdot 42q + 4q^2 + 2r^2 - 2rq - 1 \cdot 16r - 47 \cdot 98s + 4s^2 - 2rs - 4sq$$

Differentiating with respect to q, r, and s in turn and equating to zero, three equations, known as normal equations, are produced from which the values of q, r, and s may be found—

$$+ 8q - 2r - 4s + 20 \cdot 42 = 0$$
$$- 2q + 4r - 2s - 1 \cdot 16 = 0$$
$$- 4q - 2r + 8s - 47 \cdot 98 = 0$$

whence $q = 4 \cdot 33$ m, $r = 7 \cdot 47$ m, and $s = 10 \cdot 03$ m.

The probable levels are now obtained as

$$\begin{aligned} Q &\quad 138 \cdot 64 \text{ m} \\ R &\quad 141 \cdot 78 \text{ m} \\ S &\quad 144 \cdot 34 \text{ m} \end{aligned}$$

Rather than obtain the expression for E by expanding the quadratics and then differentiating to produce the normal equations, it is possible to draw up a table of coefficients of q, r, and s from the expressions for the observation errors. The separate values obtained upon squaring are—

$$E_1 = 1[q - 4 \cdot 32]^2 = q^2 - 8 \cdot 64q + 18 \cdot 66$$

Then $\quad \dfrac{\partial E_1}{\partial q} = 2q - 8 \cdot 64$

$$E_2 = 1[(r - q) - 3 \cdot 17]^2 = q^2 + r^2 - 2rq - 6 \cdot 34r + 6 \cdot 34q + 10 \cdot 05$$

$$\frac{\partial E_2}{\partial q} = 2q - 2r + 6 \cdot 34$$

and $\quad \dfrac{\partial E_2}{\partial r} = 2r - 2q - 6 \cdot 34$

$$E_3 = 1[(s - r) - 2 \cdot 59]^2 = r^2 + s^2 - 2rs + 5 \cdot 18r - 5 \cdot 18s + 6 \cdot 71$$

$$\frac{\partial E_3}{\partial r} = 2r - 2s + 5 \cdot 18$$

and $\quad \dfrac{\partial E_3}{\partial s} = 2s - 2r - 5 \cdot 18$

$$E_4 = 1[s - 10 \cdot 04]^2 = s^2 - 20 \cdot 08s + 100 \cdot 8$$

$$\frac{\partial E_4}{\partial s} = 2s - 20 \cdot 08$$

$$E_5 = 2[(s - q) - 5 \cdot 68]^2 = 2s^2 + 2q^2 - 4sq - 22 \cdot 72s + 22 \cdot 72q + 64 \cdot 52$$

$$\frac{\partial E_5}{\partial q} = 4q - 4s + 22 \cdot 72$$

and $\quad \dfrac{\partial E_5}{\partial s} = 4s - 4q - 22 \cdot 72$

Then on addition

$$\frac{\partial E}{\partial q} = (2q - 8 \cdot 64) + (2q - 2r + 6 \cdot 34) + (4q - 4s + 22 \cdot 72) = 0$$

Dividing by 2

$$(q - 4 \cdot 32) + (q - r + 3 \cdot 17) + (2q - 2s + 11 \cdot 36) = 0$$

It will be seen that the terms in brackets equal the weighted errors multiplied by the coefficients of q for those errors. Similarly

$$\frac{\partial E}{\partial r} = (r - q - 3 \cdot 17) + (r - s + 2 \cdot 59) = 0$$

and

$$\frac{\partial E}{\partial s} = (s - r - 2 \cdot 59) + (s - 10 \cdot 04) + (2s - 2q - 11 \cdot 36) = 0$$

Again the separate terms are equal to the weighted errors multiplied by the relevant coefficient. The three equations which now follow are in each case half those previously produced but give the same values for q, r, and s. The work can, however, be conveniently tabulated as follows from the expressions for E_1, E_2, etc.

Wt.	q	r	s	N
1	1	0	0	− 4·32
1	− 1	1	0	− 3·17
1	0	− 1	+ 1	− 2·59
1	0	0	+ 1	− 10·04
2	− 1	0	+ 1	− 5·68

The coefficients are entered in turn as they apply: in the last column the numerical value is entered. Now multiply each line by the coefficient of the unknown under consideration on that line, including the value of the weight.

For q which appears on three lines

$1(+ 1)(+ 1)q + 1(+ 1)(- 4 \cdot 32)$ from the first line
$1(- 1)(- 1)q + 1(- 1)(+ 1)r + 1(- 1)(- 3 \cdot 17)$ from the second line
$2(- 1)(- 1)q + 2(- 1)(+ 1)s + 2(- 1)(- 5 \cdot 68)$ from the fifth line

which give $(q - 4 \cdot 32) + (q - r + 3 \cdot 17) + (2q - 2s + 11 \cdot 36)$

$$\therefore \qquad 4q - r - 2s + 10 \cdot 21 = 0$$

For r which appears on two lines

$1(+ 1)(- 1)q + 1(+ 1)(+ 1)r + 1(+ 1)(- 3 \cdot 17)$
$1(- 1)(- 1)r + 1(- 1)(+ 1)s + 1(- 1)(- 2 \cdot 59)$

whence $\qquad q \mid 2r - s - 0 \cdot 58 = 0$

and for s which appears on three lines

$1(+ 1)(- 1)r + 1(+ 1)(+ 1)s + 1(+ 1)(- 2 \cdot 59)$
$1(+ 1)(+ 1)s + 1(+ 1)(- 10 \cdot 04)$
$2(+ 1)(- 1)q + 2(+ 1)(+ 1)s + 2(+ 1)(- 5 \cdot 68)$

whence $\qquad - 2q - r + 4s - 23 \cdot 99 = 0$

Three normal equations for the solution of q, r, and s are thus obtained.

EXAMPLE. A, B, C, and D form a round of angles at a station such that

$$(A + B + C + D) = 360°.$$

Their observed values are

$$A = 110° \; 37' \; 45'', \qquad B = 82^c \; 15' \; 35''$$
$$C = 66° \; 24' \; 40'', \qquad D = 100° \; 42' \; 10''$$

$(A + B)$ was measured separately twice and found to average $192° \; 53' \; 25''$. Find the most probable value of each angle if all six measurements were of equal reliability.

The expression $A + B + C + D = 360°$ forms an equation of condition and allows the elimination of one unknown since it is a requirement which all four probable values must satisfy. Since $(A + B)$ has been measured twice, the weight of the mean may be taken as twice that of a single observation.

If a, b, and c be the most probable values of angles A, B, and C respectively. then the errors of observation are

$$a - 110° \; 37' \; 45'', \qquad b - 82° \; 15' \; 35'', \qquad c - 66° \; 24' \; 40'',$$
$$(a + b + c) - 259° \; 17' \; 50'', \qquad (a + b) - 192° \; 53' \; 25''$$

And so, on tabulating, we get

Wt.	a	b	c	N
1	+ 1	0	0	− 110° 37′ 45″
1	0	+ 1	0	− 82° 15′ 35″
1	0	0	+ 1	− 66° 24′ 40″
1	+ 1	+ 1	+ 1	− 259° 17′ 50″
2	+ 1	+ 1	0	− 192° 53′ 25″

Thus

$$4a + 3b + c - 755° \ 42' \ 25'' = 0$$
$$3a + 4b + c - 727° \ 20' \ 15'' = 0$$
$$a + b + 2c - 325° \ 42' \ 30'' = 0$$

whence

$$a = 110° \ 37' \ 46''$$
$$b = 82° \ 15' \ 36''$$
$$c = 66° \ 24' \ 34''$$

and so

$$d = 360° - (a + b + c) = 100° \ 42' \ 04''$$

EXAMPLE. A tunnel was required to be set out between two points A and B on a line running approximately West to East. The points were not intervisible so a line CD was set out with C and D approximately South of A and B respectively.

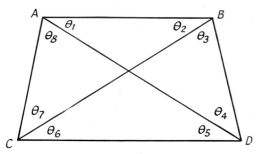

FIG. 13.4

The length of CD and the following angles were measured—

$$A\hat{C}B = 45° \ 24' \ 10'', \qquad B\hat{C}D = 37° \ 14' \ 12''$$
$$C\hat{D}A = 29° \ 38' \ 52'', \qquad A\hat{D}B = 63° \ 19' \ 35''$$
$$C\hat{A}D = 67° \ 43' \ 04'', \qquad C\hat{B}D = 49° \ 47' \ 08''$$

Adjust the angles to the nearest second assuming that those taken at A and B are of twice the weight of the other four. Explain, quoting the relevant formulae, how you would compute the length A to B and the alignment angles $C\hat{A}B$ and $A\hat{B}D$.

(*L.U., B.Sc.*)

Angle	Measured Value	Weight
$C\hat{A}D$	67° 43′ 04″	2
$A\hat{C}B$	45° 24′ 10″	1
$B\hat{C}D$	37° 14′ 12″	1
$A\hat{D}C$	29° 38′ 52″	1
$A\hat{D}B$	63° 19′ 35″	1
$C\hat{B}D$	49° 47′ 08″	2

Let the probable values of the angles be designated as shown.

Two conditions are to be satisfied—

$$\theta_5 + \theta_6 + \theta_7 + \theta_8 = 180° \ 00' \ 00''$$
$$\theta_3 + \theta_4 + \theta_5 + \theta_6 = 180° \ 00' \ 00''$$

and so

$$\theta_5 + \theta_6 + \theta_7 = 180° \ 00' \ 00'' - 67° \ 43' \ 04''$$
$$= 112° \ 16' \ 56''$$
$$\theta_4 + \theta_5 + \theta_6 = 180° \ 00' \ 00'' - 49° \ 47' \ 08''$$
$$= 130° \ 12' \ 52''$$

On tabulating we get—

Weight	θ_4	θ_5	θ_6	θ_7	N
1	+ 1	0	0	0	− 63° 19′ 35″
1	0	+ 1	0	0	− 29° 38′ 52″
1	0	0	+ 1	0	− 37° 14′ 12″
1	0	0	0	+ 1	− 45° 24′ 10″
2	0	+ 1	+ 1	+ 1	− 112° 16′ 56″
2	+ 1	+ 1	+ 1	0	− 130° 12′ 52″

Whence the normal equations are—

$$3\theta_4 + 2\theta_5 + 2\theta_6 \qquad\quad = 323° \ 45' \ 19''$$
$$2\theta_4 + 5\theta_5 + 4\theta_6 + 2\theta_7 = 514° \ 38' \ 28''$$
$$2\theta_4 + 4\theta_5 + 5\theta_6 + 2\theta_7 = 522° \ 13' \ 48''$$
$$\qquad\quad 2\theta_5 + 2\theta_6 + 3\theta_7 = 269° \ 58' \ 02''$$

And so,

$$\theta_4 = 63° \ 19' \ 45'', \qquad \theta_5 = 29° \ 38' \ 51''$$
$$\theta_6 = 37° \ 14' \ 11'', \qquad \theta_7 = 45° \ 23' \ 59''$$

Hence

$$\theta_3 = 49° \ 47' \ 13'' \quad \text{and} \quad \theta_8 = 67° \ 42' \ 59''$$

Alternative Method. The Method of Correlates may be applied to this (and the previous problems) to give a solution.

The angular errors in the two triangles ACD and BCD are $+ 18''$ and $- 13''$ respectively and these require equal and opposite total corrections.

Let $\varepsilon_3, \varepsilon_4, \ldots, \varepsilon_8$ be the respective individual corrections to the measured values.

Then

$$\varepsilon_3 + \varepsilon_4 + \varepsilon_5 + \varepsilon_6 = + 13''$$
$$\varepsilon_5 + \varepsilon_6 + \varepsilon_7 + \varepsilon_8 = - 18''$$

and for the Least Squares condition

$\Sigma w_3 \varepsilon_3{}^2$ is to be a minimum, and on differentiation,

$$w_3\varepsilon_3\delta\varepsilon_3 + w_4\varepsilon_4\delta\varepsilon_4 + w_5\varepsilon_5\delta\varepsilon_5 + w_6\varepsilon_6\delta\varepsilon_6 + w_7\varepsilon_7\delta\varepsilon_7 + w_8\varepsilon_8\delta\varepsilon_8 = 0 , \qquad (1)$$

also

$$\delta\varepsilon_3 + \delta\varepsilon_4 + \delta\varepsilon_5 + \delta\varepsilon_6 = 0 \quad . \qquad . \qquad . \qquad (2)$$

and

$$\delta\varepsilon_5 + \delta\varepsilon_6 + \delta\varepsilon_7 + \delta\varepsilon_8 = 0 \quad . \qquad . \qquad . \qquad (3)$$

since the equations for the corrections must not be invalidated by the small increments $\delta\varepsilon_3$, etc. Multiply equations (2) and (3) by $- \lambda_1$ and $- \lambda_2$ respectively and add to equation (1) to obtain:

$$\delta\varepsilon_3(w_3\varepsilon_3 - \lambda_1) + \delta\varepsilon_4(w_4\varepsilon_4 - \lambda_1) + \delta\varepsilon_5(w_5\varepsilon_5 - \lambda_1 - \lambda_2)$$
$$+ \delta\varepsilon_6(w_6\varepsilon_6 - \lambda_1 - \lambda_2) + \delta\varepsilon_7(w_7\varepsilon_7 - \lambda_2) + \delta\varepsilon_8(w_8\varepsilon_8 - \lambda_2) = 0$$

Now $\delta\varepsilon_3$, $\delta\varepsilon_4$, etc., are independent quantities and so

$$w_3\varepsilon_3 = w_4\varepsilon_4 = \lambda_1; \quad w_5\varepsilon_5 = w_6\varepsilon_6 = \lambda_1 + \lambda_2; \quad w_7\varepsilon_7 = w_8\varepsilon_8 = \lambda_2$$

But

$$w_3 = w_8 = 2w, \text{ and } w_4 = w_5 = w_6 = w_7 = w$$

therefore $\varepsilon_3 = \dfrac{\lambda_1}{2w}$, $\varepsilon_4 = \dfrac{\lambda_1}{w}$, $\varepsilon_5 = \dfrac{\lambda_1 + \lambda_2}{w}$, etc., and these values may be substituted in the correction equations to be satisfied, to give—

$$\frac{\lambda_1}{2w} + \frac{\lambda_1}{w} + \frac{\lambda_1 + \lambda_2}{w} + \frac{\lambda_1 + \lambda_2}{w} = + 13''$$

and
$$\frac{\lambda_1 + \lambda_2}{w} + \frac{\lambda_1 + \lambda_2}{w} + \frac{\lambda_2}{w} + \frac{\lambda_2}{2w} = -18''$$

i.e.
$$3 \cdot 5\lambda_1 + 2\lambda_2 = +13w$$
$$2\lambda_1 + 3 \cdot 5\lambda_2 = -18w$$

Whence $\lambda_1 = 9 \cdot 9w$ and $\lambda_2 = -10 \cdot 8w$

The values of ε_3, ε_4, ε_5, etc. (correct to the nearest second), are now deduced to be—

$$\varepsilon_3 = +5'', \qquad \varepsilon_4 = +5'', \qquad \varepsilon_5 = -1''$$
$$\varepsilon_6 = -1'', \qquad \varepsilon_7 = -11'', \qquad \varepsilon_8 = -5''$$

These corrections are applied to the measured values and the probable values of the angles found to be—

$$\theta_3 = 49° \ 47' \ 08'' + 05'' = 49° \ 47' \ 13''$$
$$\theta_4 = 63° \ 19' \ 35'' + 10'' = 63° \ 19' \ 45''$$
$$\theta_5 = 29° \ 38' \ 52'' - 01'' = 29° \ 38' \ 51''$$
$$\theta_6 = 37° \ 14' \ 12'' - 01'' = 37° \ 14' \ 11''$$
$$\theta_7 = 45° \ 24' \ 10'' - 11'' = 45° \ 23' \ 59''$$
$$\theta_8 = 67° \ 43' \ 04'' - 05'' = 67° \ 42' \ 59''$$

It will be seen that $ABCD$ forms one of the basic triangulation forms mentioned in Chapter 9, and in determining the probable values of the eight angles, the following equations of condition are to be satisfied—

$$\theta_1 + \theta_2 + \theta_3 + \theta_4 + \theta_5 + \theta_6 + \theta_7 + \theta_8 = 180°$$
$$\theta_1 + \theta_2 = \theta_5 + \theta_6$$
$$\theta_3 + \theta_4 = \theta_7 + \theta_8$$
$$\Sigma \log \sin \theta_1 = \Sigma \log \sin \theta_2$$

The first three equations are obtained by considering the whole figure or pairs of opposite triangles, and the final equation may be obtained as follows—

$$AB = BD \frac{\sin \theta_4}{\sin \theta_1} = DC \frac{\sin \theta_6}{\sin \theta_3} \frac{\sin \theta_4}{\sin \theta_1}$$
$$= AC \frac{\sin \theta_8}{\sin \theta_5} \frac{\sin \theta_6}{\sin \theta_3} \frac{\sin \theta_4}{\sin \theta_1}$$
$$= AB \frac{\sin \theta_2}{\sin \theta_7} \frac{\sin \theta_8}{\sin \theta_5} \frac{\sin \theta_6}{\sin \theta_3} \frac{\sin \theta_4}{\sin \theta_1}$$

Thus
$$\sin \theta_1 . \sin \theta_3 . \sin \theta_5 . \sin \theta_7 = \sin \theta_2 . \sin \theta_4 . \sin \theta_6 . \sin \theta_8$$
$$\therefore \qquad \Sigma \log \sin \theta_1 = \Sigma \log \sin \theta_2$$

In this example it will be seen that the values determined satisfy the condition that $\theta_3 + \theta_4 = \theta_7 + \theta_8$.

Also
$$\theta_5 + \theta_6 = 66° \ 53' \ 02'' = \theta_1 + \theta_2$$

and
$$\frac{\sin \theta_1}{\sin \theta_2} = \frac{\sin \theta_4 \sin \theta_6 \sin \theta_8}{\sin \theta_3 \sin \theta_5 \sin \theta_7} = \alpha \text{ say}$$

Then
$$\frac{\sin \theta_1 - \sin \theta_2}{\sin \theta_1 + \sin \theta_2} = \frac{\alpha - 1}{\alpha + 1} = \frac{2 \sin \dfrac{\theta_1 - \theta_2}{2} \cos \dfrac{\theta_1 + \theta_2}{2}}{2 \sin \dfrac{\theta_1 + \theta_2}{2} \cos \dfrac{\theta_1 - \theta_2}{2}}$$

$$= \frac{\tan \dfrac{\theta_1 - \theta_2}{2}}{\tan \dfrac{\theta_1 + \theta_2}{2}}$$

$(\theta_1 - \theta_2)$ may now be determined, since $(\theta_1 + \theta_2)$ is known and α can be calculated since θ_3 etc., are known. The individual values of θ_1 and θ_2 are then derived from the values of $(\theta_1 + \theta_2)$ and $(\theta_1 - \theta_2)$.

Method of Equal Shifts. The relationship between sums of log sines can be used to establish the best values of angles in triangulation figures. One such approach is by the method of equal shifts.

The method, briefly, is to first balance the angles to suit the conditions to be satisfied, such as, say, the summation of the angles in the various triangles and closing the horizon at a central station. These corrections are now included with a further correction to ensure the log sine condition is satisfied.

EXAMPLE. In the network shown (Fig. 13.5), all angles and the lengths of AB and DE have been measured. Adjust the following observed values by the method of equal shifts, if $\log DE - \log AB = 0\cdot0195859$. All values given are of equal weight.

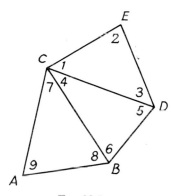

(9) $C\hat{A}B = 72° 22' 16''$

(8) $A\hat{B}C = 68° 28' 13''$

(7) $A\hat{C}B = 39° 09' 25''$

(6) $C\hat{B}D = 75° 05' 48''$

(5) $C\hat{D}B = 70° 30' 52''$

(4) $B\hat{C}D = 34° 23' 38''$

(3) $C\hat{D}E = 58° 30' 17''$

(2) $C\hat{E}D = 79° 47' 10''$

(1) $D\hat{C}E = 41° 42' 18''$

FIG. 13.5

Taking the triangles to have no spherical excess, the sums of the angles in each triangle should be $180° 00' 00''$ and the log sine condition linking AB and DE should be satisfied.

It will be seen that the errors in the three triangles are $+ 24''$, $+ 18''$ and $- 15''$ respectively, and these are eliminated by applying total corrections of $- 24''$, $- 18''$ and $+ 15''$ shared out equally to the angles of the respective triangles.

Triangle	Corrections to Each Angle		
ABC	$- 8'' (9)$	$- 8'' (8)$	$- 8'' (7)$
BCD	$- 6'' (6)$	$- 6'' (5)$	$- 6'' (4)$
CDE	$+ 5'' (3)$	$+ 5'' (2)$	$+ 5'' (1)$

Now it can be shown that

$$\frac{DE}{AB} = \frac{\sin 9 \sin 6 \sin 1}{\sin 7 \sin 5 \sin 2}$$

From the data, this relationship becomes

$$\log DE - \log AB = \log \sin 9 + \log \sin 6 + \log \sin 1 - \log \sin 7$$
$$- \log \sin 5 - \log \sin 2$$
$$= 0{\cdot}0195859$$

The measured angles give the following—

$\log \sin 9 = \bar{1}{\cdot}9791103$	$\log \sin 7 = \bar{1}{\cdot}8003367$
$\log \sin 6 = \bar{1}{\cdot}9851395$	$\log \sin 5 = \bar{1}{\cdot}9743853$
$\log \sin 1 = \bar{1}{\cdot}8230146$	$\log \sin 2 = \bar{1}{\cdot}9930625$
$\bar{1}{\cdot}7872644$	$\bar{1}{\cdot}7677845$

$\therefore \qquad \Sigma \log \sin 9 - \Sigma \log \sin 7 = 0{\cdot}0194799$

The angles measured are to be amended so that

$$\Sigma \log \sin 9 - \Sigma \log \sin 7 = 0{\cdot}0195859$$

i.e. by applying a correction of $0{\cdot}0001060$, so that the difference between the sums of the relevant log sines of the measured angles becomes that as measured directly, but at the same time not invalidating the corrections already made to the sums of the angles in the triangles.

Let corrections of $\pm \alpha''$ be further allotted to the angles, so that the final corrections will be

$(-8 + \alpha)$ and $(-8 - \alpha)$ to angles (9) and (7) respectively,
$(-6 + \alpha)$ and $(-6 - \alpha)$ to angles (6) and (5) respectively,
$(+5 + \alpha)$ and $(+5 - \alpha)$ to angles (1) and (2) respectively.

These corrections, which are in seconds of arc, are now multiplied by the values of log sin 1″ for the respective angles and applied together so that the corrected angles will satisfy the expression

$$\Sigma \log \sin 9 - \Sigma \log \sin 7 = 0{\cdot}0195859$$

i.e. $7 \times 10^{-7}(-8 + \alpha) - 26 \times 10^{-7}(-8 - \alpha) + 6 \times 10^{-7}(-6 + \alpha)$
$\qquad - 7 \times 10^{-7}(-6 - \alpha) + 24 \times 10^{-7}(5 + \alpha) - 4 \times 10^{-7}(5 - \alpha)$
$$= 1{,}060 \times 10^{-7}$$

whence $\qquad\qquad\qquad\qquad \alpha = 11''$

The corrections now become

Angle (9): $+ 3''$	Angle (7): $- 19''$	Angle (8): $- 8''$
(6): $+ 5''$	(5): $- 17''$	(4): $- 6''$
(1): $+ 16''$	(2): $- 6''$	(3): $+ 5''$

with corrected values of

$B\hat{A}C = 72°\ 22'\ 19''$	$B\hat{C}D = 75°\ 05'\ 53''$	$C\hat{D}E = 58°\ 30'\ 22''$
$A\hat{B}C = 68°\ 28'\ 35''$	$C\hat{D}B = 70°\ 30'\ 35''$	$C\hat{E}D = 79°\ 47'\ 04''$
$A\hat{C}B = 39°\ 09'\ 06''$	$B\hat{C}D = 34°\ 23'\ 32''$	$D\hat{C}E = 41°\ 42'\ 34''$

Adjustment of a Closed Precise Traverse.

If the bearing of AB, (*see* Fig. 13.6) one of the lines of a precise traverse, is to be maintained after corrections have been applied to its latitude and departure then the relationship

$$\frac{k_1 y_1}{k_1 x_1} = \frac{\text{correction to latitude}}{\text{correction to departure}} = \frac{\text{latitude}}{\text{departure}}$$

must obtain, and the correction to length AB must also be k_1l_1. Assuming the error in the measured length AB to be proportional to $\sqrt{l_1}$, then the weight of that observation may be taken as $1/l_1$.

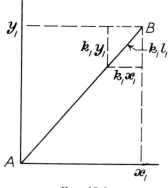

FIG. 13.6

Thus for the whole traverse, $\sum \dfrac{k_1^2 l_1^2}{l_1}$ is to be a minimum to satisfy the Least Squares condition, whilst $\Sigma\, k_1 x_1$ equals the total departure correction (D) and $\Sigma\, k_1 y_1$ equals the total latitude correction (L). Therefore we can write

$$k_1 l_1 \delta k_1 + k_2 l_2 \delta k_2 + \ldots + k_n l_n \delta k_n = 0$$
$$x_1 \delta k_1 + x_2 \delta k_2 + \ldots + x_n \delta k_n = 0$$
$$y_1 \delta k_1 + y_2 \delta k_2 + \ldots + y_n \delta k_n = 0$$

If the second and third equations are multiplied by $-\lambda_1$ and $-\lambda_2$ respectively and then added to the first, a further equation results—

$$\delta k_1 (k_1 l_1 - \lambda_1 x_1 - \lambda_2 y_1) + \delta k_2 (k_2 l_2 - \lambda_1 x_2 - \lambda_2 y_2) \\ + \delta k_n (k_n l_n - \lambda_1 x_n - \lambda_2 y_n) = 0$$

Since δk_1, δk_2 etc., are independent quantities

$$k_1 = \frac{\lambda_1 x_1 + \lambda_2 y_1}{l_1}, \qquad k_n = \frac{\lambda_1 x_n + \lambda_2 y_n}{l_n} \text{ etc.,}$$

so that

$$\lambda_1 \sum \frac{x_1^2}{l_1} + \lambda_2 \sum \frac{x_1 y_1}{l_1} = D$$

$$\lambda_1 \sum \frac{x_1 y_1}{l_1} + \lambda_2 \sum \frac{y_1^2}{l_1} = L$$

λ_1 and λ_2 can therefore be determined and values of k_1, $k_1 x_1$ and $k_1 y_1$ readily follow for each line of the traverse, allowing corrected departures and latitudes to be established.

If the traverse referred to in Table II, Chapter 4 be subjected to correction in this manner, the following values are established.

TABLE I

Line	l (m)	$\dfrac{y^2}{l}$	$\dfrac{xy}{l}$	$\dfrac{x^2}{l}$	$k = \dfrac{\lambda_1 x + \lambda_2 y}{l}$	ky (m)	kx (m)
AB	103·40	101·71	+ 13·16	1·70	− 0·000111	+ 0·01	0·00
BC	157·25	155·71	+ 15·51	1·70	− 0·000121	+ 0·02	0·00
CE	143·36	44·94	− 66·51	98·41	− 0·000356	+ 0·03	− 0·04
EG	169·08	102·82	+ 82·54	66·27	− 0·000084	− 0·01	− 0·01
GJ	176·74	161·83	+ 49·11	14·90	+ 0·000052	+ 0·01	0·00
JL	110·60	8·35	− 29·21	102·26	+ 0·000356	+ 0·01	− 0·04
LA	140·83	0·44	− 7·87	140·39	+ 0·000335	0·00	− 0·05
		+ 575·80	+ 56·73	+ 425·63		+ 0·07	− 0·14

From the Equations of Condition

$$425\!\cdot\!63\,\lambda_1 + 56\!\cdot\!73\,\lambda_2 = -0\!\cdot\!13$$
$$56\!\cdot\!73\,\lambda_1 + 575\!\cdot\!80\,\lambda_2 = +0\!\cdot\!07$$

Whence $\qquad \lambda_1 = -\dfrac{1}{3,068} \quad$ and $\quad \lambda_2 = +\dfrac{1}{6,506}$

k_1 is now determined for each line and the corrections to latitude and departure, rounded off to 0·01 m, are derived as indicated in Table I. They can be compared with those of Tables III and IV of Chapter 4.

EXERCISES 13

1. In order to establish three bench marks, two circuits of precise levels $ABCDA$ and $ACBA$ were observed with the following results.

First Circuit

Station	.	.	A		B		C		D		A
Level difference	.		+ 12·314		+ 18·198		+ 34·210		− 64·702		

Second Circuit

Station	.	.	A		C		B		A
Level difference	.		+ 30·446		− 18·176		− 12·320		

Determine, by the method of least squares, the levels which should be assigned to the bench marks at B, C and D if the reduced level of A is 192·134 m A.O.D. *Answer:* $D = 256\!\cdot\!829$ m A.O.D.

2. In a topographical survey, the difference in level between two points A and B is found by three routes, via C, D, E and F, via G and H, and via J, the distances being—

Route (1). $AC = 120$ m, $CD = 162$ m, $DE = 300$ m, $EF = 258$ m, $FB = 132$ m. Route (2). $AG = 240$ m, $GH = 306$ m, $HB = 384$ m. Route (3). $AJ = 294$ m, $JB = 462$ m.

The sections on route (1) are each levelled four times, those on route (2) eight times, and on route (3) twice; the established differences in level thus obtained being 30·81 m, 30·57 m, and 31·08 m, respectively.

If the probable error in any section at each levelling is proportional to its length, and the usual laws for combinations of readings hold, find the most probable value for the difference in level between A and B.   (*L.U., B.Sc.*)

Answer: 30·72 m.

3. The following observations were made on two adjacent angles subtended at point O, and on the total angle contained between the two outer lines OA and OC—

$$A\hat{O}B = 58° \ 56' \ 42'', \ B\hat{O}C = 17° \ 46' \ 26'', \ A\hat{O}C = 76° \ 43' \ 06''$$

Find the most probable values for $A\hat{O}B$ and $B\hat{O}C$.

Answer: $A\hat{O}B = 58° \ 56' \ 41''$, $B\hat{O}C = 17° \ 46' \ 25·3''$.

CHAPTER 14

FURTHER INSTRUMENTS AND TECHNIQUES

SINCE this book was first published a number of interesting instruments have been produced by various manufacturers, and new techniques have been developed to the stage at which they can be considered as standard practice. Opportunity has been taken in this edition to describe and discuss some of these instruments and methods, and these further notes should be read in conjunction with the relevant preceding chapters.

Levels and Levelling

The most radical development in this field since the war has been the introduction of the self-levelling principle. Chapter 3 gives details of conventional levels and various self-levelling instruments. The Zeiss Ni2 self-levelling device was introduced in about 1950 and, as shown previously, it incorporates a swinging-prism compensator. At about

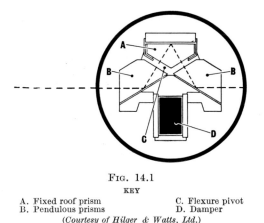

FIG. 14.1

KEY

A. Fixed roof prism   C. Flexure pivot
B. Pendulous prisms   D. Damper

(*Courtesy of Hilger & Watts, Ltd.*)

this time also the Cowley instrument was introduced into the U.K. and this used a mirror and pendulum compensating device. Their success prompted other manufacturers to develop and market self-compensating instruments and the automatic level should not now be regarded as a novelty but as an obvious aid to speedy, precise levelling. Some of the new instruments not already dealt with in Chapter 3 are described below.

Watts Autoset Level: CTS Self-aligning Level. The Hilger & Watts Autoset 2 precise level and the Vickers S700 self-aligning level, both incorporating the same design of compensator shown in Fig. 14.1, have been described in Chapter 3. This compensator is also used in the Watts Autoset 1, illustrated in Fig. 14.2, an erect staff image being produced.

It will be seen that no footscrews have been fitted to this latter instrument; the base of the instrument has a concave undersurface and this fits over the dome head of the tripod which possesses three bearing pads. By moving the level in the tripod head, it is possible to centre the circular bubble on the level which can then be fixed in position by means of a mounting bolt on the tripod. The two swinging prisms and the fixed roof prism then automatically arrange for a level line of sight, and the manufacturers give a mean closing error of ± 2 mm per kilometre of single levelling.

Fig. 14.2. Watts Autoset 1 Level
(Courtesy of Hilger & Watts, Ltd.)

Kern GK1A. Some of the automatic levels now in production have shapes somewhat different from the Watts Autoset and from the conventional levels of Chapter 3. The Kern GK1A, shown in Fig. 14.5, is a case in point. Although a horizontal line of sight is here controlled by a pendulum, this latter is not suspended by a system of wires, but swings in a magnetic bearing. Footscrews have been eliminated and the ball-joint head found in all Kern GK levels is used. The conical bearing surface of the level base fits on the spherical surface of the tripod head and a circular bubble is centred as in the Watts Autoset 1 level. Other external features are the telescope focusing screw, the horizontal slow-motion screw and, beneath the objective, the protection cap for the screw used in adjusting the horizontal line of sight.

The compensator is established at a distance $\frac{1}{2}f$ from the objective whose focal length is f. The image of points lying on a straight line in the horizontal plane, which contains the objective's principal focus on the image side, will appear at the horizontal line of the reticle, whatever the distances of these points from the objective. The horizontal plane referred to is the horizon of the compensator. Fig. 14.4 shows schematically the behaviour of a horizontal ray from the staff.

FIG. 14.3. DETAILS OF AUTOMATIC COMPENSATOR
(Courtesy of Kern & Co. and Survey & General Instrument Co., Ltd.)

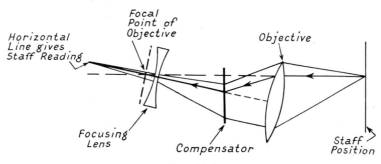

FIG. 14.4

The rays are shown as if passing through the compensator rather than being reflected back

Other rays from the point are refracted to meet the horizontal ray at the graticule.

Horizontality of the line of sight can be checked in the manner outlined in Chapter 3 (Fig. 3.23); the correct staff reading, e, is set by turning the capstan-headed adjusting screw beneath the objective.

In the instrument it will be seen that the rays of light enter through the objective and are reflected back by the roof prism of the compensator. They pass through the focusing lens and are shifted vertically through two prisms to converge on the graticule. From this they

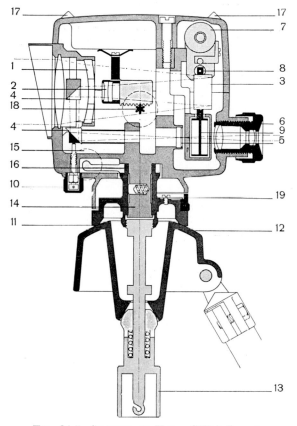

FIG. 14.5. SECTION OF KERN GK1A LEVEL

KEY

1. Objective
2. Focusing lens
3. Roof prism of compensator
4. Right-angled prisms
5. Reticule
6. Eyepiece
7. Yoke-shaped permanent magnet
8. Compensator axis
9. Damper pistons
10. Adjusting screw with protection cap
11. Conical bearing
12. Tripod head with spherical surface
13. Fastening screw
14. Cylindrical axis
15. Horizontal slow-motion screw
16. Friction coupling
17. Open sights
18. Focusing knob
19. Horizontal circle

(Courtesy of Kern & Co. and Survey & General Instrument Co., Ltd.)

emerge through the eyepiece, which is not at the same level as the objective.

Fig. 14.3 shows details of the compensating device of this instrument. The yoke-shaped permanent magnet has two conical poles covered by sapphire caps which are spring mounted. One end of the pendulum axis is in contact with one of the caps but at the other end there is a small air gap which, the manufacturers claim, helps to reduce friction to a minimum. High centring-accuracy for the pendulum can be achieved, and at the same time fatigue failure, which has been known to occur with wire suspended compensators, is obviated. The entire compensating unit, complete with its damping pistons, can be

FIG. 14.6. ZEISS NI4 LEVEL
(Courtesy of Carl Zeiss and Degenhardt & Co., Ltd.)

exchanged for repairs if necessary. The probable error over a kilometre of two-way levelling is stated to be ± 2·5 mm.

Zeiss Ni4 Level. The Zeiss No. 4 level has been devised for use in the building industry, and is said to have an accuracy of about ± 8 mm for one kilometre of two-way levelling. Fig. 14.6 shows this instrument which has its eyepiece above the objective. It has three footscrews, and the focusing knob is on top of the telescope, the whole being somewhat dissimilar to the conventional dumpy level: Fig. 14.7 shows a section of the instrument. The circular bubble used for initial levelling, with the footscrews, is viewed through the upper of the two viewing systems. After this has been achieved, there will be some residual lack of horizontality of the line of sight, and this is now compensated for by means of that circular bubble used for the initial levelling of the instrument. This bubble acts as a negative lens, positioned approximately in that part of the telescope where the focusing lens would lie, and the image

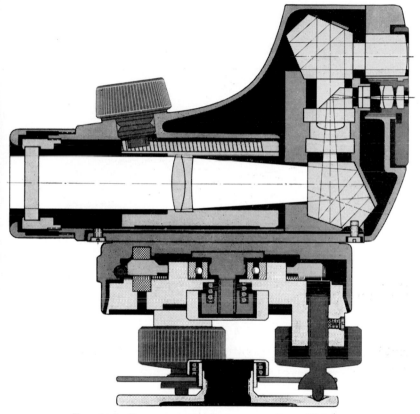

Fig. 14.7. Diagrammatic Section of Zeiss Ni4 Level
(*Courtesy of Carl Zeiss and Degenhardt & Co., Ltd.*)

formed by the objective is projected through the bubble on to the graticule. The objective is within the body of the instrument and is moved by the focusing knob backwards or forwards inside the telescope, whose forward end consists of an optical adjusting wedge. This wedge, which can be rotated about the telescope axis, allows the line of sight to be set horizontal and replaces the graticule adjustment of the conventional level.

Fig. 14.8 shows an idealized representation of the path of rays through the instrument with the bubble as though it were initially on the longitudinal axis of the telescope. The objective would form an image of a distant object at A, but the bubble would magnify and reposition this image at the graticule. When the instrument is slightly tilted through an angle α, the bubble would move through distance a and the image formed by the objective would move through distance b from

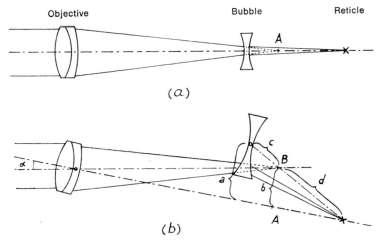

Fig. 14.8

(a) Horizontal telescope. (b) Inclined telescope.

A to B. Thus the requirement to be satisfied is that the final image will form on the graticule, and evidently

$$\frac{a}{b} = \frac{c + d}{d}$$

But if r be the radius of curvature of the bubble, then $a = r\alpha$ and also $b = f\alpha$; therefore

$$\frac{r}{f} = \frac{c + d}{d}$$

The magnification, m, due to the bubble is the ratio of image distance to object distance for this bubble, and is equal to $(c + d)/c$. Thus

$$\frac{r}{f} = \frac{m}{m - 1}$$

The ratio r/f can be taken to be constant but the magnification depends on three values, radius of curvature, refractive index and bubble thickness, which are sensitive to temperature variations. This has entailed careful proportioning of the bubble to co-ordinate the above influences so that variations due to temperature are virtually zero. Compensation must function for all allowable sighting distances,

and so
$$f = \frac{b}{\alpha} = r\frac{m-1}{m}$$

∴
$$b = \alpha r\frac{m-1}{m}$$

Distance b will remain constant whilst the objective is focused on near or far objects, and so the graticule will move on a line parallel to the optical axis with respect to the objective. It is claimed that such initial setting of the centring bubble can be readily made within 0·3 mm, and this causes a heighting error of 0·1 mm which is negligible in normal staff work.

Other Self-aligning Levels

Most automatic levels are compact in size, and one, introduced by Askania, is permanently fastened to its container base plate and

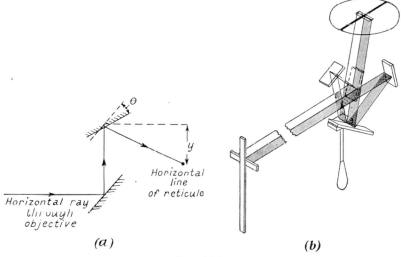

(a)

(b)

Fig. 14.9

mounted with it on its tripod. Fig. 14.9a refers to the Ertel Werke level which incorporates a pendulum-controlled mirror as does the Franke-Askania Na1 whose controlled mirror lies below a fixed deflecting prism. The two reflecting faces are parallel when the vertical axes are truly vertical.

The Cowley level, mentioned previously as one of the earliest automatic levels, contains two mirror systems, one on each side of the central vertical plane. On the left side are two fixed mirrors and on the right side are a further three, two being fixed and one mounted at the top of, and at right angles to, a metal pendulum. A horizontal bar is observed, and the two images given by the sets of mirrors are brought into coincidence by moving that bar along its staff into the horizontal plane through the optical centre of the level (Fig. 14.9*b*). Heighting readings are made with respect to the bar on the staff and an accuracy of the order of 5 mm for a 30 m sight is quoted.

Whereas the Cowley level is box-like, two other types could be said to resemble the periscope to some estent in appearance. These are the Koni 007 by Zeiss of Jena and the 5190 level by Filotecnica Salmoiraghi of Milan. The latter indeed has a height of 500 mm and a mass of over 6 kg.

Vibrational Effects in Automatic Levels

Despite the action of the dampers fitted to the compensating devices in the instruments just described, it is possible for periodic vibrations induced by wind, traffic or plant to affect the reading accuracy. The advantages of automatic levelling are then to some extent offset by the disadvantage of sensitivity.

It is essential with such instruments that the tripod be set up with the legs pushed firmly into the ground. Vibrations can then be stopped or damped by lightly restraining the tripod with the hands. Such a practice could not be tolerated in conventional spirit levelling but is permissible with automatic levels. With the levels firmly bedded, the collimation height will be unaffected and any slight tilt induced is automatically corrected for by the compensator. In some locations, however, where for example continuous vibrations due to mechanical engineering equipment are liable to be encountered, it may be preferable to use a spirit level. It is not possible to generalize, and in any given case the decision must be made having regard for the circumstances. The damping devices fitted to automatic levels are extremely efficient, and it is only occasionally that higher frequency vibrations cause resonant vibrations in the compensator of sufficient magnitude to affect accuracy.

Theodolites

Askania Precision Theodolite Tpr (after Gigas). This instrument measures horizontal and vertical angles direct to 0·2 sec, the readings being obtained visually or recorded photographically. It can be utilized for first order triangulation work, determination of latitude or longitude at Laplace stations, observation of passage of stars or artificial satellites, and flare triangulation for connecting triangulation networks across wide sea-gaps. Should there be no intermediate islands available upon which one or more stations could be sited it may be possible to intersect from stations on the main networks on a flare which has been parachuted from an aircraft situated mid-way between them. Whilst the intersections are maintained the simultaneous operation of the camera shutters is controlled by radio.

The theodolite consists basically of a tribrach, rotatable upper part or alidade, a broken telescope of magnification × 63, and a withdrawable clock case. Standard accessories include an optical micrometer and a recording camera, three footplates, two telescope eyepieces of magnification × 40 and × 80 respectively, and small tools. Electrical equipment (switch box and cables) for illumination of the reticule and circles, and camera shutter release, is also provided for operation with the normal type 12-V car battery.

It will be seen from Fig. 14.10 that the telescope differs from any of those in instruments mentioned previously, since the eyepiece is at

FIG. 14.10 PRECISION THEODOLITE, AFTER GIGAS, TYPE TPR

trunnion axis level on one of the standards, thereby allowing horizontal viewing whatever the inclination of the telescope. Further below is situated the optical micrometer (carrying the manufacturer's name) with its three fastening screws and reading microscope: the micrometer knob is on the side of the micrometer assembly. The camera and its locking device can be clearly seen; above it may be found a circular bubble for rough levelling, together with knob for setting the horizontal circle as required. Below the camera the horizontal clamp and slow motion screw are located and, at the opposite side of the camera to the micrometer, the withdrawable clock case is positioned. The instrument has an overall height of 370 mm and an approximate weight of 31 kg, but when it is packed for transit the weight is about 48 kg.

Over and above this the packed accessories and the switch box weigh some 23 kg.

For photographic recording 35 mm film is used, giving a 24 mm square picture; the magazine capacity is given as 48 pictures. The number of readings obtainable per day is increased when photographic recording is used, or alternatively, full use may be made of those periods most favourable for triangulation observations.

The manufacturer recommends that the camera be positioned even when the circles are to be observed visually, in order that the equilibrium of the rotating alidade be maintained. A tripod is not supplied with the instrument which has been designed for pillar observations only, and when setting up, the footplates are placed on a circle of 300 mm diameter, the footscrews being on a similarly sized circle. The instrument is centred using a centring pin sighted from two directions

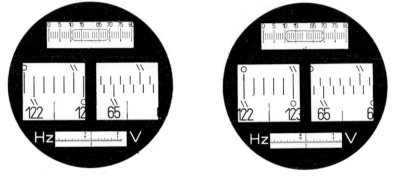

FIGS. 14.11 AND 14.12. PHOTOGRAPHIC RECORDING OF BOTH CIRCLES

at right angles by shifting the footplates as required. Preliminary levelling is possible by means of the small circular bubble and final levelling of the instrument is carried out by means of a plate level which is not shown on the illustration but which can be observed in the micrometer reading eyepiece. This level can be taken out of its housing and it is possible to adjust the bubble length to cover about 60 or 80 divisions (see the illustrations of the circle readings, Figs. 14.11 and 14.12). When replaced in the housing the level adjustment must be checked. It is also possible to adjust the movement of the alidade in azimuth about the vertical axis if, say, temperature changes have caused a change in fit of the main axis. It is desirable that the movement be easy, but if there be too much play accurate circle readings are not likely to result, and if the fit be too tight, excessive backlash is noticed in the slow-motion screw.

Circle Readings. In this instrument, as in those described earlier, there is provision for setting the circles at any required value. The position of the knob serving the horizontal circle has been indicated; that serving the vertical circle is above trunnion-axis level on the opposite standard to the one carrying the telescope eyepiece. Displacement of the circles, by accidental displacement of the knobs whilst obtaining a round of angles, is prevented by locking devices. A rough vertical circle reading can be obtained directly from a circle surrounding the telescope eyepiece. The horizontal and vertical circles, graduated at 2 double minute or 4 minute intervals, have diameters of 200 mm and 140 mm respectively, and are illuminated artificially whether observations are being made by day or night. The coincidence method of reading is adopted and two points of coincidence, one double minute or two minutes apart, can be obtained by the micrometer which has a range of 4 minutes and enables readings direct to 0·1 double seconds or 0·2 seconds to be made since one interval corresponds to 0·1 double second. In the field of view of the reading microscope the observer will see (i) the plate bubble, (ii) the horizontal and vertical circles, the graduations of the one being read and brought into coincidence by actuating the micrometer knob, and (iii) the micrometer reading. Certain marks with the whole degree numbers may be seen in the field of view as well as the major graduations at 4 minute intervals. These are 0, ` and `` indicating 0, 10 and 20 double minute respectively or 0, 20 and 40 minute respectively. If only one coincidence be made then the readings can be determined directly but with slightly less accuracy than if two coincidences be made to compensate for possible very slight graduation errors. The readings shown in Figs. 14.11 and 14.12 respectively are—

$$122° \ 48' \ 00'' + 2 \times 01' \ 55·9'' = 122° \ 51' \ 51·8''$$
or
$$122° \ 50' \ 00'' + 2 \times 00' \ 55·8'' = 122° \ 51' \ 51·6''$$

depending upon which coincidence be obtained. Since 122 represents 122° 40' and each major graduation represents 04', note the position of the index line. If both coincidences be taken, the mean obtained would be 122° 51' 51·7''. Alternatively instead of doubling minutes and seconds, the two can be added together to give the required mean, no average being required since they are both read to units of double minutes or seconds, as

	122°	24'	00''
		01'	55·9''
and		25'	00''
		00'	55·8''
	122°	51'	51·7''

The same procedure applies to the vertical circle which being moveable does not give elevations or depressions directly. An automatic self-adjusting index eliminating the need to centre the altitude-level bubble is fitted on the latest version of this instrument. Other theodolites pioneered by Askania and now manufactured by Franke & Co. of Geissen under the title "System Askania" have such an index in the form of a compensating pendulum: this is discussed later.

Photographic Recording. When the film has been inserted and other preliminaries attended to, the camera is then attached to the theodolite.

The shutter release is arranged, through the switch box, to be either manual release or to be remote control or radio set release. In the former case, where reading and delayed exposure occur, the circles are illuminated for the reading and the shutter released some one-quarter of a second later: the shutter can be locked if visual reading only be required. In the latter case, for say flare triangulation observations, continuous illumination of the circles occurs and there is no time lag in exposure. Temperature effects are minimized in this case by using a glass of low thermal conductivity and isolating the lamps from the interior, although some manipulation of the switch box may be possible

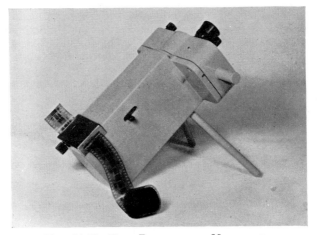

FIG. 14.13. FILM READER AND MICROMETER

if the time interval between exposures is large enough to enable switching on and off sequences. A connexion can be made from the camera to a chronograph so that the shutter release may be recorded. Also the clock is photographed at the time of pointing so that the times of the circle readings with the date and observer's name, marked on a white tablet alongside the clock, are related.

To obtain readings from the film, after developing, etc., a film reader (Fig. 14.13), which is one of the special accessories, is set up and used in conjunction with the micrometer removed from the instrument. The micrometer is used exactly as when measuring angles visually although this time it views the film directly. Two coincidences are made, the sum and difference of the micrometer readings are determined and these are related to each other on a table to obtain a correction for film shrinkage which may be either positive or negative. This correction is now applied to the circle and micrometer readings to obtain the reading for this particular pointing.

Automatic Index for Vertical Circle Readings. As mentioned previously, Askania incorporated into some instruments one of the more radical innovations of the post-war years, at least so far as theodolite design is concerned, the altitude bubble being replaced by a pendulum; the principle of the arrangement is shown in Fig. 14.14. The vertical circle is read through an optical arrangement which includes a prism attached to a pendulum, and the path of the reading rays (5–9 in the diagram) is so adjusted that, with the vertical axis BB truly vertical, vertical angles are correctly obtained. If the theodolite tilts through a from the vertical, however, then, assuming for convenience

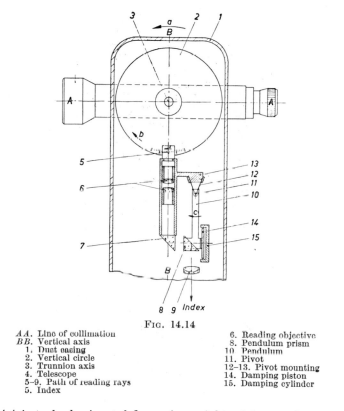

Fig. 14.14

AA. Line of collimation	6. Reading objective
BB. Vertical axis	8. Pendulum prism
1. Dust casing	10. Pendulum
2. Vertical circle	11. Pivot
3. Trunnion axis	12–13. Pivot mounting
4. Telescope	14. Damping piston
5–9. Path of reading rays	15. Damping cylinder
5. Index	

that AA is to be horizontal for a given sight, AA must be rotated in direction b in order to remain horizontal. The vertical circle is carried round with the telescope in direction d, but owing to the influence of the pendulum, the reading obtained at the index is not thereby changed. Due to the tilt of BB, the pendulum (10) carrying the prism (8) swings in direction c and the path of the reading rays is so influenced that the correct circle reading is still obtained.

With a sufficiently large range of action of the pendulum, large tilts of the vertical axis BB could be accommodated, but since preliminary levelling is carried out by means of a target bubble, which, when central, ensures that BB is plumb to within $\pm$ 3 min, then it is only essential for the pendulum to have this range. Vertical circle readings are then automatically converted by the pendulum prism arrangement. An air damping device is incorporated to damp out oscillations in the pendulum.

In addition to the Tu one-second theodolite (Fig. 14.15) Franke make the Tt and Tts tacheometer theodolites which also have automatic collimation by means of a pendulum compensator.

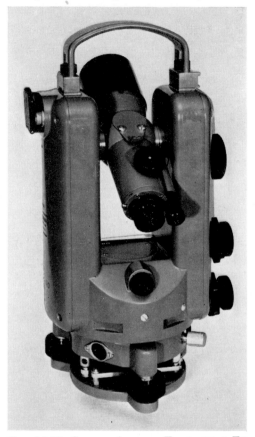

FIG. 14.15. SYSTEM ASKANIA THEODOLITE TU
(Courtesy of Franke & Co. Optik G.m.b.H.)

Telescope positions on the Tu are marked on the standard which carries the illumination mirror for circle reading, and the vertical circle gives zenith distances in position I (face left), so that when the telescope is horizontal, the vertical circle reading is 90°; in position II (face right), the equivalent reading is 270°. The other standard carries the micrometer telescope clamp and slow-motion device, and the horizontal clamp and slow-motion device, this being the lowest of the three items (Fig. 14.15).

The adjustments for the Tu are somewhat similar to those listed in Chapter 4 for the vernier theodolite, although capstan screws are not available for centring the plate bubble. This is carried out by means of an adjusting screw under the marking "II" on the standard carrying the illumination mirror. The vertical circle index can also be adjusted, if necessary, by means of a screw on the same standard. If a distant signal is observed, the sum of the angles read on the vertical circle with the telescope in positions I and II successively should be 360° 00' 00". Any discrepancy amounts to twice the correction to be applied. The angle to be expected in position I, say, is deduced and the micrometer

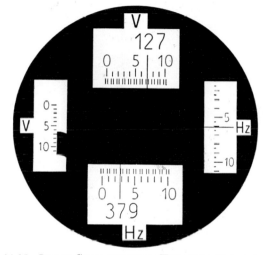

Fig. 14.16. Image Seen through Eyepiece Microscope of Zeiss Th3 Theodolite
(Courtesy of Degenhardt & Co., Ltd.)

reading set in accordance. The signal is reobserved, thereby disturbing the circle readings which are then made to coincide using the relevant screw.

Zeiss Th3: Wild T1-A: Watts ST300. Among other theodolites with automatic compensation for vertical index error are the Zeiss Th3, the Wild T1-A and the Watts ST300. A typical vertical circle reading on a Zeiss instrument having centesimal graduations is shown in Fig. 14.16. The vertical index level gives the index for the micrometer reading of the vertical circle at one end of the bubble. The vertical circle readings are indicated by V and are differently coloured to those relating to azimuth readings (Hz). The micrometer drum is used to obtain a bisection of double graduations by the division line to give 127·7 grade, and the upper edge of the index level then gives 0·064 grade to be added to the initial value. The same micrometer drum serves for both horizontal and vertical circles, but naturally the vertical index level is not associated with the former circle.

Fig. 14.17*a* relates to the Wild T1-A instrument and shows the path of a ray entering by reflection from the illuminating mirror. Rays pass through the vertical circle to a prism which transmits them downwards and back through the graduations of the vertical circle. Two further prisms then pass the rays downwards through a fluid held in a hermetically-sealed container. Any slight inclination of the rhombohedral prism above the the two-piece objective below it is compensated

FIG. 14.17*a*. WILD T1-A THEODOLITE
(Courtesy of Wild Heerbrugg, Switzerland)

by this fluid, whose surface will be horizontal when at rest. Two right-angled prisms now pass the rays through the horizontal circle graduations. Yet a further prism passes the rays outwards and upwards and through the plane parallel-plate optical micrometer which is controlled by a drum outside. The rays are next deflected horizontally inwards, images now being formed of the vertical circle again and the horizontal circle. The rays pass through the scale of the optical micrometer and the pencil of rays now carries three images which are deflected by a final prism into the microscope tube, giving readings, of the type shown in Fig. 14.17*b*, directly to twenty seconds or 0·01 grade. If a distant signal be observed, the vertical angles registered for the two telescope positions should sum to 360° 00′ 00″ in both the Zeiss Th3 and the Wild T1-A theodolites.

The pendulum prism is also used in the automatic indexing of the vertical circle of the new Watts ST 300 Microptic theodolite (Fig. 14.18), a range of ± 5′ from plumb being accommodated. This instrument has a micrometer giving circle readings direct to 10 seconds of arc (or to 0·005 grade), the setting mark being placed symmetrically about a main scale division. As in the "one second instrument" described in Chapter 4 both circles are viewed through a microscope on the right-hand upright.

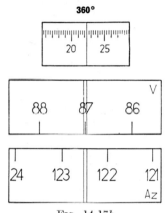

FIG. 14.17*b*
Vertical circle reading = 87° 22′ 20″
(*Courtesy of Wild Heerbrugg, Ltd.*)

The theodolite has an integral traverse base, its upper part being replaceable by targets so that the three-tripod system of surveying can be adopted. Upper and lower azimuth motions are incorporated and it will be observed that the tangent screws for the altitude and azimuth movements are concentric with their respective clamps.

Another instrument using a liquid compensator is the Kern DKM 2-A which reads direct to 1 second, and which will no doubt replace the DKM 2 (Fig. 4.24*a*) in the future. Coarse or approximate circle readings are still registered to 10 minutes of arc in the reading eyepiece, but the relevant value is enclosed in a small square on actuating the micrometer; the micrometer reading is also naturally shown separately, and is added to the coarse reading.

Adjustments to Modern Theodolites

Permanent adjustments for the glass-arc theodolites follow a similar pattern to those given in Chapter 4. The major adjustments quoted are for
(*a*) the plate bubble,

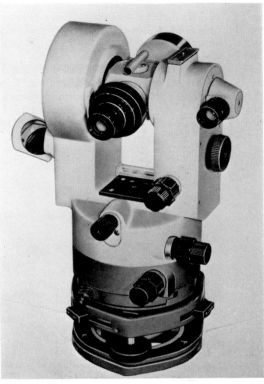

FIG. 14.18a. WATTS ST300 MICROPTIC THEODOLITE
(*Courtesy of Hilger & Watts, Ltd.*)

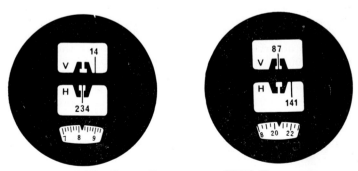

FIG. 14.18b. CIRCLE READINGS OF ST300 THEODOLITE
(*Courtesy of Hilger & Watts, Ltd.*)

(*b*) horizontal collimation (to place the line of sight at right angles to the trunnion axis), and

(*c*) adjustment of the vertical circle zero and altitude bubble.

The adjustment for horizontality of the trunnion axis (No. 3, p. 102) is discounted by Hilger & Watts, for the Microptic No. 2 theodolite, who state that precise machining ensures that this axis will remain at right angles to the vertical axis with sufficient accuracy. Cooke, Troughton & Simms give an adjustment similar to that quoted above for their Tavistock theodolite, but state that it is not likely to be needed unless rough usage or very long service is involved.

(*a*) and (*c*) have already been mentioned for the "compensated" theodolites and (*b*) can be carried out following the instructions given on pp. 99 and 102. Alternatively for (*b*), the telescope can be directed on to a sharp sighting mark some three hundred feet away, at about instrument height, and the horizontal circle reading noted on one face. After transitting and resighting, it is then noted on the other face. The two readings should differ by 180° and any discrepancy is twice the error. The circle reading which should have been noted can therefore be deduced and set by the micrometer in conjunction with the azimuth tangent screw: the graticule is then adjusted to ensure that the vertical line lies on the sighting mark. A repeat check can now be undertaken.

Adjustment (*c*), to remove the index error of glass-arc instruments such as the Watts No. 2 and Wild T2, etc., which are not "compensated," can be effected by first sighting a distant mark with the altitude bubble at the centre of its run and reading the vertical circle face left and then, after transitting, face right. In the case of the Watts instrument, the relevant angles of depression or elevation must now be deduced since the vertical circle reads zero when the telescope is horizontal on face left; but for those instruments which read zero when the telescope is directed to the zenith, the sum of the two should be 360°. In either case, the mean of the discrepancy can be allotted to the observed face-left reading to obtain a mean vertical reading. The telescope is then directed back to the sighting mark and the corrected vertical angle registered, using the micrometer in conjunction with the altitude-bubble tangent screw. All that then remains is to centre the altitude bubble by slackening one of the relevant adjusting screws and tightening the other.

Fennel FLT Code Theodolite. It will be appreciated that if the field data can be automatically registered, then booking errors are eliminated; in addition there is a likely saving in time. The TPR theodolite which is still manufactured by Askania, was one of the forerunners in providing such registration, but the whole operation is not automatic in so far as the provision of the final field data is concerned.

The FLT code theodolite (Fig. 14.19*a*) has been developed to allow the field measurements to be processed directly via the computer. Instead of normal ciphers, the circle graduations are coded numbers developed as a ring code (Fig. 14.19*b*) consisting of four information zones, representing hundreds down to tenths of grades in the centesimal

FIG. 14.19a. FENNEL FLT CODE THEODOLITE
(*Courtesy of Otto Fennel G.m.b.H.*)

FIG. 14.19b. RING CODE SECTION
(*Courtesy of Otto Fennel G.m.b.H.*)

system, with clock-tracks for positioning the black–white and white–black changes of ten complete sequences in each information zone. In addition to this digital information, each circle reading shows graduation lines, from two diametrically opposed positions, in two rectangular blocks; this is the analog part of each registration.

These circle registrations are recorded on 35 mm film (Fig. 14.19*c*), together with station and target identification data which are fed in by the surveyor using an input device. This film is casette-loaded and is contained in one of the standard casings near trunnion-axis level.

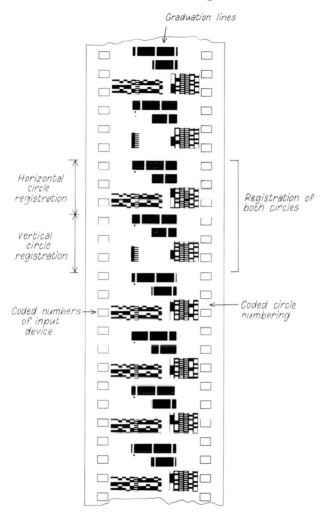

FIG. 14.19*c*. REGISTRATION OF CIRCLES
(*Courtesy of Otto Fennel G.m.b.H.*)

About 680 readings of the horizontal circle can be accepted by a film of length 10 m, but only 340 readings when vertical angles are being observed, since the horizontal circle reading is also then automatically exposed. By interpolating the line in the smaller of the two rectangular blocks of the analog portion with the lines in the longer block, readings are obtained to 0·0001 grade. One group of registrations therefore corresponds to an entry in a field book, but if the wrong characteristic number is applied to a detail point the code reading is inevitably disturbed.

For the determination of distances between traverse stations the 2 m horizontal subtense bar is used, whilst for detail points the 1 m vertical bar is used so that such points are located by length and bearing (i.e. by the polar system) and heighting information is also derived. After the film has been processed, the code is converted into a five-channel punched-tape code using the electronic converter Zuse Z24, and this tape can then be fed into a computer for the evaluation of the required information, including co-ordinates. It is understood that the accuracy of a parallactic angle measured once is $\pm$ 4·4 cc, i.e. $\pm$ 4·4 centesimal seconds (0·00044 g) or, say, $\pm$ 1·5 seconds sexagesimal.

Instruments of this nature are suitable for use in organizations such as the Ordnance Survey; for instance the West German Federal State of Hesse has a large number (over twenty) in regular use for land relocation purposes. Each instrument requires some seven people, including a surveyor and observer, in the field and although these instruments determine field data they cannot be used for setting out. There is, however, a definite saving of time in the field and in the drawing office.

Developments such as the FLT theodolite have been discussed by H. L. van Gent in a paper to the 12th F.I.G. Congress held in London during 1968. In this paper, he mentions that Messrs. Breithaupt of Kassel have produced a digital theodolite having both punched-tape recording and direct visual recording so that the pre-setting of certain angular values is permitted.

Askania Potentiometer Theodolite. Another interesting instrument, the Potentiometer Theodolite, has been developed for monitoring radio installations established for air traffic control, but amongst other possible uses are (i) the location of sounding vessels in hydrographic surveying (by intersection from two shore stations), (ii) the determination of satellite orbits, and (iii) pilot balloon observations for meteorological purposes.

The azimuth and altitude movements need to be continuous in the main function of this device; for the consequent continuous recording of the data, electronic means are employed, each conventional glass circle having been replaced by a worm-gear whose motion is transferred to a potentiometer. Precision of measurement accordingly depends upon the precision obtaining in the gears. The main body of the instrument (Fig. 14.20a), which has a "broken" telescope and also a finder telescope, contains the worm-gears, potentiometers, control units and dials for direct reading of the circles. The potentiometers are coupled to the worm-gear shafts through a changeover gear having two different transmission ratios, and to set them to zero the shaft of the change gears is coupled via a friction clutch. Fig. 14.20b shows the tracking gear, including a worm-gear drive which has a transmission ratio of 1:360, and also the drive to the relevant circle dial which has two "hands" to give direct readings. One of these "hands" gives 10° units and the other, which makes a complete revolution for each 10°, allows estimation to 0·01°.

FIG. 14.20a. ASKANIA POTENTIOMETER THEODOLITE
(*Courtesy of Askania G.m.b.H.*)

The potentiometers are connected by cable to the control box with which they form a double bridge circuit, and to the diagonals of this the recorders are connected.

Distance Measurement

Two modern distance measuring instruments have been reviewed in Chapter 9, both using the principle of modulated carrier waves. The Tellurometer MA100 and the Wild DI10 Distomat are two recent developments which are complementary to, say, the MRA101, and they are of particular interest to the civil engineer since their maximum ranges are in his general "field," i.e. 2,000 m for the MA100, and the second version of the DI10.

Both devices have light sources transmitting infra-red light, which is amplitude-modulated and returned after reflection by prism reflectors mounted at the far station. The phase of the returned signal is

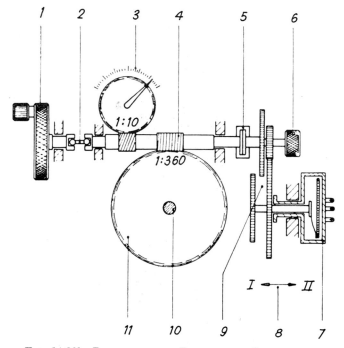

Fig. 14.20*b*. Potentiometer Theodolite: Schematic of Tracking Gear

There is one gear each for elevation and azimuth in the theodolite.

1. Control
2. Universal joint
3. Reading dial
4. Worm
5. Friction coupling
6. Potentiometer knob
7. Potentiometer
9. Range shifter
9. Gear-shift mechanism, 1:1 up to 1:10
10. Theodolite axis
11. Worm wheel

(*Courtesy of Askania G.m.b.H.*)

compared with that of the transmitted signal. Each instrument uses a gallium-arsenide diode as the light souce for the infra-red carrier beam and a photo-diode to receive the returned light, but different modulating frequencies are imposed on the respective carrier beams. For instance in the DI10 (Fig. 14.21) a frequency of 13,486,860 Hz is initially applied and the phase angle is established by a phase meter. The frequency is then raised to 14,985,400 Hz and the consequent change in phase is measured, counting the number of complete cycles in that change. It will be noted that the difference in frequencies implies a wavelength of 200 m. whilst the higher of the two frequencies implies a wavelength of 20 m. Therefore if a phase difference of $1.38\,\pi$ were detected a double distance of $1.38\,\pi \times 200/2\,\pi = 138$ m would be

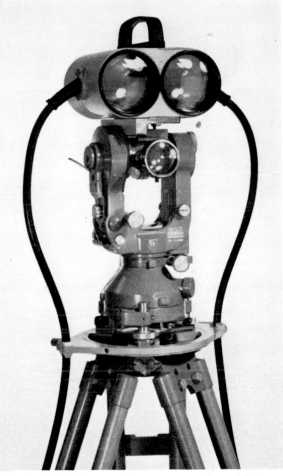

FIG. 14.21. AIMING HEAD OF THE WILD DI10
ON THE WILD T2 UNIVERSAL THEODOLITE
(Courtesy of Wild Heerbrugg, Switzerland)

being measured and this would be read out as 69 m, allowing for the outward and inward passes. In the first version of the DI10 two reading circles, set at right-angles to each other, are shown on the face of the control box; one of these registers tens and hundreds of metres, whilst the other circle is divided from zero to 9·98 m in intervals of

0·02 m and allows estimation to 0·01 m. The two circle readings are added together, the limiting registration being 1,000 m. In the second version the readout is direct to 0·01 m but 1,000 m must be added to readings taken when measuring distances between the limits of 1,000 m and 2,000 m. The panel of the control box has also been somewhat modified in the second version.

A variable-frequency oscillator produces the modulation frequencies mentioned, being synchronized with crystal oscillators of the relevant frequency. In order to read directly to 0·01 m a mean value over many cycles of the higher frequency must be obtained, since the phase shift of the reflected signal may fluctuate when subjected to air shimmer and "noise." Both the reference wave and reflected wave are transformed to a frequency of 2,400 Hz by heterodyning them with a third slightly different frequency. One cycle of this lower frequency is obviously equivalent in length to very many cycles of the higher modulation frequency, but the phase shift is still the same. This is established by a resolver, phase detector, and d.c. motor, the read-out circles being appropriately linked to the resolver.

During the required sequence of operations to establish distance, but before the direct reading is obtained, an internal calibration line is switched in, through which the radiation from the transmitting diode passes directly to the receiving diode. A calibration value is now set on the read-out circles to take into account geometrical and electronic eccentricities in the aiming head, which contains the diodes, and in the reflector. By actuating another control switch a "start" value is applied to those circles to allow for the change of electronic data of the phase-shifting elements during the change of modulation frequency so that the read-out refers to a realistic zero. When this control switch is set to "measure," distance is finally read out, the circles coming to rest after some 15 seconds.

This instrument can be mounted directly on a T2 theodolite (face right) and used directly to measure distance with bearing, but the MA100 has a better resolution since higher measuring frequencies, of the order of 75 MHz, are imposed on the carrier wave.

Laser Beams

The development of the laser (*l*ight *a*mplification through *s*imulated *e*mission of *r*adiation) has been one of the most important technological developments of the 1960's. Laser light is very intense and is essentially monochromatic. Moreover it can travel over appreciable distances in narrow beams without divergence, so that its power can be applied accurately. It also has a fourth characteristic, coherence, which can be taken here as meaning that it is composed of regular continuous waves, like those produced by a radio transmitter, and so its phase at any one instant bears some known relation to its phase after a known time interval. The laser can thus be considered analogous to the oscillator of such a transmitter except that the oscillations are produced at optical frequencies which are much higher than radio frequencies. Lasers can be categorized as pulsed solid-state, pulsed gas, continuous-wave gas and semi-conductor junction lasers. Such a vast field cannot be effectively reviewed here but the student can refer to, *A Guide to the Laser* published by Messrs. MacDonald, for general information, and to a paper *Lasers and Surveying* presented by L. F. Rentmeester to the 12th F.I.G. Congress (*op. cit.*) for background information.

As mentioned narrow laser beams can be projected and this is most advantageous since errors due to ground reflections can be eliminated. The MRA101 Tellurometer, for instance, has a beam width of only 6° but some ground reflection or traffic reflection may occur and induce slight errors in measurement. Also, like the Geodimeter Model 6, lasers only require reflectors at the far station, whereas another instrument is needed there if microwave methods are being used for the distance measurement. The advantage of the microwave system over the Model 6 is that radiation from the latter has to compete with radiation from the sun, so that its daylight range is reduced, but the laser allows improvement on this. In addition the coherent light of the laser can be modulated at very high frequencies, thereby improving the resolution of distance, and since this can be frequency modulation as in the MRA101, there is reduced sensitivity to intensity fluctuations caused by the atmosphere. Instruments based on conventional light sources have to adopt amplitude modulation, since their light is not coherent, and such devices are more sensitive to the above fluctuations.

The continuous-wave gas laser is very suitable for the modulation technique of distance measurement and Messrs. AGA have produced the Geodimeter Model 8 based on a continuous-wave helium–neon laser. It has an optical unit and control unit in one container mounted on a tripod; the optical unit, shown diagrammatically in Fig. 14.22, includes the laser, the modulation system, the transmission optics and the receiver unit.

The laser beam has a diameter of 20 mm after expansion, and its divergence is said to be 0·1 m per kilometre. Accordingly, to assist in pointing, a lens may be temporarily inserted into the beam, thereby widening it vertically. On traversing in azimuth some reflected light should be observed through a sighting telescope, whose eyepiece is mounted on the panel of the container, when the beam passes the reflector. A fine pointing can then be effected after the lens has been withdrawn from the path of the laser beam.

In the model 6 Geodimeter a Kerr cell was used for modulating the light, but the Model 8 employs crystals whose optical characteristics change when subjected to an electrical field. Four frequencies between 29,970,000 Hz and 31,467,000 Hz can be applied to the transmitted beam, and the photocell in the receiver unit is caused to vary in sensitivity in phase with another frequency, 1,500 Hz lower than the relevant modulator frequency. Thus the output of the photocell, which is subjected to the returned beam, has a frequency of 1,500 Hz and its phase shift is equal to the phase shift of the transmitted signal. This phase shift is determined by reference to a signal of the same frequency, formed by direct mixing of the two frequencies, whose phase is delayed in the resolver and then compared to the photocell signal phase in the detector. The resolver knob is turned until the null indicator on the panel indicates zero, and then the phase shift of the resolver is read out on a three-figure register. By using the various frequencies and by combining the measurements to the reflectors with measurements of a calibrated path within the Geodimeter, the distance to the reflector can be ultimately calculated. The maximum range is quoted as 50 km, with a mean square error of 1 part per million $\pm$ 5 mm.

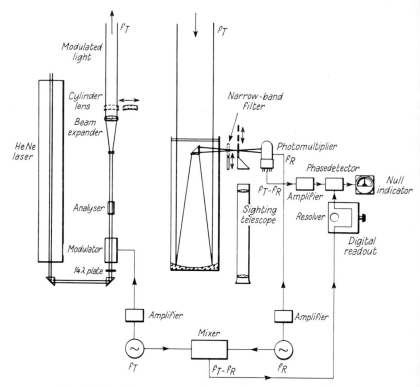

Fig. 14.22. Diagrammatic Presentation of the Optical Unit of the Geodimeter Model 8
(Courtesy of AGA (U.K.), Ltd.)

The helium–neon laser emits its radiation within a narrow band width in the red part of the spectrum where the sun's radiation is lower, and so this particular model does not suffer from the daylight restrictions previously mentioned. An optical filter is also inserted in front of the photocell and this allows the passage of most of the returned laser beam but eliminates light of other wavelengths in the range sensitive to the observer's eye and to the photocell. This particular laser is of low power and consequently Messrs. AGA state that there is virtually no risk of eye damage, particularly if normal precautions are taken. On the other hand the ruby laser, which belongs to the pulsed solid-state type, is capable of a very high power over a very short period of time, and damage will be incurred by the eye if it is struck. Ruby lasers can be used in military range-finding, since high power is needed for a good return signal from a diffusely reflecting object and where the requirements of accuracy are not so stringent.

Messrs. AGA also manufacture Geodimeter model 700 whose light source is again a helium–neon laser. This instrument will read out directly horizontal distance, if required, rather than the slope distance, on a console unit separated from the instrument unit. It has a range of 3 km and its accuracy is quoted as $\pm$ 5 mm to $\pm$ 1 mm per km. In addition however, vertical and horizontal angles can also be read out since a coded angular system has been incorporated. The following combinations are possible—

(i) As a combined instrument for the establishment of either vertical angle and slope distance or horizontal angle and horizontal distance.

(ii) As a "one-second" theodolite.

Thus, the instrument can be used as a setting-out device as well as a distance measurer.

One other interesting application of the helium–neon laser is for the control of alignment since it gives a collimated visible light which can serve as a setting-out line between a transmitter and a reflecting target. Messrs. Watts manufacture alignment equipment comprising the Spectra Physics Laser Model 120 with a beam spreader telescope and mount. A beam of red light is given by the laser as a reference line, diameters of 16 mm and 101·6 mm obtaining at the instrument and at a distance of 1·61 km respectively. One of the two forms of mount, the ST600, can be fitted into the co-ordinate stage levelling base, a component of the Watts Traverse and Base-line Measuring equipment. This unit consists of a three-screw levelling base supporting the co-ordinate stage which has movement ranges of $\pm$ 19 mm in two mutually perpendicular directions in plan: these directions can be set as required by means of a separate azimuth motion. A theodolite, if fitted with the same base flange as the ST600 mount, can also be accepted by this base for setting-out purposes, as indeed can an optical plummet and targets.

The other mount, ST603, uses a platform system, one part being permanently fixed to a Watts ST200 theodolite (*see* Chapter 4) and the other part to the laser, the two being clamped together when the laser is mounted above the theodolite telescope. Adjustments are provided so that the laser beam centre line is parallel to the line of sight of the theodolite both in azimuth and elevation.

Since a suitable laser beam can be seen on any target introduced into its path, if such a beam can be placed in the line of sight of a theodolite, then it can be used for setting-out purposes using the horizontal and vertical motions of the theodolite. The Kern DKM2-A Laser Theodolite is one such instrument using this principle. A laser is mounted on the tripod of a Kern DKM2-A theodolite and its light is passed through a condener lens whereby its diameter is reduced sufficiently for it to pass along a light-carrying fibre of diameter 0·08 mm. This fibre passes the light to a reticule in the telescope of the theodolite and this illuminated reticule indicates the line of sight. The beam is placed on the telescope axis by means of a light-dividing cube whilst a filter absorbs any laser light which might be reflected towards the eyepiece. Most commercially available lasers can be used as the light source providing

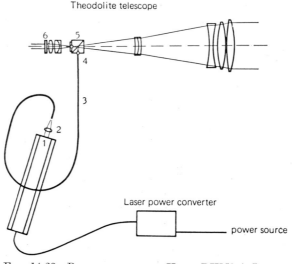

Theodolite telescope

Laser power converter

power source

FIG. 14.23. PRINCIPLE OF THE KERN DKM2-A LASER THEODOLITE

1. Laser beam	4. Reticule
2. Condenser lens	5. Light-dividing cube
3. Light-carrying fibre	6. Filter

that the necessary connectors are available; the one supplied by Kern is the Spectra-Physics Gas-laser 120T. Under favourable atmospheric conditions targets can be aligned fairly precisely to within a few millimeters within a range of 400 m; the theodolite can be used independently of the laser beam if required.

The value of the laser as a means of plumbing in deep shafts or vertical alignment on construction sites is also being studied. Fig. 14.24 indicates the principles of a device designed by Chrzanowski (op. cit.) and Masry in which a helium–neon laser is used in conjunction with an autocollimating telescope. The flat surface mirror is initially adjusted to be perpendicular to the line of sight of the telescope and then the laser tube is adjusted until a spot of light, formed by the laser beam partially reflected by the beam splitter and then the flat mirror, coincides with the centre of the telescope reticule; a blue filter is used for viewing the laser light through the telescope. This places the line of sight perpendicular to the laser beam and the whole unit is then so positioned that the beams reflected by the mercury pool and beam splitter coincide with the line of sight of the telescope and the beam

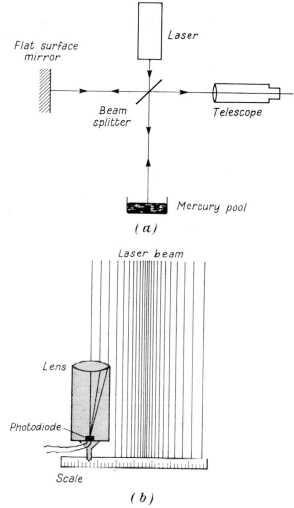

Fig. 14.24. PLUMBING BY LASER

(a) Laser optical plummet.
(b) Location of centre of intensity of laser beam.

reflected from the mirror. This last stage ensures that the line of sight will be vertical.

The centre of intensity of the laser beam has to be established for accurate work since it is known to move about the position that it would take up in a uniform and still atmosphere. Chrzanowski and

Ahmed have designed a detector for this purpose which locates the mean position of two points of equal intensity located in the beam on either side of the estimated centre line. This is repeated in a plane at right angles to the previous plane and therefore the centre of the beam can be located. This is achieved by a lens focusing the laser beam on to a photocell vertically below, whereupon a milliammeter measures the output of the cell. The marked optical centre of the lens indicates the position of the centre of the beam.

Gas lasers can also be used in holography or wave-front reconstruction photography. In normal photography the emulsion responds to intensities of light but not to relative phases of light from the object so that a two-dimensional image results. In holography the light waves scattered by the object are recorded on a plate in such a way that the amplitude and phase information is retained. This hologram can then be illuminated by laser light to reconstruct the object three-dimensionally in the observer's eye. Initially coherent light was essential in both stages, but ordinary photography has been used by Redman and Wolton* in the first stage, with laser light in the second stage producing the three-dimensional image. This technique might well be developed further in photogrammetry.

Precision Measurement

In the earlier chapters, attention has been paid to the principles involved in triangulation and topographical surveys and in some setting-out procedures. Increasing use of high precision measurement is now also made to meet the needs of present-day construction and industrial processes. Some examples of both techniques and instruments used in shafts, measurement of structural deformations in dams, autocollimation and tracking of missiles are discussed in the following paragraphs.

Vertical Alignment

Optical plumbing becomes important with the increase in height of modern framed structures, being required both in the erection of the framework and in the construction of the internal shafts. Plummets are available for upwards and downwards sighting to allow the establishment of a vertical line, and these are normally manufactured so as to be interchangeable with theodolites on their tripods.

Watts Autoplumb. This instrument has two telescopes, one of low power for sighting downwards and locating the instrument over the ground mark, and the other of high power for sighting upwards on a target (Fig. 14.25a). Coarse levelling is sufficient for locating the Autoplumb over the ground mark, since the upward-sighting telescope is fitted with the same type of compensator as is used in the Autoset levels. This automatically ensures a vertical line of sight, even though the instrument is tilted by several minutes of arc. Fig. 14.25b illustrates the optical principles.

---

* *Holographic Photogrammetry*, J. D. Redman and W. P. Wolton, 11th Congress of the International Society of Photogrammetry, Lausanne, 1968.

For external use, a target mounted on an adjustable offset arm is clamped to the structure; for floor-to-floor use, viewing is carried out through sighting holes left during construction. Two sightings are made on the target, the second one after rotating the Autoplumb through 90° in azimuth. As shown in Fig. 14.25c, a truly vertical line

FIG. 14.25a. WATTS AUTOPLUMB
(Courtesy of Hilger & Watts, Ltd.)

is established by positioning the target until it is located at the intersection of the two planes defined by a horizontal line on a graticule seen through the eyepiece of the upward-sighting telescope.

Errors in verticality are measured by means of an internal optical micrometer controlled by a graduated drum; this tilts the upward line of sight either side of the true vertical. Accuracy of a single sighting is not less than say 1 mm in 40 m, and, for two opposite sightings with the instrument turned 180° between each, this can be improved to about 0·3 mm in 40 m.

U$ SE $ AND A$ DJUSTMENT $ OF THE A$ UTOPLUMB $. The instrument is attached to a three-screw levelling base which can be mounted on a tripod, and a centring motion is available for positioning over ground points. Since the downward-sighting telescope is not compensated, it is essential that the instrument be levelled using a plate bubble, positioned parallel to the telescope axes, and a circular bubble, found at the top of the case.

The upward-sighting telescope is, as mentioned earlier, provided with a compensator of the type included in the Autoset level, and the sighting line is deflected upwards by means of a pentagonal prism which can be tilted by a micrometer movement. In the field of view, a horizontal line defining a vertical plane, together with three concentric circles, will be observed, and the line of sight is vertical when

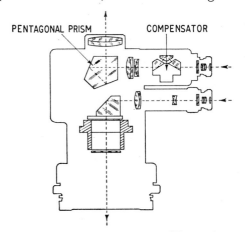

F$ IG $. 14.25$b$. O$ PTICAL $ S$ YSTEM $ OF THE W$ ATTS $ A$ UTOPLUMB $
(Courtesy of Hilger & Watts, Ltd.)

the micrometer reading is 10·00. One revolution of the micrometer drum displaces the line of sight by an angle of $\tan^{-1} 0·001$: one division of the drum scale represents one hundredth of this movement. A movement of $\tan^{-1} 0·002$ is possible in either direction from the vertical, and accordingly the displacement of the line of sight at a height H above the instrument axis, due to a micrometer reading change from, say, 10·00 to 11·52, will be $1·52H/1,000$.

A pawl engaging with vee-slot locators, positioned at 90° to each other on the bevelled fitting just above the instrument base, allows the instrument to be readily placed in two positions mutually at right angles.

Fig. 14.25d shows the plan of four ground stations in a building with holes left for the vertical sighting line in the falsework, etc., of the floors, together with a target, about 250 mm², on which a line AB has been scribed. An assistant arranges that line AB is brought into coincidence with the horizontal line in the field of view on instructions from the surveyor at the instrument. Extension lines AA_1 and BB_1 are then scribed on to the falsework or on to a suitable plate. The Autoplumb is swung through 90° and line AB is now positioned at $A'B'$ to lie on the horizontal line in the field of view. Extension lines $A'A_1'$ and $B'B_1'$ are again scribed and the target removed.

Lines can now be set on the pairs of scribed lines to give the intersection or, alternatively, if lines PO and $P'Q'$ mutually at right angles have been previously scribed on a Perspex or metal plate larger in dimension than the hole in the falsework, then these lines can be brought into coincidence with lines A_1ABB_1 and $A_1'A'B'B_1'$, and point O is

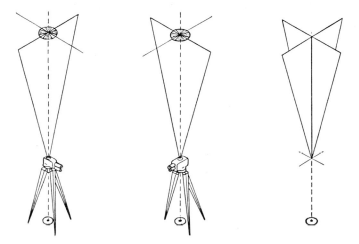

FIG. 14.25c

Vertical line defined at intersection of two planes at right angles.
(Courtesy of Hilger & Watts, Ltd.)

now located to correspond with its respective ground station. Measurements can be related to point O for setting-out purposes.

It has of course been assumed above that the instrument is in adjustment and that the line of sight is vertical with the micrometer set at 10·00. To check the adjustment of the instrument, sightings can be made near to a target line with the instrument set at 10·00. The target line is moved until it is close to the horizontal sighting line in the field of view, and then, using the micrometer drum, correct lining-in can be effected. Reading M_1 is now noted. The instrument is rotated through 180° in azimuth and a further reading M_2 of the micrometer taken for coincidence. If the instrument is in adjustment, $\frac{1}{2}(M_1 + M_2) = 10\cdot00$: if not, then $\frac{1}{2}(M_1 + M_2)$ is set as the micrometer reading and the target moved for coincidence with the line of sight which is now vertical.

The micrometer can now be set at 10·00 and the line of sight moved back into position by adjustment of the reticule. The adjustment can be rechecked now, but the manufacturer states that the error should

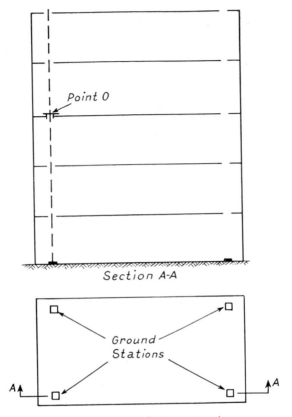

Point O

Section A-A

Ground
Stations

A ↑   ↑ A

Plan at Lowest Floor Level

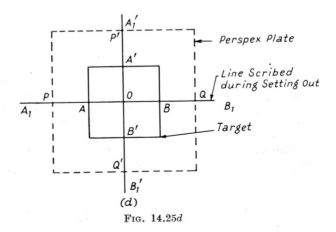

A_1'

P' ——— Perspex Plate

A'

Line Scribed
during Setting Out

P O Q B_1

A_1 A B

B' ——— Target

Q'

B_1'

(d)

FIG. 14.25d

not exceed 0·50 revolution of the drum for this correction to be carried out.

It will be appreciated that if the zeroing of the micrometer drum is correct, and the micrometer drum reads M_1 on lining onto the target, then

$$\text{Relevant displacement} = \left(\frac{10 \cdot 00 - M_1}{1,000}\right)H$$

where H is the height of the target above the instrument.

Various forms of target are available for use with structural steelwork, etc., and a more refined floor target (ST554) than that discussed is available, but it will be apparent that targets can be designed by the surveyor to suit the problem in hand.

Kern and Wild each make such an instrument, the former with separate telescopes for upward sightings and the latter with a switch-over prism and one telescope. A mean error of $\pm$ 1 mm to $\pm$ 2 mm for a single measurement at one hundred metres is quoted for the Kern OL.

Mine Tunnel Shafts

In addition to alignment it is also necessary to measure shaft depths, Common practice is to measure depth directly by steel tape if the shaft is not of great depth, or alternatively to work down the shaft in tape lengths, the engineer using the hoisting device when marking the position of the lower end of the tape against the shaft wall.

Particularly in mining engineering, where deep shafts are encountered, the transference of temporary bench marks down such shafts and the checking of them is a difficult and arduous task. Dr. D. J. Hodges* has described a method for expediting the procedure using the Geodimeter.

Fig. 14.26 illustrates the method. At station Y over the shaft is a plane mirror capable of accurate orientation in the required vertical and inclined planes. The geodimeter is set up at X and a reflector is set at Z at the bottom of the shaft or any intermediate point whose level is required. With the mirror Y in the vertical plane, the distance XY is measured by the geodimeter, after which the mirror is rotated and the distance XYZ is measured. By subtraction, the depth YZ is obtained, and if the R.L. of XY is known, the setting up of a T.B.M. at the bottom of the shaft can be carried out by reference to Z.

Deformation of Structures

The principles of triangulation can be adopted to measure the deformation undergone by a structure such as a dam. In this case, pins can be sited on the downstream face of the dam and their positions, both horizontal and vertical, can be determined with respect to observation pillars located on firm rock in the vicinity. In order to check the behaviour of these observation pillars, further pillars are constructed outside the pressure zone of the dam so that the observation pillars can be position fixed each time they are used for the "deformation

* *Colliery Guardian*, 25th July, 1963.

survey." It is essential that centring errors be a minimum, and to this end various methods are employed such as grouting a centring plate to the top of the pillar, or employing special ball-tipped bolts at the base of the instrument to fit into a hollow cylinder cast into the pillar. An instrument reading to the order of 0·5 sec or better would be used in the measurement of angles.

It is also possible to run precise traverses along the inspection galleries of the dam and along the top, and these can be linked with the observation pillars of the triangulation net at suitable points. Some plumbing will be involved to penetrate or extend to the internal galleries. The traverse sides can be measured directly by invar wires under constant tension and the angles by the type of instrument just

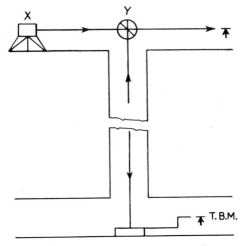

FIG. 14.26. SHAFT-MEASURING WITH THE GEODIMETER

mentioned. Many more points can be tied-in by precise traverse than by triangulation, but since the traverse is founded on a triangulation net, absolute deformations are determined. Outlines of methods and equipment used are given in Bulletins 5 and 7 published by Kern and Company, the latter containing accounts by K. Egger of work carried out on certain dams in Switzerland.

The measurement of such deformations can also be established by laser beams transmitted from stations outside the pressure zones. Messrs. Spectra-Physics claim that an accuracy of ± 1 mm is attainable by their Geodolite 3G over distances not greater than 1,000 m and that small movements applied to reflectors positioned on a dam have been accurately recorded. This device incorporates a helium–neon gas laser whose beam is amplitude modulated: five frequencies are available, the higest being 49,164,710 Hz and resolution to the larger of either one millimetre or one part in a million is quoted. Like the Geodimeter Model 8 this instrument has geodetic applications as well, whilst another version, 3A, has photogrammetric applications since it can act as a precision altimeter for vertical control.

The British Aircraft Corporation have introduced the Electrolevel to assess structural displacements. As shown in Fig. 14.27 the basic component is a spirit level vial containing an alcohol solution and three platinum electrodes. When the bubble moves the electrical resistance

Electrolevel vial

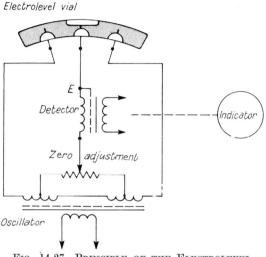

FIG. 14.27. PRINCIPLE OF THE ELECTROLEVEL
(*Courtesy of British Aircraft Corporation*)

between the inner and outer electrodes changes and a signal proportional to the angular displacement from the horizontal is given. The device is in fact equivalent to a damped pendulum.

Remote readings of the induced deflections can be taken at a conveniently placed detector and this is a considerable advantage when the structural member is difficult of access.

Autocollimation

In certain industrial processes and installations and in aspects of nuclear research a high accuracy of pointing is required, and it is possible to obtain theodolites which have had an autocollimating device incorporated. Using these a line can be established at right angles to a plane mirror target or, conversely, the mirror can be positioned at right angles to a fixed line of sight.

For a discussion of the use of autocollimation in precise setting out, the reader is referred to "Precise surveying in the construction of Nimrod," by D. W. Berry, in the *Chartered Surveyor*, April, 1961: a Watts Engineering Microptic No. 2 instrument was used in this exercise. This instrument (Figs. 14.28 and 14.29) is provided with auto-collimation and auto-reflection facilities, the conventional internal focusing system being combined with an auto-collimating eyepiece having a filament lamp and semi-reflector. When the instrument is focused to infinity and the lamp switched on, an image of the duly illuminated graticule is projected through the objective and reflected

FIG. 14.28. ENGINEERING MICROPTIC THEODOLITE
(Courtesy of Hilger & Watts, Ltd.)

back by a mirror established at the far station. The returned image is now seen through the eyepiece of the telescope, since the rays can pass through the semi-reflector, and the line of sight is perpendicular to the reflector target when the image is superimposed on the graticule viewed directly. If the sighting distance is such that a relatively small

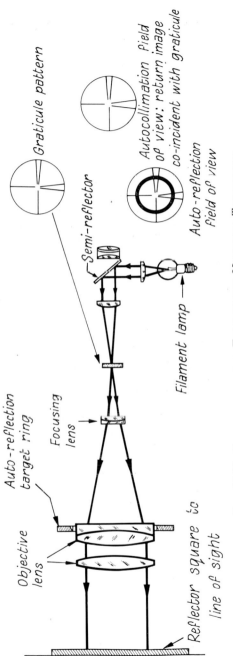

Graticule pattern

Autocollimation field of view: return image co-incident with graticule

Semi-reflector

Auto-reflection field of view

Filament lamp

Auto-reflection target ring

Focusing lens

Objective lens

Reflector square to line of sight

FIG. 14.29. PRINCIPLES OF THE ENGINEERING MICROPTIC THEODOLITE
(*Courtesy of Hilger & Watts, Ltd.*)

image is presented the auto-reflection technique may be adopted. It will be noted in Fig. 14.28 that a white ring surrounds the objective lens and, with the telescope focused to infinity, the reflection of this ring can be seen in the field of view. By adjusting the telescope pointing, or the target position as the case may be, the reflected ring image can be arranged symmetrically about the graticule. If the auto-collimation system be now switched in, assuming that the returned image is determinable, still more precise adjustment may be possible. There is also a small ring on the surface of the objective lens and this can be used in auto-reflection to achieve perpendicularity since it too can be placed symmetrically about the graticule together with the ring image.

The three-tripod system of traversing is catered for by one model of this theodolite, and versions with hollow vertical axes to allow direct downward sighting are available, and these can include modified trunnion axis bearings for mounting and using in unusual positions. A parallel-plate micrometer can be supplied for measuring alignment errors, etc., and naturally special mounting stands in lieu of tripods are available in this class of work.

Technical Reports 4 and BS7 published by Messrs. Hilger & Watts also refer to aspects of work covered by Mr. Berry. Apart from measurements of angle by theodolite and of length by invar tape in catenary, levelling operations were carried out and it was found that the Watts Autoset level previously mentioned gave a precision equal to the nominal Precise level with fewer observations (ratio 1:1·5).

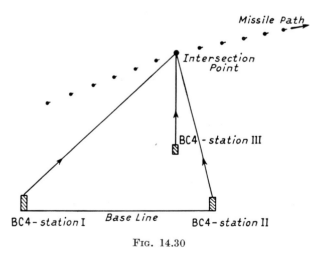

FIG. 14.30

Missile Tracking

An Askania theodolite developed for first-order triangulation surveys and for observation of the passage of artificial satellites has been discussed earlier in this chapter, and further instruments are now available for observations relating to the trajectories of missiles. Such an instrument is the Wild BC4 Ballistic Camera which is a precise photo theodolite.

The use of the photo theodolite has been discussed in Chapter 11, and it has been shown that the location of one point in plan requires one such instrument being set up in turn at each end of a base line or

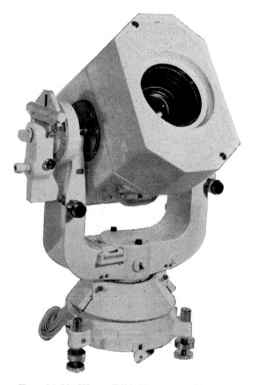

FIG. 14.31 WILD BC4 BALLISTIC CAMERA
(Courtesy of Wild Heerbrugg, Ltd.)

alternatively, of course, two instruments working simultaneously. This must be the case in the observation of missiles which are moving at high velocity, and normally at least three synchronized instruments, positioned as shown in Fig. 14.30, will be required for position fixing on their paths.

The BC4 instrument, shown in Fig. 14.31, can be said to consist of two fundamental units—the camera carrier and the camera body. The former is a modified portion of the T4 theodolite, consisting of the base and alidade plus standards. Horizontal circle readings are made direct to 0·1 sec through an eyepiece.

The camera body rests in the Y bearings of the standards and consists of the housing, lens diaphragm and shutter, and focal plane frame with various recording instruments. Five different lens cones are available, four being equipped with Wild lenses used in normal aerial cameras of focal length 115 mm or 210 mm and one being of focal length 305 mm. In all cases, a picture of size 180 mm square is obtained and

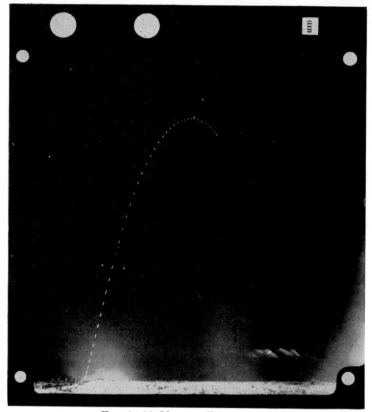

FIG. 14.32 MISSILE TRAJECTORY
(*Courtesy of Wild Heerbrugg, Ltd.*)

on it are given data including the focal length and fiducial marks. The number of exposures per second can be varied from 30 to 6.

A field adjustment frame is supplied for orientation of the instrument on terrestrial targets. This is firmly clamped to the focal plane frame and carries magnifying glasses at each corner which allow it to be centred with respect to the fiducial marks: three slow-motion screws are used for this purpose. A telescope eyepiece is also carried at its centre and this, together with the reticule and the camera lens, forms a telescope which allows the orientation of the instrument. A typical trajectory photograph is given in Fig. 14.32.

ADDITIONAL EXERCISES

1. The sketch plan, below, shows the area of a smallholding which is to be surveyed by chain survey methods. A provisional layout of survey lines is shown on the plan: AD is 94·08 m long and the wall has a height of 2·5 m.

Criticize the choice of lines and show on the plan your amended layout of survey points and lines, giving reasons for your selections.

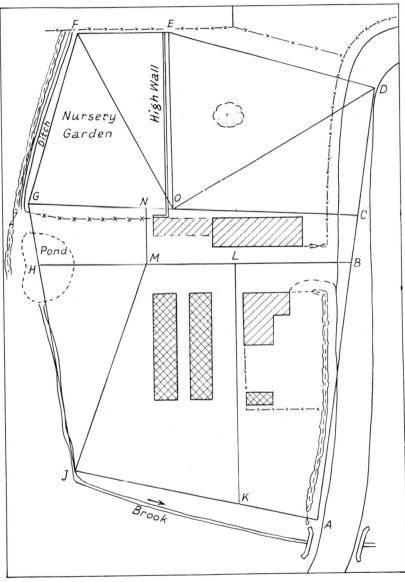

2. What are the main differences between chain and theodolite traverse surveys, and how do they affect the selection of survey stations?

What methods would you adopt to carry out the following surveys? Give reasons for your selections.

(i) A built-up area of about 10 ha, being part of a redevelopment scheme to be plotted to scale 1/200.

(ii) An area of about 4 ha in open agricultural land with little detail, to be plotted to a scale of 1/500.

(iii) An area of about 80 ha in rough country overgrown with scrub. The plan is required quickly and is to be plotted to a scale of 1/1,250, with approximate spot levels.

3. On a chain survey, the positions of two points C and D have been established by means of triangulation from points A and B. The lengths obtained are AB 600, BC 400, AC 600, AD 300 and DB 700.

Calculate the length of DC.

Answer: 506 units.

4. (*a*) Describe the features possessed by a geodetic level which are not normally associated with the engineer's level and briefly explain how these features enable precise levelling to be carried out.

(*b*) A storm-water drain is to be constructed to a gradient falling at 1 in 150 for 65 m from manhole B and at 1 in 100 for 75 m from manhole B to manhole A which is to have an invert level 91·856 A.O.D.

Levels are taken from a benchmark, value 92·450, beyond manhole A, and pegs are driven at the positions of the manholes from which sight rails for use with a 3 m long "traveller" are to be fixed.

Compute the reduced levels in the accompanying table by the Height of Instrument method and determine staff readings "a", "b" and "c", each of which will indicate when the respective peg has been driven to the stated level below the corresponding sight rail.

B.S.	I.S.	F.S.	R.L.	Staff Station	Remarks
3·176			92·450	B.M.	
	3·015			M.H.A	Ground level
	"a"			M.H.A	Peg 1 m below sight rail.
	1·064			M.H.B	Ground level
	"b"			M.H.B	Peg 1 m below sight rail.
1·469		0·832			Change point
	1·384			M.H.C	Ground level
	"c"			M.H.C	Peg 1 m below sight rail.
1·237		2·225			Change point
		2·822		B.M.	

Draw a longitudinal section to horizontal and vertical scales 1:1,000 and 1:100 respectively.

Answer: "c" = 1·198 m.

5. Two sewers each of diameter 0·3 m are to be laid from points A and C to a low point B, the gradients of AB and CB being 1 in 100 and 1 in 200 respectively. The discharge from point B is taken to a stream, and the ground levels and chainage are as follows—

Point	Chainage from A (m)	Ground Level (m A.O.D.)	Invert Level (m A.O.D.)
A	0	31·846	
	40	31·708	
	80	31·120	
B	120	30·816	28·712 Lowest point
	160	30·724	
	200	30·532	
C	240	30·498	

Plot a section of the sewers to a scale of 1:1,000 horizontal and 1:100 vertical, and show the invert levels at points A and C.

What are the maximum and minimum depths of dig?

Select a suitable length of sighting rod for each length of sewer and fix levels for sight rails at A, B and C.

Answer: Maximum dig = 2·196 m; Minimum dig = 1·166 m.

6. (*a*) What is meant by the sensitivity of a bubble tube?

(*b*) A level sight held on a staff held 80 m from the instrument reads 2·378 m and the bubble is found to be two divisions off the centre of the run towards the staff. If the level tube is in adjustment and has a sensitivity of 40 seconds, what is the true reading on the staff? Take $\sin 1'' = 1/206,265$.

Answer: 2·347 m.

7. The following staff readings were obtained when levelling for the line of a proposed sewer, the readings underlined being backsights. The bracketed values were taken beneath a footbridge, the positive reading being obtained with the staff held upside down on the soffit of the bridge.

<u>0·518</u>   (O.B.M. 86·520)   3·962   <u>0·182</u>   0·518   0·072   (0·894,   + 3·662)   1·036

(T.B.M. 82·220).

Book and reduce the levels by both methods, carry out any arithmetic checks which may be possible and state any misclosure involved in the work.

What is the clearance underneath the footbridge?

Answer: 4·556 m.

8. Four points A, B, C, D are pegged out in a straight line so that $AB = BC = CD = 30$ m. A level is set up at B and the readings taken on staffs at A and C are 1·624 m and 1·262 m respectively. With the level at D, readings on staffs at A and C are 1·788 m and 1·456 m respectively. Calculate the collimation error of the instrument.

Describe how to correct this maladjustment with the instrument at D, (i) for a tilting level, (ii) for a dumpy level.

Answer: 0·050 m in 100 m.

9. A triangulation survey has to be continued from a trigonometrical station near the coastline of a territory to an island out at sea. A station has been established on a mountain on the island at a known level of 825 m and, while ascending to the station on the mainland, it is noticed that at a height of 150 m, shown on the aneroid, the island station just appears above the sea horizon.

Considering the combined effects of curvature and refraction, estimate the distance away of the island station, and mention reasons why the result may be in error. Assume that the earth is a sphere of radius 6,367 km.

Answer: 158 km.

10. The following observations were obtained during an open theodolite traverse from A to D in order to set out a sewer tunnel in a built-up area—

Horizontal distance $AB = 150$ m	Horizontal angles
Horizontal distance $BC = 105$ m	(clockwise from back station)
Horizontal distance $CD = 90$ m	$A\widehat{B}C = 60° 20'$, $B\widehat{C}D = 120° 10'$

Calculate the plan length and bearing of the tunnel centre-line AD, given that the bearing of AB is N 59° 00' E.

Answer: Bearing N 26° 08' W.

11. An open theodolite ground traverse carried out between two points A and B on the line of a proposed tunnel resulted in the following observations being obtained—

Line	Whole Circle Bearing	Length (m)
A—(i)	086° 37′	128·88
(i)—(ii)	165° 18′	208·56
(ii)—(iii)	223° 15′	96·54
(iii)—B	159° 53′	145·05

Calculate the bearing and distance of point B from point A.

Answer: 157° 35′; 43·35 m.

12. Two stations X and Y are 510 m apart. From theodolite stations R and S on opposite sides of XY, the following angles were observed—

$$X\widehat{R}S = 54° 12′ \qquad X\widehat{S}R = 49° 18′$$
$$S\widehat{R}Y = 41° 24′ \qquad Y\widehat{S}R = 47° 12′$$

Calculate distances XR and YR to the nearest metre.

Answer: 353 m; 376 m.

13. Show that the horizontal distance (H) between a tacheometer, having a multiplying constant K and zero additive constant, and a vertically held staff is given by the expression $H = K s \cos^2 \theta$, where s is the relevant staff intercept, and θ is the angle of inclination of the line of sight.

Such a tacheometer, having a multiplying constant of 100, was set up at station A whose reduced level was 51·24 m above datum. The telescope was directed to a staff held vertically at station B, reduced level 54·67 m, and an angle of elevation of 2° 30′ was recorded on the vertical circle. If the instrument axis was 1·10 m above A and the horizontal distance from A to B was known to be 86·1 m estimate the staff readings.

Answer: 1·00; 1·43; 1·86.

14. A road 10 m wide, built on an embankment, has its centre line on a circular horizontal curve of radius 200 m, the road being level. A scheme for widening envisages increasing the formation width by 3 m on the inside of the curve, the side slope originally at 2 (horizontal): 1 (vertical) being made 3:1.

The ground can be assumed to be level transversely but rising in the longitudinal direction such that the height to formation level increases uniformly from 5 m to 7 m between chainages 2,200 and 2,300.

Calculate the quantity of filling required between the above chainages, allowing for the curvature but neglecting any effect of bulking or shrinkage, by applying the prismoidal rule to areas of cross-section at chainages 2,200, 2,250, and 2,300 m.

Answer: 3,322 m³.

15. Tabulated below are the areas within the contour lines at the site of a reservoir, obtained by planimeter from a plan of scale 1:1,000.

Contour Level (m A.O.D.)	Area (mm²)
60	85,160
58	79,355
56	70,970
54	61,835
52	56,775
50	40,290
48	16,900

If the lowest draw-off level is 48 m A.O.D. and the maximum top-water level 60 m A.O.D., estimate (a) the full storage capacity, (b) the top-water level for 50 per cent of full storage capacity.

Answer: Capacity 650,000 m³, say.

16. A trench is to be excavated in open cut for a pipe of diameter 1 m, the sides having a batter of 1 vertical to 2 horizontal.

If the width of the trench at the bottom of the excavation is to be 2·4 m and the ground surface has a slope of 1 in 10 at right angles to the pipe line, calculate the volume of excavation between two points P and R, 100 m apart. The depths of excavation at P, Q and R are 1·80 m, 2·10 m and 2·30 m respectively, and $PQ = QR$.

Answer: 143·2 m³.

17. (a) Write notes on the method of setting out in the field half-widths of embankments or cuttings.

(b) The following table is an extract from the field notes taken in setting out the half-widths of an embankment whose formation width is 20 m.

The numerators are the depths below formation level of the ground at the centre line and at the toes of the side slopes on either side, whilst the denominators are the corresponding horizontal distances from the centre line to these points.

Calculate the total volume of material between the extreme sections.

Distance	L	C	R
2,700	− 7·90	− 5·64	− 3·26
	24·94	0	15·78
2,720	− 6·34	− 5·36	− 7·13
	21·82	0	23·40
2,740	− 4·23	− 5·67	− 8·42
	19·18	0	25·92

Answer: 7,440 m³.

18. (a) Explain briefly, using neat sketches where appropriate, the function of horizontal transition curves in road alignment.

(b) A horizontal curve on a main road is to be transitional throughout. The deviation angle is 37° 46′, design speed 100 km/h., maximum centrifugal ratio permitted is 0·25, and the rate of change of radial acceleration must not exceed 0·3 m/s³.

Calculate (i) the length of the full curve,
(ii) the minimum radius of the curve,
(iii) the tangent length.

Answer: (i) 454·2 m; (ii) 314·7 m; (iii) 223·6 m.

19. Three straights AB, BC and CD have whole circle bearings of 60°, 90° and 45° respectively. AB is to be connected to CD by a continuous reverse curve formed of two circular curves of equal radius together with four transition curves. BC, which has a length of 800 m, is to be the common tangent to the two inner transition curves. Determine the radius of the circular curves if the maximum speed is restricted to 80 km/h, and a rate of change of radial acceleration of 0·3 m/s³ obtains.

Answer: 1,140 m.

20. A parabolic vertical curve connects two straights, one of slope 1 in 100 falling to the right and the other of slope 1 in 50 rising to the right at a valley. A point P on the first straight has a chainage of 4,000 and a reduced level 67·10 m above datum, whilst a point Q on the other straight has a chainage of 4,300 and a reduced level of 67·70 m. What is the length of that vertical curve which passes through a point R at chainage 4,200 and of reduced level 66·59 m? Locate the lowest point on the curve and give its level.

Answer: 312 m; 4,128 m chainage.

21. The straight portion of a proposed railway runs east to a point I, where it deflects through 45° and then runs in a direction N 45° E. The two straights are

to be connected by a circular curve of radius 400 m together with two transition curves, giving a rate of change of radial acceleration of 0·3 m/s³ when traversed at 100 km/h.

If the tangent points taken in turn round the curve are T, T_1, T_2 and U, find (a) the distance IT, and (b) the chainage of U, if that of T is 2,400.

Tabulate offsets required for setting out curve TT_1, at intervals of 20 m.

Answer: (a) 256·3 m; (b) 2,892·8 m.

22. A parabolic vertical curve is to be introduced between two gradients of + 2·5 per cent and − 1·7 per cent, allowing for a sightline distance of 300 m at a height of eye 1·05 m. The "finished road level" of the intersection point of the two gradients would be 230·56 and the chainage 1,410.

Calculate

(a) the chainages and levels at the beginning and end of the curve,
(b) the levels at each even 50 m of chainage within the vertical curve,
(c) the chainage and level of the highest point reached,
(d) the limits of chainage within which the gradient is arithmetically less than 0·5 per cent, and explain why these limits are important especially in the case of a sagging vertical curve having a "valley" point.

Answer: (a) Chainages 1,185 m, 1,635 m.

23. Design a vertical parabolic curve of length 180 m, to connect a rising gradient of 1 in 50 with a falling gradient of 1 in 100 which meet at a summit of reduced level 78·15 m above datum and chainage 1,425·00 m.

Determine the distance apart of two intervisible points which are each 1·05 m above the surface level.

Answer: 230 m.

24. A vertical curve is to be designed to connect two gradients, one of 1 in 25 and one of 1 in 33, which meet at a crest on one lane of a proposed straight dual-carriageway road. The visibility distance for a driver's eye height of 1·05 m is to be 190 m, which is less than the total length of the curve.

Deduce from first principles the length of the curve.

Answer: 318 m.

25. The length of a line was established as 900·476 m when measured by a tape of length 30 m suspended in catenary and supported at midspan. Determine the corrected length of the line if the tension applied in the field was 169 N instead of the value of 178 N intended.

The tape, which was standardized on the flat under a pull of 89 N, had a cross-sectional area of 6 mm² and its density was 7,700 kg/m³. $E = 207,000$ N/mm².

Answer: 900·464 m.

26. A nominal distance of 30 m was set out with a steel tape from a mark on the top of one peg to a mark on the top of another, the tape being in catenary under a pull of 220 N and at a mean temperature of 17°C. The top of one peg was 0·68 m below the top of the other, which was 250·00 m above mean sea level. Determine the horizontal distance between the marks on the two pegs, reduced to mean sea level. The tape, which was standardized in caternary under a pull of 178 N and at a temperature of 20°C, had a mass of 0·026 kg/m and a cross-sectional area of 3·25 mm². Take the coefficient of linear expansion as 9×10^{-7}/deg C, Young's modulus as 155 kN/mm², and the radius of the earth as 6,367 km.

Answer: 29·994 m.

27. A, B and C are three stations on a coastline used to fix the position of a borehole P being put down off shore. Find the co-ordinates of P if $A\hat{P}B$ and $B\hat{P}C$ are found to be 35° 24′ and 38° 36′ respectively.

The co-ordinates of the stations are—

Station	North	East
A	600 m	200 m
B	700 m	500 m
C	650 m	850 m

Answer: 192·6 m N; 538·1 m E.

28. *A*, *B* and *C* are three stations on a coastline and *O* is a float at sea. *A* and *C* are respectively west and east of *BO*, and *B* and *O* are respectively north and south of *AC*. $AB = 360$ m, $BC = 248$ m and $A\widehat{B}C = 153°$ 30′. If $A\widehat{O}B$ is found to be 40° 36′ and $B\widehat{O}C$ is found to be 35° 48′, calculate the distances *AO*, *BO* and *CO*.

Suggest the composition of a party carrying out such a survey, indicating the duties of each member.

Answer: $AO = 541$ m.

29. (*a*) Define any *four* of the following terms used in field astronomy, using neat sketches where appropriate; declination, azimuth, ecliptic, altitude, meridian transit, First Point of Aries.

(*b*) At a point in the northern hemisphere, the sun was observed at transit on the observer's meridian, and the corrected altitude was 57° 13′ 44″. If the sun's declination at the time of observation was south 12° 15′ 08″, what was the latitude of the observer?

What would be the altitude of the sun at the same time, for an observer at the same latitude in the southern hemisphere?

Answer: 20° 31′ 08″ N; 81° 44′ 00″.

30. In order to determine the azimuth of a survey line *AB*, a theodolite was set up at *A*, and an extra meridian observation was taken on the sun. Determine the azimuth of the line from the following data—

Corrected altitude of centre of sun $= 31°$ 16′ 20″
G.M.T. of observation $= 16$ hr 02 min 30 sec
Declination of sun at previous G.M.N. $= 6°$ 56′ 12″ N increasing at 56″ per hour
Latitude of station *A* $= 54°$ 30′ 50″ N
Horizontal angle clockwise from *AB* to centre of sun $= 50°$ 45′ 15″

Answer: 181° 53′ 57″.

31. What is local apparent time corresponding to local sidereal time 14 hr 20 min 42 sec by chronometer at a place in longitude 01 hr 41 min 30 sec E? The table gives relevant data for that day—

UT	R	E
0600	04 hr 14 min 35·8 sec	12 hr 13 min 09·5 sec
1200	04 hr 15 min 34·9 sec	12 hr 13 min 05·0 sec
1800	04 hr 16 min 34·1 sec	12 hr 13 min 00·4 sec

Answer: 10 hr 18 min 51 sec.

32. To determine the azimuth of a survey line *AB*, two stars were observed at elongation from station *A*, of unknown latitude, and the following information was obtained—

Star	Declination	Elongation	Observed Clockwise Angle from *AB*
X	64° 52′ 36″ N	E	300° 20′ 50″
Y	75° 43′ 24″ N	W	268° 34′ 20″

Determine the azimuth of *AB*.

Answer: 79° 51′ 22″.

33. Discuss the use of aerial photographs and aerial surveys in highway engineering.

34. (a) Distinguish between a plumb point and an isocentre.

(b) Show that distortions on aerial photographs due to tilt alone are radial from the isocentre.

(c) Compare the scales of photography, the areas recorded and the strip widths given by cameras X and Y at equal flying heights.

Camera X	*Camera Y*
Negative size 180 mm × 180 mm	230 mm × 230 mm
Focal length 210 mm	150 mm

How many photographs would be taken by camera X *with* 60 *per cent overlap* in covering a strip 18 km long, when the flying height is 1,500 m.
Answer: 36.

35. Give a brief appreciation of terrestrial photogrammetry.

The horizontal angle between two points A and B was measured directly at station C and found to be 21° 40′. A phototheodolite was set up at C; and on a photograph, A was found to be 26·5 mm to the left of the principal vertical line and 12.0 mm above the principal horizontal line. On the same photograph, B was found to be 36·6 mm to the right of the principal vertical line and 24·3 mm below the horizontal line. Find the focal length of the camera and the difference in level between A and B given that AC and BC are 78 m and 72 m in length respectively.
Answer: 165 mm; 15·9 m.

36. A mosaic is to be prepared at a scale of 1 in 25,000 by enlarging vertical photographs 1·2 times. What is the limiting relationship between the flying height for photography and the ground height for the displacement on the mosaic of a point on the photograph 60 mm from the principal point not to exceed 1 mm?
Answer: 1/72.

37. A tunnel is to be set out between two points A and B which are not intervisible. Two mutually visible points C and D are chosen and the following angles observed—

$C\hat{A}D$	52° 44′ 50″	$C\hat{B}D$	60° 26′ 25″
$A\hat{C}B$	43° 23′ 20″	$A\hat{D}B$	35° 41′ 20″
$B\hat{C}D$	40° 20′ 30″	$A\hat{D}C$	43° 31′ 30″

If A is west of B, and C and D are south of A and B respectively, determine the best values for the six angles assuming that they are of equal weight.
Answer: $C\hat{A}D = 52° 44′ 44″$.

38. What is meant by (a) the probable error of an observation, (b) the relative weight of an observation?

Determine the best value of an angle which was measured separately by two observers A and B, using the same instrument, to obtain the following—

Observer A	*Observer B*
62° 40′ 40″	62° 40′ 50″
40′ 10″	41′ 00″
40′ 30″	40′ 30″
40′ 00″	40′ 20″
40′ 20″	40′ 00″
41′ 00″	40′ 50″
40′ 30″	41′ 10″
40′ 40″	41′ 00″

Answer: 62° 40′ 34″.

39. In connexion with measurements of high accuracy, explain the meaning of—

(a) Compensating and non-compensating errors, using for illustration linear measurements by tape, but excluding the special effects of catenary suspension.

(b) The probable error of an observation, and find the probable error of a single observation which, when repeated on the same angle, gave values of 43° 20′ plus: 10″; 30″; 10″; 20″; 00″; 40″; 00″; 10″; 10″; 30″; 00″.

(c) The relative weights of observations, illustrating by finding the best value of a quantity which, measured four times by a method having a probable error of 3 units, gave an average value of 1,800 units and measured nine times by a method having a probable error of 6 units gave an average value of 1,808 units.

(L.U., B.Sc.)

Answer: (b) ± 9″; (c) 1,803 units.

40. A triangulation survey to establish a centre line DB across a river consists of three triangles BCD, ACD and ABD with centre point D. Adjust the polygon to close.

Triangle	Central Angle	Left-hand Angle	Right-hand Angle
ADB	142° 19′ 40″	15° 40′ 20″	22° 00′ 00″
BDC	147° 40′ 20″	17° 55′ 30″	14° 23′ 50″
CDA	69° 59′ 50″	59° 38′ 40″	50° 21′ 40″

(Angles designated left-hand are to the left of the observer when standing at the centre station and facing the outer stations.) *(L.U., B.Sc.)*

Answer: Triangle ADB: centre + 2″; LH − 4″; RH + 1″.

INDEX

The Christmas Rose

Dilly Court is a *Sunday Times* bestselling author of thirty-five novels. She grew up in North-East London and began her career in television, writing scripts for commercials. She is married with two grown-up children and four grand-children, and now lives in Dorset on the beautiful Jurassic Coast with her husband.

To find out more about Dilly, please visit her website and her Facebook page.

www.dillycourt.com
/DillyCourtAuthor